Andrew J. Frew
Martin Hitz
Peter O'Connor (eds.)

Information and
Communication Technologies
in Tourism 2003

Proceedings of the International Conference
in Helsinki, Finland,
2003

SpringerWienNewYork

Dr. Andrew J. Frew
Department of Hospitality and Tourism Management
Queen Margaret College, Edinburgh, Scotland, U.K.

Dr. Martin Hitz
Institut für Informatik-Systeme
Universität Klagenfurt, Klagenfurt, Austria

Dr. Peter O'Connor
Institut de Management Hotelier International
Ecole Supérieure des Sciences Economiques et Commerciales
Business School, Paris, France

© 2003 Springer-Verlag Wien
Printed in Austria

Typesetting: Camera ready by authors
Printing: Novographic Druck G.m.b.H., A-1230 Wien
Printed on acid-free and chlorine-free bleached paper
SPIN 10909485

With 103 Figures

ISBN 3-211-83910-0 Springer-Verlag Wien New York

Preface

The 10th ENTER Conference will continue a tradition of sharing knowledge among researchers in the fields of travel and tourism, and information and communication technologies. The overall conference theme: 'technology on the move', reflects the currency of mobile and location based services – ambient technologies, however, above all, ENTER is about inclusivity. The key objectives of this conference are to disseminate findings, to exchange experiences on the development of new research theories and methodologies and to generally encourage informed and cross-fertilized research activities. For the Research Track this has meant providing a showcase for high quality research work both complete and in progress and, with the continued upsurge in the volume and quality of submissions, regular ENTER participants will notice a few significant changes this year. These changes affect both the format of the conference and the dissemination of research outputs.

Firstly, **format**: research presentations have been streamed into three categories; the traditional research paper sessions for full papers is of course still there but has now been complemented by a set of research in progress sessions. It is anticipated that these streams will become established features of ENTER and will grow significantly in the years ahead. For those with an eye on quality, it should be remembered that regardless of stream, all papers have been subject to rigorous double blind review by our international panel of experienced researchers. A further format innovation is the EU projects stream that has been facilitated through the positive encouragement of the European IST Directorate.

Secondly, **conference dissemination**: the printed proceedings have been and will continue to be the flagship dissemination medium for full paper work, it represents a well-established corpus of knowledge with international recognition. However, we are at a watershed this year – there is simply so much good research work coming through and so much in the pipeline that we must start considering formats that over-all will be more appropriate for the next ten years in our development. The full paper Proceedings will undoubtedly have to transform to incorporate a second volume for ENTER 2004, the research in progress work will this year be available in electronic form which in itself is a great stride forward; however, the next logical step may be for us to produce a printed volume of all research in progress and perhaps to include within that the excellent work coming through from the Ph.D. workshops as a separate section. We would thus have accessible outputs for work at all stages from early and speculative through to mature and this must be a positive step towards having an increasingly informed and interactive research community. For the research track chairs this is a fundamental concept – the proceedings are much more than a sterile record of output, they are integral to sustaining what has become one of the most vibrant and friendly research domains.

Thirdly, **post-conference dissemination**: it is intended that the research presentations from ENTER will join outputs from the other tracks to form part of the growing knowledge base on the IFITT website and we would encourage all IFITT members to make full use of this facility throughout the year!

The continuing record number of submissions and papers again at the 10th ENTER Research Conference shows the vitality of research in this area and a measure of the scale and scope can be seen in the fact that over 150 researchers from more than 25 countries will present their work. ENTER itself is but the climax of a long process involving a great many researchers and reviewers and it is perhaps appropriate to re-member these efforts and express our sincerest thanks to all for their joint effort to make a success of ENTER 2003. Thanks are also due to the IFITT Board and the sponsors and partners of ENTER, all of whom provided efficient support in the hope that all participants will have an informative, enjoyable and memorable experience. Thanks also to the overall organizing team of Inkeri Starry and Dimitrios Buhalis, whose constant determination and hard work have been in large measure responsible for bringing it all together along with the unassuming but always vital contributions from Karen Hay and Helene Forcher in the IFITT engine room!

Finally, in reflecting on the evolution of the research track, this preface has departed from its predecessors in saying little about the specific content of the printed volume, but to be frank, the content speaks for itself. The presentations have been grouped under broad headings but an examination of the diversity of the work and the inter-connected nature of the domain quickly demonstrates that these groupings are at best pragmatic – so perhaps the biggest challenge is selecting your own path through this busy schedule – enjoy ENTER 2003, we will!

Helsinki, 2003   Andrew Frew, Martin Hitz and Peter O'Connor

Contents

1 Best Paper Candidates

2 Best Paper Candidates

3 Systems Architecture

4 Virtual Communities

5   Mobile Services

6   Travel Advisory and Trip Planning I

7   Marketing and Markets I

8   Destinations

9 Travel Advisory and Trip Planning II

10 Hotel Systems and Issues I

11 Consumer Issues

12 Marketing and Markets II

13 Travel

14 ICT and National Structures

15 Hotel Systems and Issues II

16 Marketing and Markets III

17 Tourism Information Systems and Sources I

18 Tourism Information Systems and Sources II

Index of Authors

Research Conference Committee

Tenth International Conference on
Information and Communication Technologies in Travel and Tourism,
Helsinki, Finland, January 29-31, 2003.

Francois Bédard, University of Quebec, Canada
Thomas Bieger, St. Gallen University, Switzerland
Dimitrios Buhalis, University of Surrey, UK
Roberto Daniele, Queen Margaret University College, UK
Daniel Fesenmaier, University of Illinois at Urbana-Champaign, USA
Andrew Frew, Queen Margaret University College, UK
Martin Hitz, University of Klagenfurt, Klagenfurt, Austria
Stefan Klein, University of Münster, Germany
Rob Law, Hong Kong Polytechnic University, Hong Kong
Panos Louvieris, University of Surrey, UK
Valeria Minghetti, CISET-University of Venice, Italy
Nina Mistilis, University of Western Sydney, Australia
Hilary Murphy, Swansea Business School, Wales, UK
Peter O'Connor, Institut de Management Hotelier International, France
Harald Pechlaner, Leopold-Franzens-University Innsbruck, Austria
Francesco Ricci, Electronic Commerce and Tourism Research Lab at ITC-irst, Italy
Walter Schertler, University of Trier, Germany
Pauline Sheldon, University of Hawaii, USA
Mariana Sigala, University of Strathclyde, UK
Oliviero Stock, University of Trento, Italy
Ingvar Tjøstheim, Norwegian Computing Center, Norway
John van der Pijl, Erasmus University, The Netherlands
Hubert van Hoof, Northern Arizona University, USA
Hannes Werthner, University of Trento, Italy
Karl Wöber, Vienna University of Economics and Business Administration, Austria

Intelligent Information Interactions for Tourism Destinations

F. Kazasis [a],
G. Anestis [a],
N. Moumoutzis [a],
S. Christodoulakis [a]

[a] Laboratory of Distributed Multimedia Information Systems and
Applications (MUSIC)
Technical University of Crete (TUC), Greece
{ fotis, ganest, nektar, stavros } @ced.tuc.gr

Abstract

This paper presents a model supporting intelligent interactions of tourists with other tourists
and locals and the tourism information of a particular Destination before, during and after the
trip. The approach tries to bridge the "Community Gap" which is the lack of interactions
among tourists and between tourists and locals at a particular Destination. Community
interactions are very important both for prospective visitors and for Destinations for many
reasons including, greater independence and self-planning in the visit's design, exploitation of
the local society knowledge about the Destination, as well as promotion of regional policies and
collective purchases of services from prospective visitors. Modern information technology has
become ubiquitous, supporting visitors with a variety of devices ranging from handy devices, to
Community Walls, to paper interfaces, to home PCs. The paper focus is the description of the
knowledge bases and their capabilities for intelligent interactions for supporting tourism
communities at Destinations.

Keywords: visitor attraction; connected community; ubiquitous computing; knowledge support

1. Introduction

The proliferation of the Web over the last few years led companies and organizations
to try to exploit the Web for e-commerce activities. Tourism is one of the most
important applications of e-commerce. Several major tourism actors but also new
comers (information technology companies mainly) have an established Web
presence, visited by many thousands of visitors every day, offering e-commerce
opportunities for business to business transactions or business to customer (tourist)
transactions. One particular class of tourism applications in the Web is Destination
Information Systems (DIS) or Destination Management Systems (DMS)
(Christodoulakis et al., 1996; Christodoulakis et al., 1997; Evans and Peacock, 1999;
Pan and Fesenmaier, 2000; Werthner and Klein, 1999). These systems typically
provide in the Web, information about the tourism offerings of a given Destination
and may promote e-commerce activities to the potential visitor. The existing DMS
however do not support advanced models of interaction between tourists (or

prospective tourists) of a Destination, nor interaction between tourists and locals. It is the authors' belief that this is a serious limitation of the existing DMS, and therefore this paper proposes an expanded functionality that provides the tourists with intelligent interactions based on a virtual community concept of tourists and locals that has a common interest theme, "Tourism at Destination". The implementation of this functionality may be in an independent system complementary to DMS or as an expanded functionality of existing DMS. Information systems that support interactions of a virtual community over the Web, which has some specific interests (the glue of this community), are usually called Community based Information Systems (CIS). Existing CIS's in the Web focus to foster social objectives like building community cohesion, enhancing community awareness in local decision making, developing economic opportunities in disadvantaged communities, and enhanced training (Schuler, 1994). Some of them have user populations of the order of tens of thousands who are repeatedly visiting the community site. However the support that the existing CIS's offer is of general purpose and they cannot be easily used to offer advanced functionality for tourism related communities. It is considered that it is very important both for tourists and for Destinations to support advanced information models enabling the interaction of tourists and locals for tourism related subjects. Such systems will bridge the "Community Gap", which is the lack of interaction among tourists and locals at a particular Destination. Some of the advantages offered to both tourists and Destinations by these information models are detailed below.

First there is a reason of providing the visitor with evaluation information that is not given by a particular organization (which may be biased or may not recognize the real visitor preferences). In a CIS, community members like visitors, may leave evaluation information about a site of interest or a service organization (hotel, etc.). Such information may be more trustable to other visitors than recommendations given by a Destination Organization or a Travel Agent. Sometimes this provision of community evaluation is convenient even for Destination Management System Organizations, which may not want to offer their own evaluation (or recommendations) in order to avoid potential conflicts with partner organizations at Destinations. It may also be convenient for the Regional Administrations, which may be interested to have objective quality of service information delivered to visitors in order to increase competitiveness of the tourism offerings at the Destination, but they do not want to be directly involved in such an evaluation.

For prospective visitors who have a desire for greater flexibility in planning their own trip, a contact with a tourism community for the Destination offers advantages of increased security in their planning which comes not only from the evaluations that other tourists have given about sites and offerings of the Destination, but also by the fact that they can get directly in contact with members of the community at Destination to ask specific questions. In fact, it is a great misconception of today's information system builders that any information system, no matter how big its databases are, can satisfy all the potential visitor questions about the Destination offerings.

Another advantage that the community based information systems for tourism offer is the greater potential for repeat visits that they create. This is not only due to the greater quality of information and better evaluation that they offer, but also because they offer increased opportunities for human connections between tourists and locals and the emotional attachment that they carry. Established contacts during the visit may also become trustable sources of information and evaluation creating further security for future visits. From the Region Management point of view virtual communities offer a great opportunity for the Region to promote Regional policies like branding. This can be done for example by promoting the creation of specific types of tourism sub-communities with specific interests. From the prospective tourists point of view, virtual communities also offer the opportunities for collective agreements of potential visitors with local or international companies. By opening a discussion forum for a visit at a specific time and with specific interests, a group of potential visitors may be found, which may be in a position to negotiate collectively better services and better rates from companies and organizations.

This paper describes the functionality needed for the provision of community based intelligent information interactions for tourism destinations. Section 2 presents the proposed approach, section 3 describes the Knowledge Base mechanism that elaborates the approach and in section 4 indicative supported services emerged by the Knowledge Base are introduced. Finally in section 5 a specific application based on this approach as well as its evaluation are outlined.

2. A new Approach to Promote Tourist Destinations

One goal of the approach is to develop an alternative, quality based tourism support system through the environmental and cultural characteristics of an area, the anthropocentric perspective and the various local communities, and to promote the concept "tourism=hospitality". The approach may be viewed as a system diffused in the territory to facilitate and support the exchange of information between different communities in a local, physical environment (such as a destination, an area of interest, a visitor attraction) with the aim of building a dynamic shared knowledge accessed by tourists and locals.

It is based on an *interaction model* that maps digital information over physical spaces. The users can access the system through a variety of as a "travel diary" to some users. Both types of information contribute to the creation of a shared knowledge (information plus personal experiences) that is based on a combination of cold data (the basic functional information) and warm data (the description of the experience of the community of users).

The *information model* is mostly based on the concept of "push information": information is mostly circulated with the model of push media (edited information is proposed to users according to their profile and to their physical position) but the system stimulates the reaction of users in the form of annotation (comments, ratings and further contributions attached to the "core" piece of information). This annotated information is re-circulated within the system giving origin to an exponential growth (with over-annotated information also progressively moving from cold data to warm data).

The *social model* of the approach allows different roles to both users and professional editors: anybody can assume roles ranging from a "passive" use (simply retrieving information), to active "contributors" (information given and retrieved), up to "cultural managers" and "moderators" (available for the editing of the information, facilitators and managers of the shared knowledge).

The *spatial model* of the approach is strongly related to the notion of mapping digital information over material physical spaces and builds on this model various interaction models. However the system also takes full advantage of the potential of remote communication and participation: access to the shared knowledge is given also to remote users. Information is connected to the physical space via distributed interfaces to access content: Space is the main structure to navigate and interact and mobile and ubiquitous computing is augmented with personalization.

3. The Knowledge Base

The cornerstone of the approach is its Knowledge Base that efficiently supports a rich set of services and diverse interaction means. The Knowledge Base is based on a schema that captures the semantics of objects handled by the system. To enhance readability, this schema is described in five parts:

Items, *Contexts* and *Person* describe the basic modeling abstractions for the description of information items, users (people) and semantic frameworks (contexts). Destination/Region/sub-Region, Tourism attractions/sites of interest, service organizations, events, are indicative examples of tourism – related concepts modelled by Items. Contexts are used in order to group together items with common semantics or typology. For example a tourism attraction (e.g. Ancient Kydonia in Chania, Crete) can be associated to the context of Minoan Palace/City, Venetian Construction, and Venetian Church. Additionally the virtual community of the Archaeologists of Crete is also represented by a context. Visitors, locals and experts (e.g. archaeologists) are described by the concept of Person.

Traces describe the interaction of users with the system and the mechanisms for recording these actions. Evaluation information (e.g. ratings, comments left by visitors for a tourism attraction) is a specific type of traces.

Filters describe the mechanisms for expressing user interests. For example a filter expresses the fact that a visitor is interested for Archaeology and specifically for Minoan sites and for restaurants with traditional dishes nearby.

Descriptions are used to model the content of information items. A historical event (e.g. the Battle of Crete) may have several descriptions in various templates, associated with the living memories of locals or the official reports of organizations or the discussions of the virtual community of historians.

Map support refers to the associations between the information items and their spatial representation. The information for a tourism attraction (e.g. the Archaeological

Museum of Heraklion) is indexed on top of map(s) and accessed through it. Moreover the community of Archaeologists of Crete has also a representation on the map.

4. Supported Services

A number of services that exploit the Knowledge Base are described next. These are grouped into four categories, creating the content, accessing the content, support for user communities and indexing content on top of maps.

4.1 Content creation

Travel Diary: The diary is a tracing mechanism, essentially a sort of memory of user actions (places visited, topics discussed, recommendations and comments made, other persons met, personal notes) to guide user's future interaction with the system. To protect privacy the system gives only aggregated statistical data based on the evaluations of users found in their travel diaries and not the diaries themselves.

Topics of discussion: A topic of discussion is an abstract place where people "meet" each other with the objective to exchange information related to the discussion. Topics are hierarchically structured from more general to more specific ones, so that it is possible for a user to narrow the context of the discussion. A topic groups together information items and the discussions correspond to posting comments or recommendations visible from all its members.

4.2 Content access

User profile: The user profile is a way of expressing the characteristics and features of a Person. It consists of a static (demographic info such as name, sex, age, country of origin etc) and a dynamic part (interests, filters, traces). User profiles are a central issue in all system functions.

Searching: Searching refers to finding items and recommendations from other users. Searching also supports ranking in order to help the users decide more easily on the items that they are interested in.

Personalization of information: In the traditional collaborative filtering (Resnick and Varian, 1997; Resnick et al., 1997; Shardanand and Maes, 1995) the system finds a set of similar users ("friends"), based on the user ratings on common information items. The system utilizes the information about the friend's actions (e.g. the evaluations that they gave about information items) in order to suggest more information items to the user. This approach has the so-called "cold start" problem, i.e. when there are no sufficient actions in the system by the "friends", the system cannot give reliable recommendations.

In the proposed approach personalization is supported via a combination of content based and collaborative filtering (Anestis, 2001). User profiles are compared to find groups of users with similar interests called friends. Friends may also be explicitly indicated by the user, or may be found not on the basis of matching the whole profile but only a part of the profile indicated by the user. In order to provide personalization for a specific user, the traces of his/her friends are examined and items with high ranks are suggested to her.

The collaborative filtering algorithm used is based on traces. A trace type may have a feedback, which expresses the degree of satisfaction or the interest of the user. The particular methodology gives a solution to the well-known "cold start" and "over-specialization" problems reported in traditional collaborative and content-based filtering literature (Babalanovic and Shoham, 1997; Pazzani, 1999).

4.3 Communities of people

Communities are groups of people having something in common. They are related to topics and physical places that are of interest to their members. Members of the same community may have access to information not seen by others. Examples of communities may be members of a family, a group of friends, archaeologists in Crete, hotel owners etc. Communities are used to establish and maintain links between people. Members of a community have a certain degree of trust between each other and this can be used for the collaboration filtering mechanisms mentioned above. It is possible to have a moderator for a community. He/she is a user (expert) with special privileges granted from the system. He is responsible for assuring the quality of information that flows into and out from the community.

4.4 Map support

The map services were designed in order to provide users with on-line maps, adaptive to user-defined criteria. These criteria apply to the information (Items) to be projected on the maps and may refer to the content itself -contexts, time etc.- or to spatial attributes. The interface also allows performing spatial queries and acquiring movement directions.

5. The Campiello Application and evaluation

During the i3 LTR Esprit Project Campiello (http://www.campiello.org, http://www.i3net.org) the above model and its implementation was followed by the partners of the project in order to develop applications and intelligent interfaces for a variety of devices that promotes the tourism product of the cities of Venice and Chania, Crete and facilitates the creation of connected communities and better connect the members of the communities (local inhabitants; past, present and future tourists; cultural managers) by local people or tourists.

Special care was taken so that Knowledge Base supports in a uniform manner diverse community interaction means with completely different characteristics regarding both the computational and the presentation capabilities. The partners of the project have investigated three new methods of communication between social teams and information repositories (Revised Campiello System, 2000). These are intelligent paper (IP), community wall (CW) and mobile-device (MD) interfaces. The Knowledge Base that was described in this paper may be accessed by users using either their regular home PCs/monitors, their MD or a CW. The user interaction models are different for each one of the above cases. Nevertheless Internet remains the linchpin allowing the access to the information model from anywhere using various user interfaces.

IP technology allows the programming of common paper with a specific behaviour by encoding non-visible information related to the actual content of the paper. Based on this a user may leave his trace on a paper and the system will recognize it and respond by presenting him the information that meets his personal interests. Two different types of applications have been implemented, personalized newspapers and travel guides. The former allows news presentation personalized according to the user preferences already known to the system. The user may then comment/rate for a topic and sees the comments/rates for this topic by other people of a community with the same interests. The later one creates personalized maps according to the user interests as well as his geographical location, allow a user to see how other visitors/locals evaluate the sites or the tourist services provided in the area and to leave his own comments/rates for these.

CW technology is based on the use of large interactive screens located in public places where people are gathered and communicate. The information presented is mainly related to warm data (news, comments, ratings, suggestions) provided by various communities, and is dynamically refreshed according to various rules such as the information inserted to the system by other sources, the interest shown or the rating of a specific topic, etc.

MD combined with GPS for identifying location; support the finding of paths of maximum interest to the user and also the presentation of multimedia information to a visitor while wandering in a city. Such information maybe the history of a building, events relevant to his profile, or even a nearby person that shares the same interests.

The provided environment is complemented by the PC user interface that integrates all the above functionality but also the traditional structuring of tourist "cold" information for the destination.

The system was thoroughly tested in a public, open experiment in order to evaluate the capability of the Campiello system to reach its own concept and objectives as well as to verify the usability of the various technologies of the system and interfaces developed during the project. The aims of the feedback collection and the evaluation were:

The acceptance of Campiello's concept: The evaluation parameters that was set for the specific objective were: demographical in order to analyze the population participated in the experiment, profiling interests in order to analyze the acceptance of the various types of information provided by the system, tourist in order to investigate the possibility -in the future- for Campiello to provide also tourist services, popular items (e.g. events, places) in order to locate the most interesting, for the visitors, activities of the local communities and finally "campiellized" parameters (i.e. contact with the locals, acceptance of Campiello, keeping relation with locals) in order to directly analyze the main ideas of Campiello.

The technology usability: The evaluation parameters that was set for the specific objective were usability, learnability, aesthetics, error prone and handling, reliability, stability, map usability, content creation.

The main evaluation methods that were adopted were the questionnaires and the logging of user actions when interacting with the Campiello system.

Questionnaires: For the purpose of the experiment two different types of questionnaires were produced. The first one aimed to the collection of feedback regarding the technology usability and the second one aimed to the collection of feedback regarding the acceptance of Campiello's concept. About 1000 questionnaires of the first type were distributed and the number of the filled questionnaires that were finally collected was about 300. More than 70% of the users evaluated the system for more than 10 minutes. 28% have only registered to the system and have visited only the introductory pages, the 33% have spent some more time browsing whereas the 39% have tested many of the services offered. More than 4000 questionnaires of the second type were distributed. The response was more than satisfactory, since the collected filled questionnaires were about 600, i.e. 15%. Based on the analysis carried out, more than the half of the visitors had contact with locals and it is also noticeable that almost 30% expressed their wish to keep in contact with the locals they met. The most interesting result is of course the wide acceptance of the Campiello concept and the willing of the visitors to use a system like Campiello in the future.

Logging user actions: This technique involved having the system automatically collect statistics and measurements about its use. The basic purpose of this mechanism was to record the users' actions in the system. User actions in the system were classified to various types. Each user action was recorded in a transparent way through the creation of a trace. The trace types, which have been defined, were the following: Rate, Comment, Visualize, Registration, Print, Read e-mail. In addition, the type of the interaction device (PC, Paper, CommunityWall) was recorded for each trace. In this way useful conclusions were extracted regarding the use of each interaction device. During the evaluation process about 5000 traces were recorded in the Knowledge Base. The number of registered users approached the 500 people.

6. Summary and Conclusions

A new approach for the promotion of tourism attractions was presented. The approach, starting from TIS functionality, attempts to bridge the "Community Gap" (i.e. the lack of any interaction between the content providers and the content consumers) and to provide support for the promotion of a visitor attraction. Our approach handles both typical tourism information about an attraction and warm knowledge, related with the experiences of other visitors/locals associated with the attraction. Its cornerstone is the Knowledge Base (KB) that efficiently supports a rich set of services and diverse interaction means. The information space consists of items (i.e. physical places, events, or more abstract topics), contexts (categories of items used for search and recommendation issues), people and traces (implicit and/or explicit user interactions with the system). Main services include: enhanced topics of discussion, support for communities of people, matchmaking, awareness, semi-automatic creation and maintenance of personal diaries, personalized maps, advanced searching, time support and content generation (comments/ratings, insertion of new information). These are augmented with an enhanced digital map service for both the indexing and the visualization of the information along with a set of services regarding the suggestion of maximum interest paths, shortest paths, etc. The KB supports in a uniform manner diverse interaction means with completely different characteristics regarding both the computational and the presentation capabilities.

References

Anestis, G. (2001). *Designing an Information System to Support Virtual Communities of Users.* Master Thesis: Department of Electronics and Computer Engineering, Technical University of Crete, Chania.

Babalanovic, M. & Shoham, Y. (1997). Fab: Content-Based, Collaborative Recommendation. *Communications of the ACM,* 40 (3), 66-73.

Christodoulakis, S., Kazasis, F., Moumoutzis, N., Servetas, A. & Petridis, P. (1996). A Software Bench for the Production of Multimedia Tourism Applications. *Conference Proceedings on Information and Communication Technologies in Tourism* (pp. 18-28). Springer-Verlag.

Christodoulakis, S., Kontogiannis, P., Petridis, P., Moumoutzis, N., Anastasiadis, M. & Margazas, T. (1997). MINOTAURUS: A Distributed Multimedia Tourism Information System. *Conference Proceedings on Information and Communication Technologies in Tourism* (pp. 295-306). Springer-Verlag.

Christodoulakis, S., Anastasiadis, M., Margazas, T., Moumoutzis, N., Kontogiannis, P., Terezakis, G. & Tsinaraki, C. (1998). A Modular Approach to Support GIS Functionality in Tourism Applications. *Conference Proceedings on Information and Communication Technologies in Tourism* (pp. 63-72). Springer-Verlag.

Evans, G. L. & Peacock, M. (1999). ICT in tourism: A comparative European survey. *Conference Proceedings on Information and Communication Technologies in Tourism* (pp. 247-258). Springer-Verlag, Wien.

Pan, B., & Fesenmaier, D. (2000). A typology of tourism-related web sites: its theoretical background and implications. *Information Technology & Tourism* 3(3/4), 155-176.

Pazzani, M. A Framework for Collaborative, Content-based and Demographic Filtering. *Artificial Intelligence Review* 13, 5-6 (1999), 393-408.

Resnick, P., & Varian, H. R. (1997). Recommender System. *Communications of the ACM,* 40 (3), 56-58.

Resnick, P., Iacovou, N., Suchak, M., Bergstrom, P. & Riedl, J. (1994). GroupLens: An Open Architecture for Collaborative Filtering of Netnews. *Conference Proceedings on Computer Supported Cooperative Work* (pp. 175-186). New York.

Schuler, D. (1994). Community Networks: Building a New Participatory Medium. *Communications of the ACM,* 37(1), 39-51.

Shardanand, U., & Maes, P. (1995). Social Information Filtering: Algorithms for Automating the 'Word of Mouth'. *Conference Proceedings on Human Factors in Computing Systems* (pp. 210-217). Denver, CO.

The Revised Campiello System (2000). Deliverable D3-4.3: EP 25572.

Werthner, H. & S. Klein (1999). *Information Technology and Tourism - A Challenging Relationship.* New York, Springer-Verlag Wien.

Acknowledgement: The Campiello project consortium was comprised by Consorzio Milano Ricerche - Dip. Di Scienze dell' Informazione, Domus Academy, Xerox Research Center - Grenoble Lab., FORTHnet, Municipality of Chania, Technical University of Crete (TUC/MUSIC)

Measuring Customer Satisfaction with Online Travel

Juline E. Mills
Alastair M. Morrison

Department of Hotel, Restaurant and Institution Management
University of Delaware, USA
juline@udel.edu

Department of Hospitality and Tourism Management
Purdue University, USA
alastair@purdue.edu

Abstract

Currently, very little research exists to guide travel businesses in understanding customer satisfaction with travel agent Websites. In particular, no current study provides a comprehensive methodology for evaluating travel Websites with a customer satisfaction focus. To this end, the purposes of this study were to identify the potential attributes of customer satisfaction with travel Websites, and subsequently test a structural base model of customer satisfaction with travel Websites. The evidence of validity (convergent, discriminant, and nomological) and reliability reported confirmed the construct validity of the 24-item E-SAT instrument for measuring customer satisfaction with travel Websites.

Keywords: customer satisfaction; E-satisfaction; E-SAT; online travel; confirmatory factor analysis; structural equation modeling.

1 Introduction

Online travel businesses are gaining unprecedented economies of scale. This is evidenced by the growth of online travel where some travel Websites (TWs) now offer information on more than 60,000 hotels in over 200 countries (Export Today's Global Business, 2000). Consumers are indeed making use of TWs with record spending of over $6.9 billion in the first quarter of 2002 (News Bytes News Network, 2002). Consumers are also demanding services when they need and want them, gravitating more toward those TWs that provide the closest to instantaneous delivery making it more important for businesses to build TWs with the consumer's interests in mind (Eclipse Magazine, 2001). This research effort therefore examines the issue of customer satisfaction (e-consumer satisfaction or E-SAT) with TWs. Specifically the study first, identified the attributes of E-SAT with TWs; second, developed a model to depict the relationship of the proposed attributes of E-SAT; third, developed a validated measuring instrument for examining E-SAT with TWs; and fourth, tested a structural base model of E-SAT with TWs. For the purposes of this research customer satisfaction is defined as the fulfillment or gratification of the customer's wishes (Oliver, 1997). While, TWs include such sites as Priceline.com, Expedia.com, and Travelocity.com. Understanding TW satisfaction may aid businesses with improving their online marketing efforts, providing better services, and in the long run, retaining loyal customers. Findings from this study can provide insights and understanding for various segments of the travel industry to develop tailor-made and customized Websites that cater more appropriately to the needs of today's travelers. The resultant model and instrument for measuring customer satisfaction with travel Websites may provide a base for businesses to use in evaluating and refining the services offered at their Websites

2   Defining the Potential Attributes of the E-SAT Model

Research methodology proposed by Boshoff (1999) served as the guiding framework for measuring E-SAT with TWs. Strauss and Corbin (1998) grounded theory technique, a qualitative data analysis methodology, was used to catalog the attributes as they emerged from the literature review. The definitions for the first- and second-order constructs which are proposed as being attributes of E-SAT with TWs are shown in Table 1.

Table 1. Model Constructs from E-SAT for Travel Website Model

CONSTRUCTS	DEFINITIONS
TW Interface	The means by which the consumer interacts with the TW.
Access	How consumers visit the TW via search engines etcetera.
Loading	The speed with which the TW loads onto the consumer's PC.
Appearance	The ability of graphical content and layout to grasp users attention.
Navigation	How customers surf the TW using convenient linking options.
Interactivity	The TW users speed when using the Web- fast/efficient or slow.
Search	The consumer's ability to quickly locate information at the TW.
Security	The safety of personal information during transactions at the TW.
Perceived Quality	The consumer's opinion of the overall excellence TW services.
Incentives	Gifts and information that encourages use of the TW.
Feedback	Visible/auditory information regarding user actions at the TW.
Info. Reliability	The delivery of timely product and service information by the TW.
Perceived Value	The consumer's overall assessment of the utility of TW services.
Involvement	The importance of the TW in searching/purchasing travel services.
Shopping Convenience	Whether the customer views the TW as saving on time and effort.
Transaction utility	The ability of consumers to use the TW to carry out various operations.
Price	The acceptability of the price paid or payable for services at TWs.

3   The Domain of E-SAT with Travel Websites

From the literature review the model shown in Figure 1 was proposed and used to guide the remainder of the study. The model proposes that E-SAT is a direct result of the customer's experience (overall perception) at the TW. Customer experience at TW (CETW) in turn, is a multi-dimensional latent construct impacted by travel Website interface (TWI), perceived quality of the services and products offered by the travel Website (PQTW), and the perceived value of the travel Website to the customer (PVTW). A positive experience at the TW is expected to lead to higher levels of E-SAT with the TW. The model also proposed that the three constructs – TWI, PQTW, and PVTW are second-order factors and collectively, are a function of 14 first-order factors as shown in figure 1.

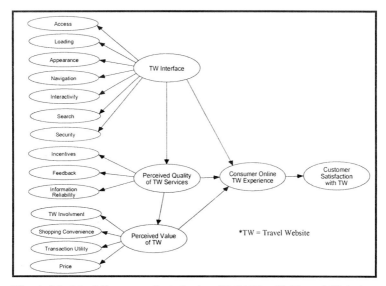

Fig. 1. Model of Customer Satisfaction (E-SAT) with Travel Websites

4   Generating Questionnaire Items

To test the proposed E-SAT for TW model 68 questionnaire items were utilized. Thirty-nine items generated from the literature review and 29 items adapted from previously validated instruments. Appendix 1 presents the original items coded (Q1-Q68). Items were adapted from Bitner & Hubbert (1994); Novak, Hoffman & Yung (2000); Yoo & Donthu (2001), or developed by the researchers. All items were measured on a 5-point Likert-type scale where 1=disagree strongly and 5=agree strongly. The E-SAT instrument was administered in two phases to college student populations. In support of the use of a student sample, Emory (1980) contended that convenience samples, such as student populations, are useful for theory testing, as was the purpose of this study. Additionally, college students use the Internet more than any other medium to research and purchase travel (Perez, 2000).

5   Exploratory Data Analyses -The First Survey

In the first phase of the research study the sample population used consisted of 250 undergraduate students at a large Midwestern U.S. university. A minimum sample size of 100 with no more than 200 based on Guadagnoli & Velicer (1988) was used. The online survey received 213 hits with 107 completing the survey, for a 50% cooperation rate (Mills, 2002). On examining the data for normality no extreme departures were noted. In checking the strength of the relationship between the independent variables significant problems of multicollinearity was found. Items (Q19, Q30, Q29, Q56, and Q64) were found to have Pearson's (r) greater than .70 and were correlating with multiple other items and as such were removed from the data set leaving 63 items. Cronbach's alpha as a measure of internal consistency was .94. Exploratory data analyses (EDA) using principal components analysis (PCA) was conducted to begin the initial reduction process of the final number of items that were to be used to represent each construct (Kim & Mueller, 1978). First, PCA (1) was conducted separately on each of the factors. Second, in PCA (2) the factors were then tested, as they are proposed as representing their respective second-order factor, to determine if any items loaded on more than one factor. Based on the results of the PCA (1) (Q42, Q43, Q44, Q53, Q54, and Q59) were removed from the dataset due to poor loadings or communalities (less than .50) on their respective first-order constructs. PCA was then conducted with varimax rotation and Kaiser normalization. The PCA (2) results for TWI revealed that some factors were related, such as navigation and search as well as loading and interactivity. Six items (Q3, Q6, Q7, Q13, Q14, and Q15) were removed and the original seven factors were reduced to five first-order factors. For PQTW (Q36 and Q33) and for PVTW (Q45, Q48, and Q52) were removed. Shopping convenience, price, and transaction utility clustered together to form one factor, renamed as TW resource maximization. Table 2 gives a brief synopsis of the PCA results.

Table 2. Summary of the Exploratory Data Results for E-SAT for TW Model

	Factor Loadings	Percentage of Total Variance Explained	Eigenvalues
PCA (1)	.52-.89	35.79-76.97	1.43-3.47
PCA (2)	.51-.88	46.02-63.92	1.43-3.47

From the PCA of the first survey 46 items remained for use in the second survey. Cronbach's Alpha for PVTW was low (.54). In an attempt to improve the reliability of this factor, two additional items (Q69 and Q70) were added to the items used to measure TW involvement for a total of 48 items to be used in the second administration of the survey.

6   Confirmatory Factor Analyses - The Second Survey

Prior to the second data collection, sample size was determined based on Chin (1998) contention that the question of appropriate sample size depends to a large extent on the power analysis and not on the number of parameters being estimated. Statistical power analysis was conducted using the G*Power statistical power analyses software. Power calculations yielded a recommended sample size of 330. The sample population used in the second survey consisted of 500 undergraduate students enrolled in an online class at another large Midwestern U.S. university. The online survey received 475 hits with 361 students completing the survey, for a 76% cooperation rate. Items (Q18, Q23, and Q47) were removed due to problems of multicollinearity. Cronbach's alpha for the remaining 45 items was (.96). The steps used in the first survey were again employed and (Q21, Q38, Q63, and Q61) were dropped due to poor communalities or factor loadings in the second survey PCA (1). While in second survey PCA (2) (Q11 and Q70) did not load well on their respective factors and were removed leaving 39 items for use in the CFA process. CFA was then performed to determine the "goodness-of-fit" of the proposed E-SAT with TW model. First-and second-order CFA was conducted. As the data are continuous maximum likelihood estimation procedure with covariance matrix was used. Modeling testing was conducted using Analysis of Moment Structures (AMOS 4.0). In the CFA models all standardized estimates were substantively reasonable and statistically significant at the p<. 0001 level. From the CFA analysis 28 items remained. The results of the final second-order CFA models are shown in table 3.

Table 3. Second-order CFA Model Results for E-SAT for Travel Websites

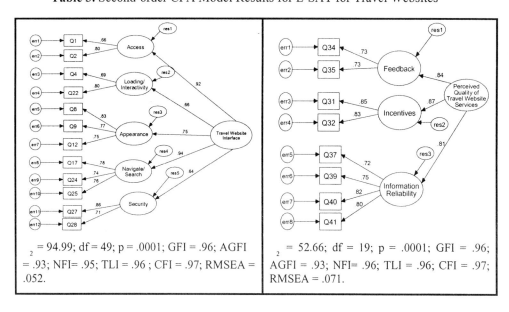

$_2$ = 94.99; df = 49; p = .0001; GFI = .96; AGFI = .93; NFI= .95; TLI = .96 ; CFI = .97; RMSEA = .052.

$_2$ = 52.66; df = 19; p = .0001; GFI = .96; AGFI = .93; NFI= .96; TLI = .96; CFI = .97; RMSEA = .071.

Table 3 Continued. Second-order CFA Model Results for E-SAT for Travel Websites

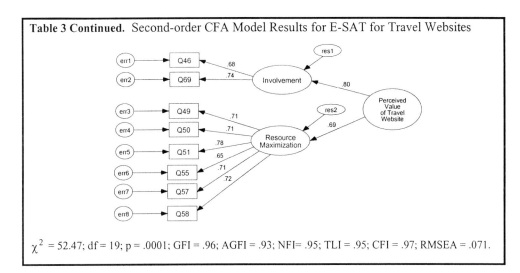

χ^2 = 52.47; df = 19; p = .0001; GFI = .96; AGFI = .93; NFI= .95; TLI = .95; CFI = .97; RMSEA = .071.

7 Fitting the Measurement Model

The measurement model, which consisted of the CFA models with two paths added, a direct effect from TWI to PQTW, and from PQTW to PVTW was then tested. After modification and the systematic removal of (Q37, Q49, Q24, Q57, and Q50), model fit was achieved with χ^2 = 384.58; df = 220; p = .0001; GFI = .91; AGFI = .89; NFI= .90; TLI = .95; CFI = .96; RMSEA = .046. All standardized estimates for the measurement model were substantively reasonable and statistically significant at the p <.0001 level. In examining the two added paths TWI impacted PQTW (.96), while PQTW impacted PVTW(.96).

8 Fitting the Structural Model

The structural model for the E-SAT for TWs consisted of the measurement model with the addition of two endogenous variables CETW and E-SAT with the TW. Direct paths from TWI, PQTW, and PQTW were added as directly impacting CETW, which in turn impacted E-SAT with the TW. The initial model was poor, however, the systematic removal of (Q12, Q40, Q55, Q62 and Q68) yielded acceptable model fit with χ^2 = 457.22; df = 242; p = .0001; GFI = .90; AGFI = .88; NFI= .89; TLI = .94; CFI = .94; RMSEA = .050. Standardized estimates for the model were substantively reasonable and statistically significant at the p<.0001 level. Within the structural model CETW had a positive effect on E-SAT with the TW (.99). Moving from measurement to structural model, TWI (.69) and PVTW (.78) had a positive effect on CETW. PQTW had a significant negative effect (-.47) on CETW. In the measurement portion of the model factor loadings ranged from .66-.86. While structural paths ranged from .62-.97. The final model with standardized estimates is shown in Figure 2.

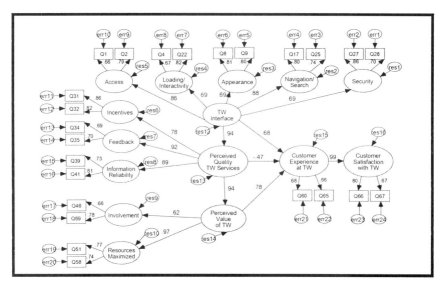

Fig. 2. Final E-SAT for Travel Website Model

9 Alternative Model Solution

One of the drawbacks to SEM is that a variety of models may fit the data. As such, additional paths were hypothesized where TWI, PQTW, and PVTW would have direct positive effects on customer satisfaction. In testing this alternate structural model, goodness-of-fit was achieved with χ^2 = 409.79; df = 239; p = .0001; GFI = .91; AGFI = .88; NFI= .90; TLI = .95; CFI = .96; RMSEA = .045, PGFI = .72. The model at first glance provided a better fit. However, the standardized estimate between CETW and E-SAT with the TW were not reasonable exhibiting a value greater than 1.00, as was the direct effect of TWI on E-SAT indicating that the alternate model was not an improvement over the original model.

10 Validity and Reliability Assessment of the E-SAT Instrument

To be able to claim construct validity, an instrument must exhibit convergent, discriminant, and nomological validity and must prove to be reliable. *Convergent Validity* - To possess convergent validity, in tandem with adequate model fit, a large percentage of the paths in the model must be (.70) and above and the number of first-order factors four or more (Chin, 1998). Using these guidelines the criteria for convergent validity was met as six of the ten paths between the three second-order factors and their respective first-order factors was greater than (.70). *Discriminant Validity* - A measure possesses discriminant validity if it does not correlate with theoretically unrelated constructs. One accepted test of discriminant validity is to determine whether the covariance and two standard errors add to less than 1.00 (Dabholkar, Thorpe & Rentz, 1996). This procedure was used on all possible pairs of the various factors. Values ranging from .17-.73 were found. Thus all factors are statistically distinct and possess discriminant validity. *Nomological Validity* - A measure has nomological validity if it "behaves as expected with respect to some other construct to which it is theoretically related" (Churchill, 1995). All measures were positively and significantly correlated (.38-.99) in the hypothesized direction as such the instrument did possess nomological validity.

Reliability for the final items in the E-SAT survey instrument were as follows: Access (.68), Loading/Interactivity (.71), Appearance (.79), Navigation/Search (.74), Security (.75), Incentives (.83), Feedback (.65), Information Reliability (.73), Involvement (.66), Resource Maximization (.70), CETW (.62), and E-SAT with TW (.78). While four items displayed lower reliability (below .70) the instrument as a whole showed a high level of reliability. Overall reliability for the entire E-SAT for TW instrument was (.93).

11 Discussion and Conclusion

Of the three proposed higher-order constructs PVTW was the most significant followed by TWI. The study results indicate that customer experience at the TW does directly affect customer satisfaction with the TW. This suggests that TW businesses must ensure that visitors to their Website have a positive experience and are satisfied with the site. To create a positive experience at the TW, these findings suggest that businesses must first aim at creating value in the mind of consumers as well as examine their Website interface. PQTW was important to the creation of value, but had a direct negative impact on customer experience contrary to the proposed model but is supported by the concept of valence theory (Reidel, Nebeker, & Cooper, 1988). Indicating that the more the user interfaces with the TW, the more varied the experience and the higher the expectations thereby decreasing the perception of quality. This makes sense intuitively as the user now has a larger framework from which to formulate and judge the quality of the TW.

The proposed E-SAT measurement instrument may be well suited for studying TWs and could serve as a diagnostic tool for travel businesses to determine what aspects of their Websites are weak and in need of improvements. While travel practitioners may wish to test the overall level of customer satisfaction at the TW, academia could view the instrument as the basis for future and improved research in the area of E-SAT with TWs. The E-SAT instrument and model was developed using a sample of college students and as such, may not be generalizable to other populations. Future research may examine the relationship demographics on the E-SAT for TW model. As the online travel industry moves towards a more customer-centric focus of operation, clearly understanding what makes a customer "satisfied" with a TW will be critical. To this end, the E-SAT for TW model offers starting point and through continued testing a viable solution for the analysis of online customer satisfaction.

Appendix 1 - Original E-SAT Questionnaire Items

TW Access-Q1.I can access TWs from a variety of other related Websites;**Q2.** TWs are easily accessed via search engines;**Q3.** I can log on to TWs anytime.
TW Loading-Q4.TWs **upload** quickly on to my computer screen;**Q5.** Information from TWs **download** quickly to my machine; **Q6.** It is easy to download information from TWs.
TW Appearance-Q7.TWs are attractive; **Q8.**TWs are creative in design; **Q9.**TWs are colorful; **Q10.**TWs show good pictures of products; **Q11.**TWs use appropriate interactive aids; **Q12.**TWs are impressive in appearance; **Q13.**TWs have a professional appearance.
TW Navigation-Q14.TWs are uncluttered (not too many links and information);**Q15.** TWs have easy step-by-step instructions;**Q16.** TWs are easy to use;**Q17.** I can easily find what I need on TWs; **Q18.**It is easy to search for information on TWs; **Q19.**TWs provide easy-to-follow paths or links.
TW Interactivity-Q20. There is very little waiting time between my actions and the TWs response;**Q21.** Interacting with TWs is slow and tedious;**Q22.** Pages on the TWs I visit usually load quickly.
TW Search-Q23.TWs search engine give me the results I need; **Q24.** The search engines on TWs are easy to use; **Q25.** Search engines on TWs provide accurate results.
TW Security-Q26.TWs protect my personal security; **Q27.** TWs provide adequate security to protect my personal information; **Q28.**I believe credit card security on TWs is great;**Q29.** I am confident in the security offered on TWs; **Q30.** I feel safe providing my credit card information on TWs.
TW Incentives-Q31.TWs offer more deals than their offline counterparts;**Q32.** TWs provide cheaper products than their offline counterparts; **Q33.** TWs offer realistic deals.
TW Feedback-Q34.TWs provide 7-day/24-hr. worldwide customer service;**Q35.**TWs notify me quickly if there is a problem with my order; **Q36.** TWs allow me to track my purchases.
TW Information Reliability-Q37. TWs provide the information I was looking for; **Q38.**I find all the travel related services I need on TWs; **Q39.**TWs provide accurate information; **Q40.**TWs provide information that is easy to understand; **Q41.**I find it is easy to obtain information from TWs; **Q42.**The services offered on TWs are tough to find elsewhere;**Q43.** On TWs I can always find the information that I need; **Q44.**TWs offer better information their offline counterparts.
TW Involvement-In your opinion, the use and availability of travel Websites are: Q45.Irrelevant/ relevant;**Q46.** Means a lot to me/means nothing to me; **Q69.Rewarding**/disappointing; **47.**Matters to me / doesn't matter; **Q48.** Of no concern / of concern to me;**Q70.** TWs are appealing to me.
Shopping Convenience-Q49.I find it convenient to use TWs;**Q50.**TWs provide a variety of products/services; **Q51.**TWs provide fast delivery for the information I desire.
Transaction Utility-Q52.I can be sure of the quality of products/services on TWs;**Q53.**I can actively interact with services on TWs;**Q54.**I can actively interact with the information on TWs;**Q55.**TWs provide access to familiar travel brands (such as American Airlines etc.)
Price-Q56.TWs offer good prices; **Q57.**TWs offer accurate prices; **Q58.**TWs offer a good variety of prices for the service I am interested in; **Q59.**TWs offer similar prices to their offline counterparts.
Online experience-Q60. My TW experience is often enjoyable; **Q61.**TWs represent a significant percentage of my online experience; **Q62.**TWs are valuable to my overall online experiences;**Q63.** So far, how has your travel Website experience met your expectation?;**Q64.**I often tell my friends about my TW experiences; **Q65.** I recommend useful TWs that I find useful to my friends.
General Satisfaction Measures-Q66. Compared to Websites that offer other travel services how would you rate your satisfaction with TW?;**Q67.**How satisfied are you with using online TW vs. traditional offline travel services?;**Q68.** How satisfied overall are you with travel on the Web?

Reference

Bitner, M., & Hubbert, A. (1994). Encounter satisfaction versus overall satisfaction versus quality:The customer's voice. *In service quality: New directions in theory and practice.* Roland T. Rust and Richard L. Oliver (eds), Thousand Oaks, CA: Sage Publications, pp. 72-94.

Boshoff, C. (1999). RECOVSAT: An instrument to measure satisfaction with transaction-specific service recovery. *Journal of Service Research, 1*(3), 236-249.

Chin, W. (1998). Issues and opinion on structural equation modeling. *Management Information Systems Quarterly, 22*(1). http://www.misq.org/ [May 20, 2002].

Churchill, G. (1995). *Marketing research: Methodological foundations, 6th edition.* Chicago: The Dryden Press.

Dabholkar, P., Thorpe, D., & Rentz, J. (1996). A measure of service quality for retail stores: Scale development and validation. *Journal of the Academy of Marketing Science, 24*(1), 3-16.

Eclipse Magazine. (2001). Destination marketing & the Internet - Part II, pp. 1.

Emory, C. (1980). *Business research methods revised edition.* Homewood, Illinois: Irwin-Dorsey Limited.

Export Today's Global Business. (2000). Hotel online bookings. Vol. 16, Issue 1,pp. 20.

Guadagnoli, E., & Velicer, W. (1988). Relation of sample size to the stability of component patterns. *Psychological Bulletin, 103*(2), 265-275.

Kim, J., & Mueller, C. (1978). *Factor analysis statistical methods and practical issues.* Thousand Oaks, CA: Sage Publications, Quantitative Applications in the Social Sciences Series, No. 14.

Mills, J. E. (2002). *An analysis, instrument development, and structural equation modeling of customer satisfaction with online travel services.* Unpublished doctoral dissertation, Purdue University, West Lafayette, IN.

News Bytes News Network, (2002, April 17). Online travel lifts e-commerce to all-time high - study. Newspaper Source Database. Item No. CX2002107U7225.

Novak, T., Hoffman, D., & Yung, Y. (2000). Measuring the customer experience in online environments: a structural modeling approach. *Marketing Science, 19*(1), 22-43.

Oliver, R. (1997). *Satisfaction a behavioral perspective on the consumer.* Boston, Massachusetts: Irwin MacGraw-Hill.

Perez, J. (2000). Internet agent. *Travel Agent, 301*(8), 36-38.

Reidel, J., Nebeker, D., & Cooper, B. (1988). The influence of monetary incentives on goal choice, goal commitment, and task performance. *Organizational Behavior and Human Decision Processes, 42*, 155-180.

Strauss, A., & Corbin, J. (1998). *Basics of qualitative research: Grounded theory, procedures, and techniques.* Newbury Park, CA: Sage.

Yoo, B., & Donthu, N. (2001). Developing a scale to measure the perceived quality of an Internet shopping site (SITEQUAL). *Quarterly Journal of Electronic Commerce, 2*(1), 31-45.

E-Mail Customer Service by Upscale International Hotels

Richard Leuenberger [a]
Roland Schegg [a]
Jamie Murphy [b]

[a] Lausanne Institute for Hospitality Research / LIHR
Ecole hôtelière de Lausanne (EHL), Switzerland
{research; roland.schegg} @ehl.ch

[b] University of Western Australia / Information Management and
Marketing Department, Australia
jmurphy@ecel.uwa.edu.au

Abstract

This research used a typical e-mail query to explore customer service by 491 properties from 13 international upscale hotel chains. The results show that even luxury hotels have difficulties providing prompt and accurate and e-mail responses to their customers. Relationships between hotel location and responsiveness suggest significant regional differences in e-mail customer service. This study illustrates that better e-mail policies and training would give hotels an immediate competitive advantage via improved eService.

Keywords: e-mails, e-service, hospitality, diffusion of innovation

1   Hospitality in a wired world

Internet technologies are a double-edged sword. Their benefits for hotels can include increased sales, customer satisfaction, lower expenses, etc. In reality, these technologies challenge an industry whose history is built on customer service.

The Internet has created a global marketplace (Kotler et al. 2002). Coupled with de-regulation, this globalism has increased tourism industry competition by giving consumers more information and increased expectations for specialized trips (Bloch and Segev, 1996). Technology has become a strategic weapon and competitive advantage in tourism (Buhalis and Main, 1998), especially as online consumers are big spenders. A study of 80,000 Australian tourists found that wired tourists spent double that of their offline colleagues (Bolin, 2002).

The Internet also offers new customer service channels (Zemke and Connellan, 2001). Grönroos' Net Offer Model (2000) highlights the importance of Internet 'Customer Interfaces', which manage customer-company interactions and service. Websites and e-mail exemplify this new 'Customer Interface' for one-to-one dialogues with current and potential customers (Grönroos, 2000; Zemke and Connellan, 2001).

How have hotels implemented the most basic of these technologies, e-mail? Building on previous hotel e-mail customer service research (Pechlaner et al. 2002; Frey et al. 2002; Gherissi et al. 2002), this study benchmarks e-mail responses by luxury hotels and investigates their relationships with hotel characteristics.

2 The challenge of e-mail customer service

E-mail is the most prevalent Internet technology among global hotels (Wei et al. 2001) and consumers' most popular Internet activity (Ramsey 2001). Just as telephones and toll free numbers pioneered new customer service delivery, e-mail adds a ubiquitous, cheap and virtual customer service channel. As customers shift from phone to e-mail communication, companies shift towards customer service via e-mail (Strauss and Hill, 2001; Zemke and Connellan, 2001).

Customers increasingly gather information from the Internet, forcing organizations to negotiate the power shift towards customers (Strauss and Frost, 2001; Kotler et al. 2002). Correspondingly, customer satisfaction has become a measure of organizational performance along with traditional metrics such as net profit and return on assets (Lovelock et al. 2001). Hotels gain a powerful customer relationship tool (Zemke and Connelan, 2001; Marinova et al. 2002) but must change in order to manage e-mail's cost, time and interactivity advantages.

Each e-mail requires unique handling; customers expect a personal response (Yang, 2001). Furthermore, 55 percent of customers expect accurate responses within six hours (Alford, 2000), yet only 20 percent of companies met this standard (Bankier, 2000). Self-service technologies automate customer employee interaction in order to accelerate response times (Meuter et al. 2000), but people like to deal with people not machines (Zemke and Connellan, 2001).

3 Conceptual development and hypotheses

Previous research has shown poor e-mail customer service by hotels (Frey et al. 2002; Gherissi et al. 2002; Pechlaner et al. 2002). Diffusion of innovations helps explain this poor service. Organizations adopt innovations in stages (Rogers, 1995) and e-mail use is probably shifting from the *initiation* stage of having a website and email address to the *implementation* stage of answering e-mail messages properly.

Researchers call for more investigation into organizational characteristics related to adoption (Fichman, 2000; Rogers, 1995). Studies of hotel characteristics have generally shown that larger, higher-rated and affiliated hotels adopted Internet technologies faster (Frey et al. 2002; Gherissi et al. 2002; Pechlaner et al. 2002; Siguaw et al. 2000).

Hypothesis 1: Based on the number of rooms, large hotels have better responsiveness (H1a) and response quality (H1b) to e-mail queries than small hotels.

Connolly (2000) however, noted that large hotel companies cited fragmented ownership as a major obstacle to IT adoption. Franchising and management contracts limit ownership and control of IT assets at the property-level, making it difficult to implement chain-wide programs and maintain uniform service quality. In an industry plagued by high turnover, staff skills also inhibit IT adoption (Connolly, 2000). Large international chains with complex ownership structure and international workforces should have greater problems implementing Internet technologies.

Hypothesis 2: Based on the hotel chain size, small chains have better responsiveness (H2a) and response quality (H2b) to e-mail queries than large chains.

Factors that influence country Internet diffusion include infrastructure, government policies, economic development, culture, language and IT penetration (Bayarmaa and Boalch, 1997). Adding culture to these variables though, may provide a more robust understanding (Maitland, 1998). Past studies have shown diffusion differences based on socio-cultural factors (Helsen et al. 1993) such as individual beliefs, value systems and attitudes to information sharing (Bayarmaa and Boalch, 1997).

Hypothesis 3: Responsiveness (H3a) and response quality (H3b) to e-mail queries will differ based on the geographical location of the hotel.

4   Methodology

Hotels and restaurants use mystery shoppers to analyze their service. A similar methodology avoids a common limitation of marketing (Lee et al. 2000) and innovation (Rogers, 1995) research, relying upon reported behavior rather than actual behavior. This mystery shopper methodology uses replies to customer e-mails as a measure of Internet technology implementation.

4.1   Data collection and experimental design

The population was hotels from the 300 largest hotel chains as listed by www.hotelsmag.com. Today's hotel chains, ranging from financially dependent to managed companies, complicate defining a luxury chain. Judgmental sampling selected 13 international chains (see Table 3) representing 491 luxury properties with an individual e-mail address. Chains with centralized e-mail systems such as Marriott, Hilton or Four Seasons were not included in this study of e-service by individual properties.

A test e-mail to 35 Austrian five star hotels not in the sample refined the final e-mail asking about room availability for a honeymoon weekend, special events for that date and nearby hospitals or medical facilities. E-mails were sent individually to override filtering programs that protect hotels from information overload and spamming (Pechlaner et al. 2002). Sending the e-mails according to properties' local time, 8 April 2002 at 8 a.m., ensured similar timing. To further reduce bias, four different sender names and e-mail addresses changed along with time zones to prevent hotels filtering e-mails with identical sender information.

4.2 Responsiveness and quality elements

Pre-testing and previous research (Gherissi et al. 2002; Frey et al. 2002; Pechlaner et al. 2002; Yang, 2001; Zemke and Connellan, 2001) led to six responsiveness and five quality variables. Six binomial elements, along with their sum as an ordinal variable, measured responsiveness: *responded*, did the hotels reply; *day response*, reply in less than 24 hours; answer the *room availability* question and answer other questions (*honeymoon*, *special events*, *medical facilities*), were follow-up e-mails necessary?

Five binomial elements, along with their sum as an ordinal variable, measured eService quality: Potential guests should be addressed politely with dear and personally, "Dear Mr. Smith" (*salutation*). The e-mail should give the hotel's identity and electronic receptionist's identity (*contact information*). As each e-mail query gives hotels a marketing opportunity, does the hotel provide information on the property (*marketing of hotel*)? For the guest's convenience, some hotels block rooms until a release date (*provisional booking*). Some hotels update customers on their reservation or provide more information about the property or location (*follow-up*).

5 Results

5.1 Responsiveness and quality

About seven out of ten hotels (71%) responded within a week's time, 30 properties (6%) never received the e-mail due to technical errors (wrong e-mail address, server time-out, etc.) and 114 properties (23%) failed to answer the query (or did not receive it). The 75.3% response rate is similar to 74% by 1-5 star Swiss hotels (Frey et al. 2002), slightly higher than the 71% of 3-4 star Austrian and Italian hotels (Pechlaner et al. 2002) and higher than 45% by Tunisian 1-5 star hotels (Gherissi et al, 2002).

The 347 respondents answered in average of 37.3 hours after receiving the e-mail room query. More than half the sample responded within a day and 259 properties (76%) responded by the second day. The last reply trickled in after 12 days. Compared to eight out of ten Swiss (Frey et al. 2002) and five out of ten Austrian/Italian hotels (Pechlaner et al. 2002) replying in one working day, international upscale hotels are not setting industry benchmarks for prompt replies.

Important for both customer and hotelier is information about the core service, the room. Of the 347 responding hotels, 328 (94%) gave the potential customer room availability information and prices for the dates requested. Almost three out of four hotels gave information on medical facilities. The willingness to pay for a honeymoon weekend is probably higher than for an ordinary weekend, yet just three in ten hotels took advantage of this sales opportunity by providing information on special events.

Most hotels acknowledged the customer but failed to acknowledge themselves. Almost nine out of ten hotels addressed the customer personally, but just over half included their contact information (name, phone number) and about one third marketed their property by adding a short description of the property) in the e-mail. Contact information is important as customers often gather information via e-mail but book through traditional media (telephone, letter or fax). Less than one in twenty hotels provisionally blocked rooms or sent a follow-up e-mail.

5.2 Hypotheses testing

Pearson's chi-square for nominal, Kruskal-Wallis (K-W) for ordinal and one-way ANOVA tests for scale-level data evaluated significant ($p<0.05$) differences based upon hotel size and location.

No significant differences in response rate, response time, and answering the questions across hotel size failed to support hypotheses 1a (see Table 1). As expected (Hyp 2a) hotels in large chains responded significantly less within 24 hours (50%) than hotels in smaller chains (65%) and answered questions related to special events (25%) significantly less than hotels in smaller chains (40%).

Table 1. Hypothesis tests for responsiveness variables

	N	Responded (N=461)	Response time (h)	Response within 24h	Room request	Honey-moon	Medic. facil.	Special event	Global resp. factor
Property Size (H1a)									
0-150 rooms	97	73.3%	36.2	51.0%	95.0%	28.0%	70.0%	30.0%	3.73
150-300 rooms	105	75.0%	34.2	60.0%	90.0%	28.0%	70.0%	24.0%	3.71
>300 rooms	146	76.8%	32.9	58.0%	97.0%	36.0%	73.0%	34.0%	3.97
CHI2(K-W) F		0.536	0.225	1.988	4.269	2.461	0.510	3.166	5.382
sig.		0.765	0.799	0.370	0.118	0.292	0.775	0.205	0.068
Size of Hotel Chain (H2a)									
1-30'000 rooms	91	77.8%	30.6	65.0%	91.0%	34.0%	70.0%	40.0%	4.00
30'000-100'000 rooms	62	72.9%	30.3	65.0%	95.0%	29.0%	74.0%	31.0%	3.94
>100'000 rooms	195	74.9%	37.1	50.0%	95.0%	30.0%	71.0%	25.0%	3.71
CHI2(K-W) F		0.662	1.370	7.781	2.106	0.560	0.321	6.171	4.646
sig.		0.718	0.256	**0.020**	0.349	0.756	0.852	**0.046**	0.098
Location (H3a)									
North America	78	72.9%	32.6	64.0%	87.0%	44.0%	72.0%	44.0%	4.10
South America	36	54.5%	45.7	36.0%	92.0%	39.0%	83.0%	19.0%	3.69
Europe	163	79.8%	27.5	63.0%	97.0%	21.0%	68.0%	28.0%	3.77
Asia/Pacific	71	83.5%	45.6	42.0%	97.0%	37.0%	72.0%	24.0%	3.72
CHI2 F		20.909	5.381	16.735	10.935	15.703	3.380	10.276	2.376
sig.		**0.000**	**0.001**	**0.001**	**0.012**	**0.001**	0.337	**0.016**	0.070
population (average)	348	**75.3%**	**34.2**	**56.0%**	**94.0%**	**31.0%**	**71.0%**	**30.0%**	**3.83**

With the exception of medical facilities and global responsiveness, all other elements showed significant location differences (Hyp 3a). Response rates in North America (71%), Europe (77%) and Pacific/Asia (76%) were significantly higher than in South America (48%). The average response time in Europe (27.5h) and North America (32.6h) was significantly faster than in South America (45.7h) and Asia Pacific (45.6h). Nearly one out of two North American hotels (44%) provided information on special events compared to about one out of four hotels in other continents.

Most reply quality elements showed no significant differences across the independent variables. Significant differences on contact information partially supported hypotheses 2b-3b (see Table 2). Hotels in large chains provided contact information less frequently (50%) than hotels in smaller chains (68%) and South America underperformed compared to the other continents.

Table 2. Hypothesis tests for response quality variables

	N	salutation formula	marketing of hotel	provisional booking	follow up	contact information	global quality factor
Property Size (H1b)							
0-150 rooms	97	87.0%	43%	7.0%	1.0%	54.0%	1.92
150-300 rooms	105	83.0%	33%	3.0%	1.0%	56.0%	1.76
>300 rooms	146	88.0%	37%	3.0%	4.0%	56.0%	1.88
CHI2 (K-W)		1.566	2.173	3.544	3.663	0.186	1.465
sig.		0.457	0.337	0.170	0.160	0.911	0.481
Size of Hotel Chain (H2b)							
1-30'000 rooms	91	89.0%	41%	4.0%	4.0%	68.0%	2.07
30'000-100'000	62	87.0%	37%	5.0%	0.0%	55.0%	1.84
>100'000 rooms	195	85.0%	36%	4.0%	2.0%	50.0%	1.76
CHI2 (K-W)		1.055	0.485	0.234	3.284	8.480	5.599
sig.		0.590	0.784	0.890	0.194	**0.014**	0.061
Location (H3b)							
North America	78	83.0%	35%	0.0%	5.0%	71.0%	1.94
South America	36	81.0%	28%	0.0%	3.0%	33.0%	1.44
Europe	163	88.0%	40%	7.0%	1.0%	53.0%	1.89
Asia/Pacific	71	89.0%	41%	3.0%	1.0%	56.0%	1.90
CHI2		2.207	2.454	9.752	3.901	14.793	2.422
sig.		0.530	0.484	**0.021***	0.272	**0.002**	0.066
population (average)	348	70.6%	34.2	56.0%	94.0%	31.0%	71.0%

* 3 cells had expected count of less than 5.

6　Discussion

This e-service research of response and quality variables stems from a judgmental sample from the 300 largest hotel chains listed by www.hotelsmag.com and therefore fails to generalize to all luxury or international hotels. Furthermore, although the authors double-checked the coding of e-mail responses, there was no reliability testing as Krippendorf (1980) suggests for content analysis. Regardless, the results reflect e-mail customer service by almost four hundred luxury hotels in five continents.

That these results resemble similar studies of lower category and independent hotels (Gherissi et al. 2002; Frey et al. 2002; Pechlaner et al. 2002) suggests an equality of opportunity. E-mail, with few resource and knowledge restrictions, meets Roger's (1995) criteria for rapid adoption (e.g. relative advantage, compatibility, complexity, triability and observability) and is perhaps too easy to adopt.

Hotels showed a similar level of quality, independent of their location. This could suggest two stages of organizational adoption, *initiation* followed by *implementation* (Wolfe 1994; Rogers 1995). South American hotels have yet to pass the *initiation* stage, demonstrated by their significantly lower response rate. Similar results in reply quality though, could suggest that neither North nor South American hotels have moved to the *implementation* stage of e-mail customer service. Depending on their location, hotels handled the e-mail request differently. Norris (2001) found that company resources and country development determine technology access and diffusion; a North-South divide holds back poorer countries. This research partly confirms a digital divide, with South America hotels on the poorer side.

7   Managerial implications

This study shows that e-mail use, even by upscale hotels, has vast potential for improvement. Only two out of three customers received a reply that answered their initial questions. Business use of email is an implicit promise to customers; *"if firms provide email addresses, they must answer incoming mail"* (Strauss and Frost, 2001: 309-10). It would be goofy to list numbers in the phone book and fail to answer the phone. Yet hotels list e-mail addresses on their website and fail to answer e-mails.

E-mail is as important as a phone call, fax or letter. Hotels should establish, and train staff on, e-Service policies (Strauss and Hill, 2001; Zemke and Connellan, 2001). Based on customers' frequently asked e-mail questions, hotels should develop a Frequently Asked Questions section on their web sites and craft template e-mail answers. These templates should use basic business communication procedures such as polite greetings, thanking the recipient addressing them by name, answering the questions, and identifying the hotel -- name, postal address, phone and fax numbers and web site address -- as well as sender.

Better e-mail policies and training would give hotels an immediate competitive advantage via improved e-Service. Depending upon hotel characteristics, some hotels provided markedly better e-mail responses than their competitors. As e-mail programs are relatively simple and inexpensive, hotels may realize a better Internet investment return by focusing on this basic technology in lieu of additional website features.

References

Alford, P. (2000). E-business in the travel industry. *Travel and Tourism Intelligence*. London, Corporate IntelligenceGroup.

Bankier, J.B. (2000). E-Mail Customer Service. *Jupiter Vision Report* 01/08.

Bayarmaa, B. & G. Boalch (1997). A Preliminary Model of Internet Diffusion within Developing Countries. *AusWeb97 - Third Australian World Wide Web Conference.* Lismore, Australia, Southern Cross University.

Bloch, M. & A. Segev (1996). *The impact of electronic commerce on the travel industry.* University of California, more information on http://groups.haas.berkeley.edu/citm [21 August 2002].

Bolin, R. (2002). Domestic tourism Internet usage: Pinning down the E-tourist. *12th International Research Conference of the Council for Australian University Tourism and Hospitality Education «Tourism and Hospitality on the Edge».* Fremantle, Australia.

Buhalis, D. & H. Main (1998). Information technology in peripheral small and medium hospitality enterprises: strategic analysis and critical factors. *International Journal of Contemporary Hospitality Management* 10(5): 198-202.

Connolly, D. J. (2000). Strategic Investment in Hotel Global Distribution Systems. *Trends 2000 - Proceedings of the 5th Outdoor Recreation & Tourism Trend Symposium «Shaping the Future».* Lansing, Michigan

Fichman, R. (2000). The Diffusion and Assimilation of Information Technology Innovations. Framing the Domains of IT Management. In R. Zmud (ed.), *Projecting the Future Through the Past* (pp. 105-128). Cincinnati: Pinnaflex Publishing.

Frey, S., Schegg, R. & J. Murphy (2002). Guten Tag? Bonjour? Buon Giorno? Bun di? Electronic Customer Service in the Swiss Hotel Industry. *12th International Research Conference of CAUTHE «Tourism and Hospitality on the Edge».* Fremantle, Australia.

Gherissi, T. L., Schegg, R. & J. Murphy (2002). The State of Electronic Customer Service in the Tunisian Hotel Industry. *Proceeding of the 7th Association Information Management Conference.* Hammamet, Tunisia.

Grönroos, C. (2000). The NetOffer model: a case example from the virtual market space. *Management Decision* 38(4): 243-252.

Helsen, K., Kamel, J. & S. Wayne (1993). A new approach to country segmentation utilising multinational diffusion patterns. *Journal of Marketing* 57(October): 60-71.

Kotler, P., Jain, D. C. & S. Maesincee (2002). *Marketing Moves.* Boston: Harvard Business School Press.

Krippendorff, K. (1980). *Content analysis: An Introduction to Its Methodology.* Beverly Hills, CA: Sage Publications.

Lovelock, C.H., Patterson, P.G. & R.H. Walker (2001). *Services Marketing: An Asia-Pacific Perspective.* Sydney: Pearson Education Australia.

Lee, E., Hu, S.M. & R. Toh (2000). Are consumer survey results distorted? Systematic impact of behavioral frequency and duration on survey response errors. *Journal of Marketing Research* 37(February): 125-133.

Maitland, C. (1998). Global Diffusion of Interactive Networks: The Impact of Culture. In C. Ess & F. Sudweeks (eds), *Proceedings Cultural Attitudes Towards Communication and Technology 1998* (pp. 268-286). University of Sydney, Australia.

Marinova, A., Murphy, J. & B.L. Massey (2002). Permission e-mail marketing as a means of targeted promotion. *Cornell Hotel and Restaurant Administrative Quarterly* 43(1): 1-9.

Meuter, M.L., Ostrom, A.L., Roundtree, R.I. & M.J. Bitner (2000). Self-Service Technologies: Understanding Customer Satisfaction with Technology-Based Service Encounters. *Journal of Marketing* 64(3): 50-64.

Norris, P. (2001). *Digital Divide: Civic Engagement, Information poverty and the Internet in democratic societies.* New York: Cambridge University Press.

Pechlaner, H., Rienzner, H., Matzler, K. & L. Osti (2002). Response Attitudes and behavior of Hotel Industry to Electronic Info Requests. In K.W. Wöber, A.J. Frew, & M. Hitz (Eds.), *Information and Communication Technologies in Tourism 2002* (pp. 177-186). Wien: Springer.

Ramsey, G. (2001). *The e-mail Marketing Report, Executive Summary.* New York: eMarketer Inc.

Rogers, E. (1995). *Diffusion of Innovations.* New York: The Free Press.

Siguaw, J.A., Enz, C.A. & K. Namiasivayam (2000). Adoption of Information Technology in U.S. hotels: Strategically driven objectives. *Journal of Travel Research* 39(November): 192-201.

Strauss, J & R. Frost (2001). *E-Marketing.* New Jersey: Prentice-Hall.

Strauss, J. & D.J. Hill (2001). Consumer complaints by Email: An exploratory investigation of corporate responses and customer reactions. *Journal of Interactive Marketing* 15(1): 63-73.

Wei, S., Ruys, H.F., vanHoof, H.B., & T.E. Combrink (2001). Uses of the Internet in the global hotel industry. *Journal of Business Research* 54(3): 235-241.

Wolfe, R.A. (1994). Organizational Innovation: Review, Critique and Suggested Research Directions. *Journal of Management Studies* 31(3): 405-431.

Yang, Z. (2001). Consumers' perceptions of service quality in Internet commerce: Strategic Implications. *Proceedings of the American Marketing Association Summer Educator's Conference.* Washington D.C., 76-77.

Zemke, R. & T. Connellan (2001). *e-service.* New York: Amacom.

Electronic Travel Markets: Elusive Effects on Consumer Behavior

Anssi Öörni[a] and
Stefan Klein[b]

[a] Department of Information Systems
Helsinki School of Economics, Finland
oorni@hkkk.fi

[b] Department of Information Systems
University of Muenster, Germany
stefan.klein@uni-muenster.de

Abstract

This paper examines the effect of Internet-based electronic markets on consumer search in the travel and tourism industry. Two experiments provide the empirical basis for the paper. Which addresses the question of whether consumer search in electronic markets is different form search in conventional markets. In this work we will refine the efficiency of consumer search and its effects with the following specific questions: 'Is consumer search in electronic markets more productive?' and 'Is consumer search in electronic markets more efficient?' Based on our analyses, we present propositions about major obstacles that have prevented current travel markets from reaching the high efficiency related to the hypothesized electronic markets. Our paper ends with a critical outlook of the market development. Are there ways to overcome some of the observed obstacles and are there indications that the industry has understood the problems and is moving?

Keywords: Electronic travel markets, consumer search, search costs.

1   Introduction

The conventional wisdom concerning Internet competition is that the information-related inefficiencies largely vanish in electronic consumer markets (Brynjolfsson and Smith 2000). In the extreme version of this view, the Internet is expected to render retailer location irrelevant (Rayport and Sviokla, 1994), and consumers may become fully informed about product characteristics and prices (Bakos, 1997).

Even as interorganizational information systems facilitate data exchange between interested parties, limited evidence exists to support the hypothesized increase in market efficiency. Largely, the expected benefits rest on reports of the domestic air travel markets in the United States in the 1980s (Copeland and McKenney, 1988; Hopper, 1990; Bakos, 1991). Higher levels of price competition were reportedly observed and these observations were attributed to the development of computerized reservation systems (Bakos, 1991). However, it has also been noted that many other factors have probably affected the markets simultaneously; e.g. deregulation of the US air travel markets took effect in the same time frame (Copeland and McKenney, 1988). Furthermore, a number of studies have revealed increasing market concentration and rising airfares since deregulation (e.g. Bailey and Williams 1988).

Our work follows several other studies concerned with the effects of electronic markets on consumer search for differentiated products. Degratu et al. (2000) examined online grocery markets and found mixed support for the hypothesis of price sensitivity being lower in online than in conventional channels. Bailey (1998) analyzed prices for books, compact discs, and software in the Internet and conventional outlets and found evidence that prices on the Internet were, on average, higher. Brynjolfsson and Smith (2000) demonstrated that substantial price dispersion existed in electronic markets for books and compact discs while the average prices were somewhat lower. They have also provided evidence suggesting that even the price-sensitive book buyers have preferences with regard to online sellers (2001). Clay et al. (2001) respectively found that many on-line book buyers may not be engaging in search despite its potential benefits. Lee (1998) found that prices of used cars sold via electronic auction (AucNet) tended to be higher than comparable prices in conventional auction markets. Clemons et al. (2002) found substantial price dispersion for domestic airline tickets offered by online travel agents in the US. Anckar and Walden (2002) observed that on-line reservation of travel services can be a complex task, often exceeding the capabilities of even an educated customer.

Previous research has employed measures such as average prices, price dispersion, and price sensitivity to study the efficiency of information search in electronic consumer markets. While these studies provide valuable information on the efficiency of the Internet as a source of product information, they also have limitations that should be addressed to increase our confidence in the results. Market prices may reflect events other than seller reactions to consumer search.

We will discuss how consumer behavior in electronic markets deviates from what is expected in electronic markets literature. Seller actions are also scrutinized to infer what changes should be made to the current markets to better market efficiency.

2   Theoretical background

Changing identity of sellers and buyers in markets and also fluctuations in supply and demand result in uncertainty, since information becomes obsolete (Stigler, 1961). Both sellers and buyers must therefore update their information, and there is often no better means to do that than search. Search is not, however, without costs. Search costs prevent consumers from acquiring perfect information, which is reflected in multiple prices in most consumer markets.

Stigler (1961) proposed, that high search costs will lead value maximizing consumers to limit their pre-purchase search, which results in less than perfectly informed purchase decisions. Since consumers vary on their market knowledge and search costs, relatively wide price dispersion persists in many consumer markets. While a consensus prevails over the key measures of search costs – 1) the amount of search and 2) price dispersion for products of comparable quality – multiple diverging views have been adopted over the possible welfare effects of search costs and over the most likely markets to benefit from decrease in search costs. Two basic types of welfare effects may result from declining search costs (Bakos, 1997): 1) Consumers will be able to more accurately find products meeting with their tastes and needs. 2) The equilibrium price may decline from the monopolistic level towards the level of marginal production costs. Realization of these effects may, however, depend crucially on the level and distribution of search costs. If search costs were zero, all consumers would be able to acquire full market knowledge and make the perfect choice. The result would be, in effect, a market of pure competition given that conditions related to number and characteristics of market parties were satisfied. However, as Stigler (1961) pointed out, search costs are unlikely to ever vanish totally from consumer markets. And the effects of even arbitrarily low, yet positive, search costs can be quite dramatic as suggested by e.g. Diamond (1971).

To capture the idea of consumers searching for products they like, it is necessary to introduce heterogeneity across products. Anderson and Renault (1999) propose that some product heterogeneity is even desirable to motivate consumers to search more extensively than they would search, if finding better prices were the only incentive. Initially, an increase in heterogeneity motivates consumers to extend their search. However, once a taste for variety is high enough so that a sufficient number of consumers search, the situation is close to the case of perfectly informed consumers, and thereafter the equilibrium price rises with taste for variety since markets become increasingly segmented.

To summarize, both the amount of consumer search and the dispersion of prices, when quality differences are accounted for, are widely accepted indications of search costs. Thus, these are the central measures by which we will attempt to determine the efficiency of electronic consumers markets relative to conventional markets. The relation of search costs and welfare effects is less clear, since search costs are likely to remain positive. However, we determine that low price levels are more likely associated with low than high search costs, and thus can be used as an additional, if not very reliable, measure to assess the effects of electronic markets on pre-purchase consumer search. Markets for differentiated products with a degree of heterogeneity in consumer tastes are likely candidates to benefit from electronic commerce.

3 Methodology and Results

Consumer behavior is known to have both its rational and habitual side. It has been demonstrated that consumers may not commit themselves to pre-purchase search due to lacking attention (Simon, 1986) or motivation (Schmidt and Spreng, 1996). Hence, the decision of whether to search is not necessarily affected by the expected costs and benefits of search alone. Furthermore, consumers may lack the cognitive or informational resources needed to formulate the problem at hand. Since electronic markets are expected to facilitate pre-purchase search with prior preferences (see e.g. Bakos, 1997), prior preferences and motivation should be controlled in tests.

We arranged two experiments to compare the costs and benefits of information search and product comparison. Both experiments were optional assignments of a university level course on electronic commerce. The first experiment, carried out in September 1999, consisted of two tasks: the subjects were asked to 1) arrange a conference journey to Hawaii in March 2000 and 2) to design a winter vacation for the season 1999-2000. The participants in the study had one week to complete these tasks, and they were free to schedule their work. The subjects were also allowed to use all the information sources they wished. The subjects were instructed that they had at their disposal a grant of FIM 10,000 (approximately € 1,700), which they were allowed to exceed and "pay" the excess costs themselves. The second task of the assignment was designed for motivating the subjects. They were instructed to design a winter vacation with a total budget of FIM 4,000 (approximately € 700). The vacation was to be arranged for the winter season 1999-2000. Every subject completing the assignment was awarded 10/100 points for the course grade. These points were not tied to the performance of the subjects and the assignment was optional. To further encourage the subjects, we announced that roughly half of them would be selected into a lottery based on their performance in the assignment. The prize of the lottery was one winter vacation with a budget of FIM 4,000, i.e. the winner was awarded the vacation designed in the second task of the assignment.

The second experiment, carried out in September 2000, comprised two tasks in which the subjects were asked to arrange a conference voyage to 1) Hawaii and to 2) Brisbane. They had one week to complete the tasks. The subjects were randomly assigned to two groups. The first group used the information sources of electronic travel markets for the first task while the second group acted as the control group. For the second task, the roles of the groups were switched.

We instructed our subjects to arrange 1) flights to and from the conference destination, 2) accommodation for the duration of the conference, and 3) local transportation. The journeys were to be designed for one person and had to cover at least the conference dates. The subjects were further instructed that they should try to minimize the travel costs while simultaneously satisfying some goals (presence in the conference destination during the conference schedule, accommodation in single rooms). Every subject completing the assignment was awarded 10/100 points for the course grade. To further motivate the subjects, they were awarded additional points for every FIM 1,000 (€ 168) they could save starting from the total approximate price of FIM 15,000 for Hawaii and FIM 17,000 for Brisbane.

The subjects of the experiments were undergraduate or graduate business students. The assignments were optional. We screened the data for completeness and for outliers. After that, we had 122 observations from the first experiment and 92 from the second.

3.1   Productivity of Search

In this work, productivity of search is operationalized as the number of alternatives worth consideration found per unit of time. If electronic markets are more efficient environments for consumer search, consumers should be able to locate prospective sellers and products more easily than in conventional markets. This topic has been addressed by setting up a field experiment and examining the search behavior of correspondent subjects. The subjects were asked to report the number of alternatives they considered and the time they used for various phases of the search process. The productivity measure was constructed by dividing the number of alternatives found by time used for search of products belonging to the given product class. We formulate the hypothesis, consumers using electronic markets find more alternatives per time unit than consumers using conventional markets, as follows:

$$H_0 : \mu_E = \mu_C$$

The α-risk is controlled at 0.05 when $\mu_E = \mu_C$.

$$H_1 : \mu_E \neq \mu_C$$

Table 1. Productivity of search (number of alternatives considered / hours searched).

Travel service	Market	n	Mean	Std. Dev.	t-value	df	p-value (two-sided)
Flights to Hawaii	Electronic	53	3.982	4.080	1.088	82	0.280
	Conventional	31	3.117	2.225			
Flights to Brisbane	Electronic	30	13.102	54.235	-0.989	29 [a]	0.331
	Conventional	54	3.301	2.681			
Accommodation in Hawaii	Electronic	53	9.929	12.256	2.130	77 [a]	0.036
	Conventional	29	5.775	5.297			
Accommodation in Brisbane	Electronic	31	7.769	8.502	-1.350	41 [a]	0.184
	Conventional	53	5.522	4.819			

a) *Degrees of freedom are decreased since homogeneity of population variances is not assumed because of the results of Levene's test of homogeneity-of-variance ($\alpha = 0.05$).*

Search in the electronic markets produced only slightly more options per time unit (see Table 1). The differences are very small; the only statistically significant difference observed was obtained for the search for accommodation in Hawaii that was more efficient in electronic markets. We conclude that hypothesis H1 found only limited support in our data.

The differences observed in productivity between electronic and conventional channels were on the whole not statistically significant. Hence, we conclude that search in electronic markets does not yet seem to be more productive than it is in conventional markets. The productivity measure is a linear approximation of the benefits of search. The marginal benefits of search are often found to be sharply decreasing. Thus, the productivity measure is likely to be biased in favor of the electronic markets, unless the prices are more dispersed there than in the conventional markets.

3.2   Efficiency of Search

The relative efficiency of information dissemination in electronic and conventional markets should be reflected in the quality of procurement decisions. In particular, the prices in the electronic market are expected to decline and become less dispersed, as consumers are better able to compare alternatives available to them. Efficiency of search is operationalized by the price dispersion our subjects reported while controlling for the quality of the products. Price dispersion is directly related to search costs, although, some dispersion typically results from differences in product quality and from varying cost structures of sellers. We formulate the hypothesis, electronic markets are more efficient than conventional markets in terms of price dispersion, as follows:

$$H_0 : \sigma^2{}_E = \sigma^2{}_C$$
$$H_1 : \sigma^2{}_E \neq \sigma^2{}_C$$

The α-risk is controlled at 0.05 when $\sigma^2{}_E = \sigma^2{}_C$.

We conducted Levene's test of homogeneity of variances to test whether the sample variances deviated substantially. To control for the duration of staying at the destination, we used the accommodation costs per night in the test. These figures are shown in table 2. The airline flights are considerably differentiated. Their prices vary as a function of flight class, time spent at the destination, seller, and time of buying the ticket. To ensure that variation in price dispersion and price level do not reflect quality differences in the services offered through electronic and conventional markets, we screened the flights to control the quality of air travel. All reported flights were in economy class, and there was little deviation in most well known quality factors such as the number of legs in flight. Time spent at the destination is one of the most important factors for flight prices and the reported flights diverged on this factor. However, length of stay is not a cost factor to the airlines. Rather, it reflects airliners' ability to price discriminate against different customer segments. Further, we instructed our subjects to search for economical prices and set a fairly strict travel budget, for we wanted them to minimize the travel costs. Hence we did not include control for length of stay in our analysis.

The mean prices and standard deviations of prices, shown in table 2, suggest that electronic air travel markets have not yet gained in efficiency compared to conventional markets. The results concerning the markets for accommodation leave more space for speculation. However, in summary, tests for price dispersion suggest that no statistically significant differences were found in the efficiency of electronic and conventional markets for flights and accommodation services.

Table 2. Price dispersion in electronic and conventional markets.

Travel service	Market	n	Mean (Price)	Std. Dev. (Price)	F-value	p-value (two-sided)
Flights to Hawaii	Electronic	55	9142	2230	1.140	0.289
	Conventional	32	8615	2180		
Flights to Brisbane	Electronic	32	12779	4980	2.015	0.160
	Conventional	53	9925	4787		
Accommodation in Hawaii	Electronic	55	653	385	0.368	0.546
	Conventional	28	763	313		
Accommodation in Brisbane	Electronic	22	222	74	0.799	0.376
	Conventional	33	273	87		

3.3 Barriers of Search

We have attempted to identify the sources of uncertainty that hindered search in electronic markets during the experiment. These problems were gathered from the diaries that subjects had to keep during the experiment. Of the total of 122 subjects, 76 reported difficulties related to locating and evaluating flights. Availability of information was the most frequent source for complaints; 48 (39%) subjects reported that they experienced difficulties finding information relevant to choosing a flight. Of these subjects, 32 (26%) reported that they were unable to find flight schedules, prices, details, and availability of seats all in any one of the electronic sources they used. Ten subjects (8%) were more specific, reporting that comprehensive flight information was available through search engines (such as Travelocity.com), yet, only for expensive business flights. Further 6 (5%) subjects complained that flight information was often disclosed only after registration or reservation.

A few subjects (14, 11%) experienced difficulty locating prospective sellers. Most of them reported that they could locate a number of sellers, but faced difficulties in evaluating them. Consumers with little prior knowledge about the market can not a priori tell apart sellers providing ample product information and those that are not able to meet their information needs. Similarly it is difficult if not impossible to know in advance which sellers have economical offers.

Some subjects reported frustration emanating from technical problems (8, 7%) and lacking search engine or interface design (6, 5%). Technical problems dealt mostly with unreachable servers while design issues comprised low usability of electronic storefronts and lacking search engine implementation causing difficulty in constructing multi-leg flights.

In summary, the majority of reported problems related to information content rather than technical issues. Lack of cohesive flight information can be traced to at least the high cost of systems integration. Law and Leung (2000), while investigating airlines' online reservation services on the Internet, reported that a higher number of airlines provided on-line flight schedule information than flight availability information. They also observed some regional differences in provision of flight availability information, which they attributed to the high cost of integrating online Web services to airlines' central reservation systems

4   Conclusions

Stigler (1961) attributed the need to search largely to a desire to update information that has become obsolete. The present electronic markets provide few means to substantially better the conventional markets in this regard. It seems that many authors have missed the nature of the Internet or networking technology in general while contemplating the effects of electronic data exchange on consumer search. The existing infrastructure for electronic markets does little to address those shortcomings, which Stigler (1961) cited as sources of market inefficiency. A high number of sellers have developed their own retail outlets in the World Wide Web. From the technological point of view, it is evident that retrieval of product information is likely to be fast once relevant information is located. However, little proof exists to support beliefs regarding easy locating of sellers, products or product attributes. The recent success of online travel supermarkets, such as Expedia, Travelocity, Orbitz or Opodo, suggests that one-stop-shopping and significantly improved usability and interaction design on the Web is appreciated by consumers (Klein 2002). However, differential pricing, price volatility and price dispersion, which result primarily from the airline's yield management strategies, combined with a lack of insight into the market structure, still pose a major obstacle for consumers to search for flights efficiently without expert assistance.

To summarize the findings presented in this work, our conclusion is that there is still little evidence of electronic markets leading invariably to lower search costs, more extensive pre-purchase search, and increased price competition. Even if electronic markets undoubtedly enhance information exchange between the market parties, they are less efficient in alleviating the market imperfections addressed by Stigler (1961). Identifying prospective sellers with suitable offerings in the Internet seem to be less efficient than previously expected. It is not uncommon to observe multiple retailers located near to each other and still prospering, while the information search costs are effectively zero (see e.g. Slade 1986). Furthermore, consumers are exposed to market information while they move about in their natural habitat, even if they are not searching for it. Certainly, there exist a number of geographical areas where markets for many goods are marginal and local markets have not emerged. Consumers having to choose their information means from international calls, faxes, conventional mail, and the Internet services are likely to benefit from the introduction of electronic markets. However, these markets have to be relatively marginal, otherwise local markets would have emerged.

In effect our research suggests that the impact of ICT on search behavior is moderated by numerous domain specific effects, such as product and market structure, which are not yet fully understood.

References

Anckar, B., & Walden, P. (2002). Self-Booking of High- and Low-Complexity Travel Products: Exploratory Findings. *Information Technology & Tourism,* 4 (3/4), 151-165.

Anderson, S. P., & Renault, R. (1999). Pricing, product diversity, and search costs: a Bertrand-Chamberlin-Diamond model. *RAND Journal of Economics,* 30 (4), 719-735.

Bailey, E., & Williams, J. R. (1988). Sources of economic rent in the deregulated airline industry. *Journal of Law and Economics,* 31, 173-202.

Bailey, J. P. (1998). *Intermediation and electronic markets: Aggregation and pricing in Internet commerce.* Technology, Management and Policy. Cambridge, MA: Massachusetts Institute of Technology.

Bakos, J. Y. (1991). A Strategic Analysis of Electronic Marketplaces. *MIS Quarterly,* 15 (3), 295-311.

Bakos, J. Y. (1997). Reducing buyer search costs: Implications for electronic marketplaces. *Management Science,* 43 (12), 1676-1692.

Brynjolfsson, E., & Smith, M. D. (2000). Frictionless Commerce? A comparison of Internet and conventional retailers. *Management Science,* 46 (4), 563-586.

Clay, K., Krishnan, R., & Wolff, E. (2001). Prices and price dispersion on the Web: Evidence from the online book industry. *The Journal of Industrial Economics,* 49 (4), 521-539.

Clemons, E., Hann, I., & Hitt, L. (2002). The Nature of Competition in Electronic Markets: An Empirical Investigation of Online Travel Agent Offerings. *Management Science,* 48 (4), 534-549.

Copeland, D. G., & McKenney, J. L. (1988). Airline Reservations Systems: Lessons From History. *MIS Quarterly,* 12, 353-370.

Degratu, A., Rangaswamy, A., & Wu, J. (2000). Consumer Choice Behavior in Online and Traditional Supermarkets: The Effects of Brand Name, Price, and Other Search Attributes. *International Journal of Research in Marketing,* 17 (1), 55-78.

Diamond, P. A. (1971). A Model of Price Adjustment. *Journal of Economic Theory,* (June), 158-168.

Hopper, M. G. (1990). Rattling SABRE - New Ways to Compete on Information. *Harvard Business Review,* 68 (May-June), 118-125.

Klein, S. (2002). Web Impact on the Distribution Structure for Flight Tickets. In K. Wöber, A. J. Frew, & M. Hitz (Eds.), *Information and Communication Technologies in Tourism 2002* (pp. 219 - 228) Wien; New York: Springer.

Law, R., & Leung, R. (2000). A study of airline's online reservation services on the Internet. *Journal of Travel Research,* 39 (2), 202-211.

Lee, H. G. (1998). Do Electronic Marketplaces Lower the Price of Goods? *Communications of the ACM,* 41 (1), 73-80.

Rayport, J. F., & Sviokla, J. J. (1994). Managing in the Marketspace. *Harvard Business Review,* 72 (6), 141-150.

Schmidt, J. B., & Spreng, R. A. (1996). A proposed model of external consumer information search. *Academy of Marketing Science,* 24 (3), 246-256.

Simon, H. A. (1986). Rationality in Psychology and Economics. *Journal of Business,* 59 (4), 209-24.

Slade, M. E. (1986). Conjectures, Firm Characteristics, and Market Structure: An Empirical Assessment. *International Journal of Industrial Organization,* 4 (4), 347-369.

Smith, M. D., & Brynjolfsson, E. (2001). Consumer decision-making at an Internet shopbot: Brand still matters. *The Journal of Industrial Economics*, 49 (4), 541-558.

Stigler, G. J. (1961). The Economics of Information. *The Journal of Political Economy,* 69 (3), 213-225.

eTourism Developments in Greece

Ourania Deimezi
Dimitrios Buhalis

Center for eTourism Research (CeTR), School of Management
University of Surrey, Guildford, GU2 7XH, UK
{d.Buhalis; rdeimezi} @acn.gr

Abstract

It is widely accepted that technological developments have had a major impact on the travel industry as a whole. However, despite the major opportunities and challenges provided by its exploitation, levels of eTourism developments are not comparable to all regions. The key objective of this research has been to examine the eTourism developments in Greece. Results indicate that the online travel market is at the early stages with great potential for development. However, the low level of cooperation and co-ordination at the destination level makes the prospect of DMS development in Greece quite distant. It is clear though that the competitiveness of Greek Tourism will depend on its ability to integrate eTourism tools at the micro and macro level.

Keywords: eTourism; Greece; ICTs, SMTEs, Destination Management Systems

1   Introduction

The implications of the Internet and other growing interactive multimedia platforms for tourism promotion are far reaching. The change that the tourism industry is experiencing presents an opportunity for countries to improve their relative position in the international market if they embrace new business models and electronic commerce (eCommerce). The use of eCommerce and the Internet in various countries varies tremendously. This study aimed to examine a number of issues related to eTourism developments in Greece, including levels of tourism establishments Internet penetration and eCommerce usage; degree of online cooperation for tourism industry players and the intention of key industry players towards the development of a Destination Management System.

2   Tourism in Greece

Greece is one of the most popular tourist destinations worldwide in terms of arrivals and significant economic resources are generated by this activity. However, the Greek tourism industry needs to seek new markets and to adjust to new tourism demand trends (Buhalis, 2001). An estimated 12.5 million visitors arrived in 2000, rating Greece in the 15th place in the world classification of tourist destinations. Most visitors (94%) originate from Europe and the United Kingdom and Germany jointly contribute for almost 50% of all arrivals (WTO, 2001). Foreign tour operators mainly distribute the Greek tourism product to incoming markets. The vast majority of Greek suppliers are SMEs and most suffer from functional and structural weaknesses (Ioannides et al. 2001). Therefore, they depend almost entirely on tour operators for their communication with customers and visibility in their major markets (Briassoulis, 1993). Domestic tourism is also very important as Kappa Research (2002) found that 82% of Greeks are taking at least one domestic holiday yearly. The Greek tourism industry lacks strategic vision (Buhalis, 2001). Tourism might have been incorporated into various government bodies, however it has not yet achieved the status of a separate Ministry. Lack of strategic directions and lack of public-private sector collaboration has resulted in deficiencies on general planning and coordination among the bodies concerned, which in turn has prevented a more harmonious development in this industry (Kakos, 2002).

3   ICT Developments and eCommerce in Greece

Electronic Commerce has emerged as a global phenomenon. However, several countries straggle in terms of Internet and electronic commerce utilisation. Indeed, the vast majority of the world population does not have access to the Internet and lacks the technological skills and knowledge to do so (Papazafeiropoulou et. al., 2001). The Internet and electronic commerce adoption in Greece is not very encouraging. During 2001, 50% of all Greeks used mobile phones, 20% used a personal computer (PC) and only 10% used the Internet. The percentage of Internet penetration for the first time passed 10% of the population (over 15 years of age) at the end of the first half of 2001 and it was expected to exceed 12% by the end of the same year (CORDIS, 2002). A large proportion of new Internet users emerge, as 75% started using the network during the last two years (Karounos and Goussiou, 2001).

In Greece, eCommerce is found in its infancy stages of growth and presents a delay of five years in relation to the main European markets (Braliev and Yatromalakis, 2002). Consumer eCommerce "remains largely embryonic throughout the region as only 12% of Internet users or 3.5% of the Greek population buy products online (Stat Bank, 2002). The importance of SMEs to the Greek economy is comparable to the one of the European Union. Although SMEs represent more than 99% of enterprises in most EU Member States, their importance in the economy is not matched by their use of e-business tools. According to a benchmarking project by the European Commission the SMEs utilisation of ICT in Greece retains at the lowest levels as demonstrated in Table 1 (EU, 2002).

There is not explicit information to indicate whether Greek tourist enterprises are connected to the Internet or whether they provide their products and services online. The growing popularity of using Internet technology in destination marketing by many national tourism organisations is already evident (Au, 2001; Buhalis and Spada,2000). In 2000, the GNTO presented its site 'Discover Greece'. GNTO's site offers information about a wide field of tourism activities, broken down into two district categories: editorial content and data about specific products, services and providers. The former helps to entice and inform the potential customer and includes photographs, maps and posters. Editorial information is supported by data on tourism suppliers, attractions, activities events and amenities (Trifona, 2001). Nevertheless, it does not offer online reservation facilities.

Table 1. SME e-business adoption rates in 2001

% of SMEs	Using ICT	Having web access	Own web site	Third party web site	Making eCommerce purchases	Making eCommerce sales
Austria	92	83	53	26	14	11
Denmark	95	86	62	N/A	36	27
Spain	91	66	6	28	9	6
Finland	98	91	58	N/A	34	13
Greece	84	54	28	8	5	6
Sweden	96	90	67	N/A	31	11
UK	92	62	49	11	32	16
Germany	96	82	65	21	35	29
Luxembourg	90	54	39	13	18	9
Netherlands'	87	62	31	N/A	23	22
Italy	86	71	9	26	10	3
Norway	93	73	47	N/A	43	10

Source: European Commission, 2002: 5

4   Research Methodology

For the purpose of this research, the use of both primary and secondary data was necessary. Firstly a comprehensive secondary research was undertaken on issues related to eCommerce, eTourism, tourism in Greece and SMEs. Given the exploratory nature of this subject area, primary research included both qualitative and quantitative methods. The 1st stage of the primary research consisted of a critical review of the tourism enterprises that include details of their online presence in the Greek Travel Pages (GTP), which is considered the most comprehensive directory of Greek tourism. Although some enterprises may not include their online details on the GTP, this is the closest estimate available. This aimed to establish the level of Greek tourism eCommerce and online representation.

The 2nd stage involved semi-structured interviews aimed to elicit critical issues and develop a set of variables. *Personal semi-structured in -depth interviews* were employed to explore further parts of the subject. Interviewees were selected to be pioneers in eTourism in Greece as demonstrated by their online presence. Content analysis employed to determine the presence of concepts within the interview results. Data was sorted into thematic frameworks summarised in tabular form and was analysed at four dimensions including *not discussed, discussed, discussed in detail, emphasised.*

Finally, the 3rd part of the research included a quantitative survey being conducted to compliment the web survey and interview findings in remote regions. Following the competition of the semi-structured interviews, a *structured questionnaire* was drafted and was administrated via email. The GTP Greek monthly travel and tourism guide provided a mailing list of totalling 750 email addresses. After a period of six weeks, a total of 54 questionnaires were completed, which produced an 8% response rate, a relatively low response rate, but acceptable given the high season for the industry.

The study then examined in total 348 websites in a one-month period to assess the level of eCommerce. The data obtained from the questionnaires were analysed using SPSS. In addition, semi-structured answers included in the interview protocol were also input in the SPSS. Frequencies were thus computed and analysed jointly with the questionnaires, as they examined the same variables. Table 2 indicates the response rate for each part of the research.

Table 2 Primary research summary

Stage	Instrument	Targeted	Achieved	Response rate
1st Stage	Web survey	348	348	-
2nd Stage	In-depth interviews	55	28	51%
3rd Stage	Questionnaires	654	54	8%

5   Analysis of findings

Qualitative research indicated that the generic state of eCommerce in Greece is poor. The majority of the expert panel agreed that worldwide, eCommerce ranks very highly for the tourism industry. However, they felt strongly that uptake is disappointingly slow in Greece at the moment. Participants felt strongly that although they would like eCommerce to have a central role in their strategy, at that moment it was of neutral importance. This was attributed to the low Internet access and inertia towards eCommerce by the overwhelming majority of the Greek population. They argued that although the international market is very important, they are quite dependent on the domestic market as well. International markets predominantly use Web based travel agencies and portals at the countries of origin to make their bookings. An interviewee from the hotel sector further commented that problems may arise with the establishment's own site pricing strategies as the websites of international tour operators' offer favourable deals to foreign tourists, since hotels are imposed by wholesalers to publish official rack rates on the website. Therefore, respondents felt that eCommerce was not their main priority. Table 3 indicates the percentage of enterprises that promote their website and email address at the GTP (2002). Although this is not a formal representation, (as the establishments that are either not included or do not advertise their online presence in GTP are excluded) it should be viewed as a close approximation of eTourism in Greece.

Independent Internet presence referring to establishments having their own Internet website, was one of the remarkable findings of this research. Due to the sampling criterion, that determined that participants had an online presence, all 28 of the interviewed tourism establishments and 98.1% of the questionnaire respondents were founded to have an independent Internet presence. However, the overall picture from the 348 total websites assessed was that the vast majority were at a very early stage, in terms of dimensions included in the measurement of website evaluation. The layout was rather simple with plain information and with low volume of graphical information. The most common form of interactivity was email and comment forms.

However, some websites provided a guest book facility and mailing lists for further contacts. In terms of language coverage, it is very encouraging that 90% of the websites examined offered their content in more than one language. The most frequent languages cited were Greek and English. This was considered as a positive sign of development as the number of available languages determines the extent to which international customers can actually understand and use the system.

Table 3 Frequency of Internet representations in GTP

	Total number of establishments included in GTP	Number of establishments having website presence *and/or* email	Having Website presence *and/or* email as % of total
Alternative Tourism establishments	15	15	100
Airlines (Greek carriers)	2	2	100
Conference and Meetings	6	5	84
Yacht Brokers	35	22	63
Hotel Chains	51	29	57
Travel Agencies	1706	868	51
Rent a Car	121	50	41
Motorcoach companies	111	23	21
Shipping Companies	72	15	21
Cruise Companies	42	5	12
Hotels and other lodgements	5,156	587	11.4
Campsites	48	2	4.2
Museums, Sites and Monuments	577	8	1.4
Total	**7942**	**1631**	**20.5**

In an attempt to explore the technological levels and their involvement, both interviewees and respondents had to state who had the responsibility for their online presence. The findings of both methods confirmed that outsourcing is the most preferable method for web site development. According to interviewees, the time pressures to launch their site and the lack of internal knowledge/ expertise force tourism enterprises to assign the development, hosting and management of their online presence to web design companies. It is often that these companies have initiated the process and have sold an eSolution to tourism enterprises. As most people do not have the technical expertise and tools to deploy and manage a website management they feel that outsourcing is the only realistic option.

The Internet as discussed in the literature revolutionises the tourism distribution channels, as it enables customers to place bookings directly. The most commonly available reservation feature in the Greek tourism Internet arena is email provision and booking forms (Table 4). It is surprising that even when they have web page few companies do not provide any online booking facility but only display telephone and fax numbers. Interviewees justified that at this stage their website has been developed simply for information provision and they have found the procedure complex. One interviewee commented; "if we provide them with email address, then we will have to arrange money transfers and other issues. Therefore why should we spend the time to read emails and reply when we are just one phone call away? If we are going to do it someday, we will provide the complete booking solution." This demonstrates the potential benefits of eCommerce are not fully explored in the Greek tourism context.

Table 4. Online booking facilities for Greek tourism

Company Category	Total surveyed	Online booking capabilities offered				
		Email	Booking Form	Real time	Other Channel	Not any kind of online booking
Accommodation	238	104	141	15	12	9
Travel Agency	119	63	58	3	2	4
Tour Operator	4	4	4	-	-	-
Car rental	4	4	3	-	-	-
Museum	18	18	-	-	-	-
eTravel Portal	6	4	4	2	-	-
Cruise Company	9	6	3	1	-	-
Coach	2	2	-	-	-	-
Tourism Associations	3	N/A	N/A	N/A	N/A	N/A
Airlines	2	1	-	1	-	-
Mainland Tras.	5	5	-	-	-	-
Shipping Comp	10	3	4	3	-	-
Total	**420**	**214**	**217**	**25**	**14**	**13**
Percentage	**100%**	**51%**	**52%**	**6%**	**3%**	**3%**

As the discussion so far suggested eCommerce is in its infancy. The majority of respondents and interviewees stated that Internet bookings count from 1 to 5% of their reservations. Lack of knowledge as to how to manage and market their website was identified as a limitation. The possibilities highlighted in the literature for online promotions such as discounts to encourage Internet bookings, upgrades and other methods used, were neither mentioned by interviewees nor observed in the web survey. Moreover, a significant difference was found between small and larger companies, with the latter achieving higher numbers of online bookings. This was attributable to the better financial resources available, the gradual appreciation of eCommerce potential and the ability to employ the technical expertise in order to promote online bookings.

A number of prime drivers were identified through the interviews (Table 5) and questionnaires (Table 6). It is clearly demonstrated that customer service, direct communication with customers, market demand and modernisation are key factors, illustrating that eTourism in Greece is demand/consumer driven. These results are in accordance with the existing literature. The lowest ranked business driver was 'fear from competitors'. This demonstrates that the low, compared to other countries level of ICT and the Internet by the tourism industry, is not a concern for already ICT advanced businesses. Therefore, competition is not perceived yet as threat. This might have important implications however to the advancement of eTourism in Greece, as competition encourages innovation and it's absence might proved a serious drawback. In addition Greek tourism enterprises underestimate the global competition emerging through the Internet.

Table 5   Content analysis grid- Main drivers towards eCommerce

	Not discussed		Discussed		Discussed in detail		Emphasised	
	Cases (n)	%	Cases (n)	%	Cases	(n)%	Cases (n)	%
Direct communication/access to consumers	5	18	2	7	0	0	21	75
Reduce their dependence on Intermediaries	9	32	0	0	1	4	18	64
More efficient and accurate customer service	0	0	0	0	10	36	18	64
Market demand	10	36	2	7	2	7	14	50
Modernisation	18	64	2	7	2	7	6	22
Cost reduction with partners and other suppliers	0	0	8	28	15	54	5	18
Fear of innovative competitors	25	89	3	11	0	0	0	0

Table 6   Qualitative research results on Main drivers to eCommerce (n=54)

	1-Very Unimportant	2- Unimportant	3- Neutral	4- Important	5- Very important	No of valid replies	Mean	STD
More efficient and accurate customer service	2 3.7%	0 0%	4 7.4%	14 25.9%	34 63%	54	4.48	0.79
Direct communication/access to consumers	2 3.7%	2 3.7%	3 5.6%	12 22.2%	35 64.8%	54	4.41	1.01
Modernisation	2 3.8%	0 0%	4 7.5%	17 32.1%	30 56.6%	53	4.38	0.92
Market demand	0 0%	2 3.7%	3 5.6%	24 44.4%	25 46.3%	54	4.33	0.75
Cost reduction with partners and other suppliers	2 3.8%	5 9.4%	13 24%	14 26%	19 35.8%	53	3.81	1.14
Reduce their dependence on Intermediaries	4 7.7%	3 5.8%	15 28%	13 25%	17 32%	52	3.69	1.21
Fear of innovative competitors	6 11%	7 13%	14 26%	13 24%	13 24%	53	3.38	1.30

Cooperation as much as competition is a feature of the emerging online tourism services sector, where commercial suppliers provide access to each other's resources. The existence of hypertext links among travel related sites is considered to be an obvious sign of cooperation. Since there is no trust to public sector organisations, networks emerge to develop comprehensive value chains. However, out of total 348 Greek tourism related websites examined, less than half (45.6%) were found to offer a hypertext link of any kind. In the web survey, several types of links were identified.

Firstly, type A the lesser frequent link (5% of all offering links) observed between sites that, ostensibly, are competitors i.e. a hotel offering link to another hotel. Nevertheless, in all observations of this type the linked competing establishments were situated in different geographical areas of Greece. Secondly, type B in which establishments offered external links to other complementary tourism service providers including hotel to travel agency, taxis, and transfer companies. This type enjoyed higher levels of occurrence as 28% of all websites offered complimentary services links. Thirdly, type C wherein establishments offer external links to other related service providers such as currency converter, weather, travel books/guides and embassies were the most frequent (52%). This indicates that the majority of businesses which went online have realised the benefits offered by offering value added services to the customer. Finally, type D, with less obviously related services such as media, journals and real estate agencies were additionally monitored (10%). At a much lesser extent (2%) links were observed which were totally unrelated with tourism services such as butchery, motor garage and car manufacturing company! The results of the survey in relation to links did not demonstrate the existence of an organised network among tourism establishments. The diversity of links indicates that close personal relationships might be a leading driving force in some cases. Indeed, Greek tourism websites research confirms yet again, the low levels of cooperation, as low percentages of hypertext links illustrate little electronic networking.

The overwhelming majority of respondents stressed their discomfort with the Greek National Tourism Organisation's (GNTO) representation of Greek tourism. The GNTO representation on the Web was assessed as inadequate. Only one of the interviewees stated that his company website had a link with the GNTO website. The cooperation was often perceived unnecessary, as respondents claimed that there was nothing to benefit from such cooperation. However, the majority agreed that the situation calls for rapid action in order Greece to remain competitive.

Both interviewees and qualitative respondents were invited to answer about *future prospects of a Destination Management System (DMS)* in Greece. As there is no DMS operating currently in Greece, interviewees were first questioned about their knowledge on the subject. Only 14% had heard or read about DMSs. This reflects that Greek tourism players are not aware of emerging trends and electronic tools. The lack of knowledge of such a well-researched and so frequently mentioned topic on the international tourism literature implies the essential need to raise levels of information provided to the industry. A detailed definition of DMSs was provided, with the intention that respondents would be in a position to state their views on the potential DMS development in Greece. The vast majority (80%) was positive towards the development of a DMS for Greece. Although the majority of participants agreed that cooperation is quite important for the competitive position of Greek tourism, they once again raised doubts towards the achievement of cooperation.

6   Conclusion

The IT penetration and the level of eCommerce both of the supply and demand sides in Greece is in embryonic stage. The results from this study indicate that the tourism business sector is lacking behind the opportunities with the rapidly changing environment. The lack of strategic business planning and the limited levels of understanding of the eTourism potential implies that the Greek tourism sector needs to rethink their position in order to compete at an international level with other destinations. This conclusion is further reinforced by the fact that the sample was made of already users of the Internet and yet their technological advancement is limited. The need for change is obvious when one considers that Greece's main international tourism markets (British and Germans) are among the top European Internet users. One of the most important inhibitors for the advancement of eCommerce revealed to be the Greek populations' inertia towards online transactions. That influences the decision makers in the tourism industry as well. Some of the reasons behind that were found to be the personal nature of business to customers' relationships, the informality of information and the low usage of the Internet by the Greek population. Greece's challenge during the next decades is to capitalise on new technology developments and to use emerging eTourism tools in order to better promote and market the tourist product.

Collaborative networking should be established through exchanging linkages with other complimentary suppliers, country information and all travel related websites. The views of the Greek tourism industry towards DMS were positive at a general level. However, the majority of respondents raised serious barriers to the potential introduction of a DMS in Greece and declared that the GNTO is the main barrier to a successful networking/cooperation of tourism operations. This highlights the need for improved relationships in the sector, an issue that impacts on tourism networks as a whole, before moving on a highly cooperative and demanding system such as a DMS. The development of a DMS can only flourish in the Greek tourism context if trust between tourism stakeholders is built. A long-term strategy and coordination at a macro level, involving both public and private sectors is a key prerequisite. It is evident however that unless the public and private sector capitalises on the emerging eTourism tools, Greek tourism will face major competitive disadvantage in the future.

References

Au, N., 2001, Destination marketing on the Internet: impact of Hong Kong tourism association web site on international travellers to Hong Kong. In Sheldon, P.J., Wober, K.W., and Fesenmaier, D.R., eds., *Information and Communications Technologies in Tourism*. 303-308. Vienna: Springer.

Braliev, A. and Yatromalakis, N., 2002, Global Competitiveness Report 2001-2002, *Harvard Business School*. Retrieved June25, 2002 from the World Wide Web: http://www.cid.harvard.edu/cr/profiles/Greece.pdf

Briassoulis, H., (1993) Tourism in Greece. In: Pompl, W. and Lavery, P. eds, *Tourism in Europe, Structures and Developments*, Wallingford: Cab International, pp. 285-301

Buhalis, D., 2001, Tourism in Greece: strategic analysis and challenges, *Current Issues in Tourism*, 4(5), 440-480.

Buhalis, D., and Spada, A., 2000, Destination management systems: criteria for success-an exploratory study. *Information technology in Tourism*. 3(1). 41-48.

CORDIS, 2002, The results of panhellenic research into the use of computers, the Internet and mobile telephones, *European Commission*, Retrieved June 4, 2002 from the World Wide Web: http://www.cordis.lu/greece/rd.htm

European Commission (EU), 2002, *Benchmarking National and Regional E Business-Stage 1 - Synthesis report*. Retrieved June10, 2002 from the World Wide Web: http://europa.eu.int/information_society/topics/ebusiness/godigital/Docs/20207_ Final_synthesis_report.doc

Greek Travel Pages (GTP), 2002 Greece's Monthly Travel-Tourism Guide, March, Athens: International Publications. [in Greek].

Ioannides, D., Apostolopoulos, Y. and Sonmez, eds. S.F., 2001,*Mediterranean Islands and Sustainable Development:practices, management and policies*. London: Continuum.

Kakos, T., 2002, Public and private sector: are there any collaboration possibilities for the tourism's benefit? *Tourism Market*, February, (145), 88-94, [in Greek].

Kappa Research, 2002, Domestic tourism trends. *Tourism and Economy*. April, (273). 14-21. [in Greek].

Karounos, T. and Goussiou, L., 2001, National survey on the use of personal computers, internet and mobile telephony in Greece, *Ministry of Development-Greek Research and Technology Network*, Retrieved June 2, 2002 from the World Wide Web. http://www.goonline.gr/English/docs/survey.doc.

Marcussen, C., 2002, *Trends in European Internet distribution -of travel and tourism services* Updated. Retrieved 30 July, 2002, from the World Wide Web: http://www.crt.dk/uk/staff/chm/trends.htm.

Papazafeiropoulou, A., Pouloudi, A. and Doukidis, G., 2001, Electronic Commerce Policy Making in Greece, BIT World 2001 conference paper. Retrieved June 4, 2002 from the World Wide Web: http://www.eltrun.aueb.gr/papers/bit.pdf.

Stat Bank. 2002, *eCommerce and Internet research*. Retrived June10, 2002 from the World Wide Web: http://www.statbank.gr/sbstudies.asp [in Greek].

Trifona, 2001, New technologies in service of tourism. *Trade with Greece*. Retrieved June 15, 2002 from the World Wide Web: http://www.acci.gr/trade/No18/69.pdf

World Tourism Organisation (WTO), 2001, *Tourism Highlights -Updated*. Retrieved June 2, 2002 from the World Wide Web: http://www.worldtourism.org/market_research/data/pdf/highlightsupdatedengl.p df.

Experience-based Internet Marketing: An Exploratory Study of Sensory Experiences Associated with Pleasure Travel to the Midwest United States

Ulrike Gretzel
Daniel R. Fesenmaier

National Laboratory for Tourism and eCommerce
University of Illinois at Urbana-Champaign, USA
{gretzel, drfez}@email.edu

Abstract

Destination marketing Websites remain largely focused on the communication of functional attributes such as price, distances, and room availability. However, it is argued that a model of embodied cognition, which assumes that consumers derive experiences largely from patterns of sensory input, better reflects the actual construction of tourism experiences in the minds of consumers than these purely cognitive approaches. Acknowledging the importance of senses for human cognition has implications for marketing real world experiences as well as creating experiences online. The paper presents an exploratory analysis that investigates whether information about colors, scents, and sounds expected during vacations in the Midwest United States can be categorized and bundled in meaningful ways, and consequently used in the context of online destination marketing.

Keywords: tourism experience; embodied cognition; sensory information; Internet marketing

1   Introduction

Marketing researchers and economists alike have drawn attention to the experiential nature of goods and services (Holbrook, 2000; Pine and Gilmore, 1999). Their findings related to hedonic values and emotional responses to consumption situations are especially relevant for marketing tourism given the inherently experiential nature of tourism products and services. Both consumption and decision making processes related to tourism are to a large extent driven by hedonic and emotional aspects (Vogt and Fesenmaier, 1998; Vogt et al., 1993) and memories of trips are a function of trip-related experiences and the stories we construct from them (Fesenmaier and Gretzel, 2002). This recognition of the experiential nature of tourism and of new consumer trends calls for marketing approaches that make use of innovative ways for communicating tourism experiences (Schmitt, 1999).

Emerging technologies increasingly afford sensory information to be communicated to users in online environments. Pictures with vivid colors and high resolution as well as high quality sounds are widely integrated into today's Websites. Technology that conveys tactile and olfactory experiences is currently being developed and promises even more immersive experiences to be possible on the Web. Nevertheless, destination marketing Websites remain largely driven by database structures (Manovich, 2001), focusing on communicating lists comprised of functional attributes such as price, distances, and room availability. Their design is based on a model of a rational and information seeking consumer which often results in simple activity-based descriptions that reflect the supply side (reflected by computer programmers and/or marketing managers) rather than an actual consumer's perception of tourism experiences. It is argued that this lack of an experiential mindset within the tourism industry is due largely to a lack of understanding of the nature of tourism experiences. Purely cognitive models of consumers can provide only limited explanations for holistic and often largely hedonic consumption experiences. What is needed is an extended model that takes sensory experiences and emotions into account. Such a model is often referred to as a model of "embodied cognition" (Malter and Rosa, 2001) and assumes that consumers construct experiences largely from patterns of sensory input (Biocca, 1997). It is the aim of this paper to explore whether people actually create coherent structures/themes out of sensory experiences that can be used for the purpose of experience-based destination marketing on the Internet.

2   The Role of Sensory Information in Communicating Tourism Experiences

Contrary to the current simplified, activity and/or amenity-based representations of tourism experiences on the Web, real world vacations are complex experience structures that involve cognitive *and* sensory stimulations as well as affective responses to certain events. Due to this multi-sensory nature of tourism products, textual and pictorial ways of describing vacations are very limited in terms of conveying a complete picture that can be used to formulate correct product expectations. Even word-of-mouth, the dominant informational strategy in the tourism domain, faces constraints with respect to the communication of first-hand embodied knowledge. Yet, creating an informed opinion based on these forms of information is essential to the potential consumers of tourism experiences. In this respect, tourism constitutes an interesting exception on the search versus experience good continuum (Nelson, 1970) as most of its attributes are not searchable and at the same time not accessible/assessable through actual product trials. Product trials have been identified as extremely powerful sources of information for the formation of brand beliefs and attitudes since they involve an experience of the product through multiple senses. Kempf (1999) found that the affective response to such trials is especially influential in the case of hedonic products.

Whereas marketing strategies for traditional consumer goods such as cars, cosmetics, and food products have long acknowledged the importance of actually touching and, if applicable, smelling and tasting a product, tourism marketing has widely neglected the significant role of olfactory, haptic, gustatory, and auditory sensations during consumption experiences. When sensory information is communicated, it is usually presented in isolated form, i.e. either smell, taste, touch, *or* sound, and is often simply translated into visual cues. The dominance of visual (and to some extent auditory) information in tourism marketing appears to be largely driven by the limited affordances of the traditional media used. However, other marketers have found very creative solutions to overcome these limitations, e.g. through scented pages in fashion magazines. Another reason for the lack of a broader range of sensory cues in tourism marketing appears to be the subtleness with which embodied cognition works. Sensory information is usually processed on a subconscious level, and thus is often less prominently mentioned by consumers. However, its influence on decision-making processes has been widely shown, especially in research related to retail environments (Morrin and Ratneshwar, 2000). Whatever the reasons are for the current absence of sensory information in the communication related to tourism products, its inclusion will become more and more important as tourism marketing increasingly uses experiences as the foundation for defining its products.

New technological developments promise inexpensive and readily-available substitutes for real world product trials (Klein, 1998). Virtual tours, for example, have become a popular means of communicating tourism related information in a way that resembles an actual visual experience (Cho and Fesenmaier, 2000). Further, streaming video allows for a quick and cost-efficient representation of the visual as well as auditory aspects of a vacation experience. However, in terms of their experiential content, these current modes of communicating tourism information are only marginally more "real" and effective than their traditional counterparts. Compelling and engaging digital experiences require immersion in a rich set of data that captures all of the human senses; consequently, it is expected that an increasing amount of research will be devoted towards the development of sensors for taste, smell, and touch (Jain, 2001). These efforts will eventually result in commercially available products waiting to be taken advantage of by consumers and marketers alike. Whether sensory tourism information is communicated through these emerging technologies or using traditional forms such as metaphors and narratives, a richer understanding of the various sensory domains and the way in which they are bundled to form holistic tourism experiences is needed.

This lack of information regarding the consumers' perception of the sensory dimensions of vacations was addressed in a recent study conducted by the National Laboratory for Tourism and eCommerce. The following presents the results of an exploratory analysis of certain sensory aspects associated with travel experiences in the Midwest United States.

3 Methodology

The findings presented in this paper are based upon a survey of 3,525 randomly selected persons who had requested travel information from a Northern Indiana tourism office during Summer and Fall, 2001. Data collection was conducted during a 2 month period (November, 2001 – December, 2001). The survey was administered following a three step process: 1) An initial survey kit, which included a cover letter, a 12-page survey, an insert with an invitation to enter a drawing and a description of the prizes offered, and a postage-paid return envelope, was sent to each person in the sample on November 6, 2001; 2) One week later, each person was mailed a post card to remind them to complete and return the survey and/or to thank them for participating in the effort; 3) Three weeks later, a second survey kit with an invitation to enter into a second drawing was mailed to all non-respondents. This effort resulted in 1,436 completed responses (an additional 111 were returned with bad addresses or insufficient responses) for a 42.1 percent response rate.

The survey included a section that asked respondents to write down the feelings, experiences, tastes, colors, scents, and sounds they associate with desired experiences when traveling to a Midwest destination. More specifically, respondents were asked to imagine a trip to a destination in the Midwest United States and report the colors that dominate their mental image, the scents they would like to smell, and the sounds they expect to hear. The survey respondents used a variety of words to describe the different sensory inputs they expect from a vacation in the Midwest. Although most descriptions were single-word concepts, some respondents used combinations of words such as "maple leaf red", "smell of home cooking", or "sound of water lapping onto lake shore" to describe favorite or expected colors, scents, and sounds. This complexity of the descriptions and the large amount of text to be transformed into numeric values, as well as the idiosyncrasies in the usage and spelling of words turned the development of a classification scheme into a rather time-consuming task. In an effort to make the classification process more efficient, a preliminary analysis of the text data was conducted using CATPAC II (Woelfel and Stoyanoff, 1993). The resulting neural network output was used to identify the words that were most frequently mentioned to portray desired sensory experiences. Frequent co-occurrences of words as displayed through short distances in the graphical representation of the network were analyzed to determine whether certain sensory concepts warranted a separate category. For instance, "lake" and "breeze" emerged as very frequently combined words; thus, a "lake breeze" concept was included among the scent categories.

A total of 11 color, 10 scent, and 14 sound categories were identified (see Table 1). The small number of categories reflects the fact that the concepts are rather broadly defined; as a consequence, the color categories are comprised of various shades of a specific color; and scent and sound concepts include several related yet not identical olfactory and acoustic experiences. The "Green" category, for example, includes shades of bright grass green as well as forest or pine green. The "Blue" category is dominated by sky/light blue and lake or water blue but also includes more unique shades such as azure and medium blue. The "Food" scent category is comprised of barbeque smells, baked bread, festival/fair food scents, and special scents such as fish boils, scotch, and fudge. Animal sounds include mainly bird calls and cricket chirps; however, other animal sounds such as dogs barking and frogs croaking were also assigned to this category. Although broader definitions lead to a loss of detail and therefore potential misinterpretations, this categorization strategy was selected because subsequent analyses were expected to provide implications as to whether such global classifications can, nevertheless, lead to useful definitions of specific vacation experiences. The underlying assumption is that experience is a function of the *combination* of certain sensory inputs, and that finer distinctions become visible and interpretable when several sense categories are bundled together.

Descriptive analyses, factor analysis, and cluster analysis were used to investigate the relationship among the different sense categories and to explore how respondents combined sensory experiences into coherent bundles.

4 Findings

Several interesting findings emerged from the analysis of the sense categories that were derived from the initial text data (see Table 1). On average, 1.6 color types, 1.8 scent categories, and 2.1 sound concepts were mentioned, implying a rather uniformly colored imagery with more complex combinations of smells and especially sounds. Green was the most frequently mentioned color category (49.5 percent of the respondents reported green to dominate their imagined trip), followed by blue with 44.2 percent (see Table 1). The most frequently reported smell was "lake breeze" (47.2 percent), and autumn leaves were mentioned by 23.4 percent of the respondents as being a scent associated with a pleasure trip to the Midwest. Animals (66.1 percent) and music (28.4 percent) were the most frequently mentioned sounds.

Table 1. Frequencies of Sense Categories

Color		Scent		Sound	
Category	%	Category	%	Category	%
Green	49.5	Lake breeze	47.2	Animal	66.1
Blue	44.2	Autumn leaves	23.4	Music	28.4
Yellow	16.7	Grass/hay	19.2	Water	25.6
Red	14.9	Fresh air	18.7	Fire	22.7
Orange	11.0	Pine trees	17.3	Wind	13.4
Brown	6.1	Food	16.8	People	11.8
Gold	5.3	Flowers	13.6	Quiet	11.2
White	4.8	Country	10.0	Children	9.4
Sand	2.1	Campfire/Smoke	9.6	Traffic	9.4
Purple/Pink	1.6	Antiques	1.0	Boats	3.4
Colorful	0.9			Horses	2.3
				Money	1.5
				Trains	1.2
				Walking	0.8

Factor analysis with a varimax rotation was used to investigate whether certain patterns emerged from the data with respect to "sensory domains", i.e. colors, sounds, and smells that were frequently mentioned together. A nine factor solution was identified as fitting the data best and at the same time leading to meaningful combinations of sensory experiences (see Table 2). Factor 1 includes sensory experiences related to the fall season, with red, yellow and orange displaying the highest factor loadings. Factor 2 is defined by colors, sounds, and smells related to water, with blue and lake breeze contributing the most to this domain. Factor 3 describes a noise theme with country smells which could be interpreted as a country music/fair experience. Factor 4 is comprised of outdoor/camping sensory experiences and is mainly defined by autumn leaves scent and campfire sounds. Factor 5 may be best described as an "indulge" factor as it is comprised of food smells, people sounds, gambling and shopping sounds, the smell of antiques, music, and the color gold. Factor 6 refers to country-related experiences. The color brown, which was several times used in the sense of earth or weathered wood, largely defines this factor. Factor 7 is defined by very "happy" sensory experiences such as colorful and pink settings, flower, grass, and food smells, as well as children and horse and buggy sounds. Factor 8 describes a more scenic nature experience and Factor 9 a peaceful experience with almost no sensory input and fresh air being the dominant sensory cue. The factor loadings are generally rather low – a finding that is expected given the broad definitions of the sense categories and the resulting overlap between certain sensory domains. However, the findings of the factor analysis indicate that associations between colors, sounds, and smells exist and, most importantly, that meaningful sense packages can be derived from the rather broad classifications.

Table 2. Description of Sensory Domains Derived from Factor Analysis

	Colors	**Scents**	**Sounds**
Factor # 1 Autumn	Red, Yellow, Orange, Gold, Brown	Autumn leaves, Campfire & smoky smells	
Factor # 2 Water	Blue, White, Not Green	Lake breeze	Water, Children, Traffic
Factor # 3 Noise		Grass, Country	Traffic, Music, People
Factor # 4 Outdoors	Green	Lake breeze, Country, Autumn leaves	Animals, Campfire, Music
Factor # 5 Indulge	Gold	Food, Antiques	Music, People, Money
Factor # 6 Country	Gold, Brown, Sand	Country	Wind, Animals
Factor # 7 Happy	Pink, Colorful	Flowers, Food, Grass	Children, Animals, Horses
Factor # 8 Scenic/Nature	Green, Blue, Sand	Pine trees, Flowers	Animals, Trains
Factor # 9 Peaceful	White	Fresh air	Quiet, Wind

The nine sensory domains identified through factor analysis were then used as the basis for a cluster analysis to investigate the diversity of "bundles" created from the sensory domains involved when thinking about pleasure trip experiences in the Midwest United States. A K-means cluster analysis was conducted using the factor scores of the nine sensory domains. A solution with seven clusters was identified as the most suitable as it resulted in meaningful cluster sizes, a high within-group coherence and between-group difference. A discriminant analysis with cluster membership as the grouping variable and the nine factors as independent variables was conducted to test the quality of the clustering solution. The results were highly significant and showed a high discriminating power of the cluster groups (with 94.5 percent of the cases being correctly classified).

Table 3 provides a short description of the factors defining the seven "experience bundles". Only those domains are presented that display relatively large positive values for the respective factor scores. For example, Cluster 1 describes a group of respondents that construct their trip theme entirely based on sensory information related to water. In contrast, respondents in Cluster 2 combine water related sense categories with the colorful, intensive smells, and animal and children sounds of the "Happy" sense domain. It is important to note that four of the seven groups combine different sensory packages when thinking about a Midwest travel experience but that only certain combinations occur and that the one-factor groups are consistently larger except for Cluster 7, which basically describes respondents that prefer minimal sensory inputs (the final cluster centers indicate that Peaceful, Scenic, and Outdoors are the only, and yet not very strongly sought after sensory domains for this group).

Table 3. Factors Defining Experience Bundles

Cluster 1 (n=261)	Cluster 2 (n=110)	Cluster 3 (n=47)	Cluster 4 (n=202)	Cluster 5 (n=32)	Cluster 6 (n=110)	Cluster 7 (n=452)
Water	Water Happy	Autumn Country	Autumn	Outdoors Country	Indulge	Not Autumn Not Water Peaceful Scenic Outdoors

5   Discussion

The findings of this exploratory study suggest that certain sensory domains exist and are often bundled following specific patterns of associations. It is important to note that all sensory domains are comprised of at least two different senses and that, consequently, no single sense seems to dominate. More specifically, the existence of cohesive sensory domains implies that the different senses are supportive of each other, and that their combination represents a meaningful way of revealing nuances in the structure of broad sense categories. Further, it appears that these sensory domains, or combinations thereof, can be used to define coherent experiences sought after by certain groups of travelers.

An important limitation of the study is that information about the specific colors, scents, and sounds imagined by the survey respondents can not be generalized as it is to a great extent influenced by the context of the study (i.e., the survey referred specifically to travel within the Midwest United States and was conducted during the fall). The analysis presented in this paper was only concerned with structural questions rather than the actual content of the sensory domains, and thus was not constrained by these limitations. However, research in the area of embodied cognition as it relates to tourism should ultimately be concerned with the construction of sensory-based experience typologies. The latter requires the collection of sensory information independent from the setting of a specific study. Further, the current study only took three types of sensory experiences into account. There is a need for expanding the scope to tactile and gustatory experiences and to establish linkages to emotional components of trip experiences.

Since consumers seem to associate certain sensory themes with certain trip experiences, it appears to be important for destination marketing to integrate sensory domains into its communication strategies. It follows from the current analyses that sensory experiences are complex but not idiosyncratic and can be used to communicate certain tourism experiences to specific groups of travelers. The results of the current exploratory study have important implications for user modelling strategies, especially in the context of tourism recommendation systems. Although a myriad of visual, olfactory, and auditory sensations are potentially associated with a vacation, the use of broad sense categories seems to be sufficient in describing desired experiences as long as different senses are taken into account and sense categories are combined into larger domains. It is argued that questions related to these sense categories could easily be integrated into current user profiles or query structures and could greatly enhance the traditional activity-based approaches.

Especially for Internet-based tourism marketing, these initial findings indicate that the integration of sensory information on Websites may play an important role in supporting information search and decision-making processes by providing the sensory cues essential for the conceptualization and evaluation of vacation experiences. Thus, future studies should focus on the relative impact/role of sensory information as compared to functional attributes in tourism-related information search and decision-making. It is posited that the current and future research will help create the foundations for building alternative and potentially more effective approaches for Internet-based communication of trip experiences.

References

Biocca, F. (1997). Cyborg's dilemma: Progressive Embodiment in Virtual Environments. *Journal of Computer Mediated-Communication*, 3(2).

Cho, Y. & D. R. Fesenmaier (2000). A conceptual framework for evaluating the effects of a virtual tour. In Fesenmaier, D. R., Klein, S. & D. Buhalis (Eds.), *Information and Communication Technologies in Tourism 2000*, pp. 314-323. Wien: Springer Verlag.

Fesenmaier, D. R. & U. Gretzel (2002). Searching For Experiences: The Future Role of the Consumer in the Leisure Experience, *Proceedings of the Leisure Futures Conference*, Innsbruck, Austria, forthcoming.

Holbrook, M. B. (2000). The millennial consumer in the texts of our time: Experience and entertainment. *Journal of Macromarketing*, 20(2): 178-192.

Jain, R. (2001). Digital Experience. *Communications of the ACM*, 44(3): 38-40.

Kempf, D. S. (1999). Attitude Formation from Product Trial: Distinct Roles of Cognition and Affect for Hedonic and Functional Products. *Psychology & Marketing*, 16(1): 35-50.

Klein, L. R. (1998). Evaluating the Potential of Interactive Media through a New Lens: Search versus Experience Goods. *Journal of Business Research*, 41: 195-203.

Malter, A. J. & J. A. Rosa (2001). *E-(Embodied) Cognition and E-Commerce: Challenges and Opportunities.* Paper presented at the Experiential E-Commerce Conference at Michigan State University, East Lansing, September 27-29, 2001.

Manovich, L. (2001). *The Language of New Media*. Cambridge, MA: MIT Press.

Morrin, M. & S. Ratneshwar (2000). The Impact of Ambient Scent on Evaluation, Attention, and Memory for Familiar and Unfamiliar Brands. *Journal of Business Research*, 49: 157-165.

Nelson, P. J. (1970). Information and Consumer Behavior. *Journal of Political Economy*, 78(2): 311-329.

Pine, J. & J. Gilmore (1999). *The experience economy*. Boston, MA: Harvard Business School Press.

Schmitt, B. (1999). *Experiential Marketing*. New York: The Free Press.

Vogt, C., Fesenmaier, D. R., & K. MacKay (1993). Functional and aesthetic information needs underlying the pleasure travel experience. *Journal of Travel and Tourism Marketing*, 2(2): 133-146.

Vogt, C. & D. R. Fesenmaier (1998). Expanding the Functional Tourism Information Search Model: Incorporating Aesthetic, Hedonic, Innovation and Sign Dimensions. *Annals of Tourism Research*, 25(3): 551-579.

Woelfel, J. & N. J. Stoyanoff (1993). *CATPAC: A Neural Network for qualitative analysis of text*. Paper presented at the annual meeting of the Australian Marketing Association, Melbourne, Australia, September 1993.

Harmonise – Towards Interoperability in the Tourism Domain

Michele Missikoff[a], Hannes Werthner[b], Wolfram Höpken[c], Mirella Dell'Erba[b], Oliver Fodor[b], Anna Formica[a], Francesco Taglino[a]

[a] Laboratory for Enterprise Knowledge and Systems
IASI-CNR, Italy
{missikoff, formica, taglino}@iasi.rm.cnr.it

[b] eCommerce and Tourism Research Laboratory
ITC-irst, Italy
{werthner, dellerba, fodor}@itc.it

[c] IFITT, Int
wolfram.hoepken@start.de

Abstract

Harmonise tackles the interoperability problem in the tourism domain, where organisations following different standards should be enabled to exchange information in a seamless manner. The project puts a strong emphasis on the combination of a social consensus process with the application of new technologies. Carefully analysing and observing the requirements of the tourist domain and its current state of the art, a comprehensive methodology has been elaborated to achieve the desired harmonisation effect. Harmonise combines suitable tools into a platform supporting an ontology-based mediation process. In this paper we describe three fundamentals of the harmonisation effort: interoperability, ontologies and mediators. And we draw a vision of a future electronic tourism market based on these fundamentals.

Keywords: interoperability; harmonisation; ontologies; mediation; semantic web

1   Introduction

Harmonise is an EU project aiming at the interoperability problem in tourism, putting a strong emphasis on the combination of a social consensus process with the application of new technologies. Carefully analysing and observing the requirements of the tourism domain and its current state of the art Harmonise proposes a comprehensive methodology and identifies best practices to achieve the desired harmonisation effect.

The project comprises two major activities reflecting the two previously mentioned aspects: the primary aim of the project is to establish an open international consortium – *Tourism Harmonisation Network (THN)* – including major tourism stakeholders, domain experts and IT professionals. The THN serves as an open forum for treating interoperability issues and coordinating the related activities within the tourism domain. The second activity focuses on the technical issues to solve the interoperability problem by means of the so-called *Harmonise Platform*.

The Harmonise Platform relies on technologies such as ontologies, mediation, semantic annotation of data sources and semantic reconciliation. The goal is to allow participating tourism organisations to keep their proprietary data format while cooperating with each other, exchanging information in a seamless manner. Since nowadays most of the e-commerce activities depend on the Internet and the World Wide Web in particular, Harmonise follows closely its evolution towards Semantic Web – the web of meaningful data. The strategy is also to maintain compatibility with emerging but already widely accepted e-business frameworks and infrastructures such as ebXML and Web Services.

Harmonise doesn't necessarily intend to invent new technologies. Rather it analyses existing and emerging approaches to the interoperability problem and seeks an ideal combination of tools and technologies to apply them in the tourism domain. This paper will introduce the basic technologies identified as relevant for our purpose and to set the fundamentals for "integrating" fragmented systems and services within a global electronic tourism marketplace. The benefits of these technologies will be discussed and put in contrast with current mainly ad-hoc proprietary approaches. The paper also designs a vision how the future electronic tourism market could look like using the results of our harmonisation initiative.

2   Direct Approaches to the Interoperability Problem

The current situation of the electronic tourist market is characterised by heterogeneity. There are global standards, only used by big companies. And there exist standards considering the needs of small tourism suppliers, but used only on a national level. Since information systems are used within the tourism domain to support market processes, there have been many initiatives for harmonising such electronic markets and reaching global interoperability. Past and most of the current approaches are mainly based on the idea of fixed, obligatory standards, defining all details of the exchanged messages. Depending on the used communication mechanism, these standards are defined as fix-structured messages, e.g. UN/EDIFACT TT&L (UN/EDIFACT, 2002) or ANSI ASC X12I TG08 (ASCX12, 2002), XML-based messages, e.g. OTA (OTA, 2001), or function- or object-oriented application programming interfaces, e.g. HITIS (HITIS, 2002) or TIN (TIN, 1992).

2.1   International Standards

The most important reason for the failure of past standardisation initiatives is their lack of flexibility. Committing to fixed standards with all message details among all communication partners on a global level is too work-intensive and local or national particularities cannot be considered. All details of the exchanged messages, including all *technical* details depending on the communication mechanism, must be committed among all communication partners. This leads to a high effort for defining and maintaining such standards, especially in the case of a global coverage. These standards tend to be very complex and therefore are almost exclusively used by large companies, e.g. hotel chains, airline companies, global distribution systems (GDS/CRS), etc. The most famous examples for international standards are UN/EDIFACT TT&L and ANSI ASC X12I TG08.

New standardisation approaches try to eliminate these shortcomings by the usage of XML (eXtensible Markup Language). XML-based standards are easier to use and enable at least a limited flexibility and extendibility. However, the usage of XML will not solve the problem that all message details have to be committed among all communication partners. An example for an international XML-based standard is the specification of the OTA.

2.2 National Standards

On a national level the drawbacks of fixed standards are less problematic than on an international level and fixed standards have proved to be suitable at least for reaching interoperability among NTOs (National Tourist Organisations), their regional DMOs (Destination Management Organisations) as well as local suppliers. Examples for such standards are the specifications of the Portuguese destination information system SIGRT, the French destination information system TourinFrance or the Finish Tourist Board. However, this interoperability is strictly limited to a national level. Those standards enable to integrate the offer of local suppliers of one country, but there is no integration of the offer of different countries nor is the offer of local suppliers integrated into global distribution systems.

Given the heterogeneity of the Web and the missing of a central power instance we foresee that one global, all-embracing standard is very unlikely to be achieved. Instead, different standards for different market segments or participant types will coexist.

3   Moving to a Conceptual Level

Existing standardisation initiatives solve the harmonisation issues directly on the level of the technical solution, i.e. the defined standards include all technical details (e.g., XML details). However, an agreement among all market participants on this level is unlikely. Instead, a harmonisation should be independent of the technical solution and should take place on a more abstract, *conceptual* level.

Within the IT domain the separation between a technical level and a conceptual level is an important fundamental principle, for example, when modelling software systems, databases or data structures in general. Conceptual aspects, e.g. the meaning of entities of a problem domain, their characteristics and relationships to other entities, are specified independently of their technical representation, e.g. their storage structure. The same *concepts* can then be used for different technical solutions and changes on the technical level can be done independent of the conceptual level.

Applying the principle of separating conceptual and technical aspects to the problem of harmonising electronic tourism markets leads to new opportunities for reaching interoperability. The agreement of all market participants can now be restricted to the conceptual level, i.e. to the concepts behind exchanged messages, and can thus be reached much easier. The exact structure and format of the messages can be flexibly defined. The common conceptual level enables a mapping among different specific message formats. The following examples will demonstrate the principle of separating conceptual and technical aspects for reaching interoperability.

3.1   IFITT RMSIG

The IFITT RMSIG (IFITT Reference Model Special Interest Group, RMSIG 2002) is an initiative within IFITT intending to harmonise electronic tourism markets. The IFITT RMSIG provides a framework for modelling electronic tourism markets, a *reference model*. Instead of a fixed standardisation, the reference model enables the flexible description of specific models understandable by other market participants, based on a common modelling language. Following the fundamental principle of separating conceptual and technical aspects, the modelling language enables the conceptual modelling of electronic tourism markets on an abstract level, based on the *Unified Modeling Language* (UML). Concrete models for specific technical solutions and communication mechanisms can then be derived from a conceptual model. In this way, the reference model enables the description of specific models for specific standards or data exchange formats and supports interoperability among different technical solutions.

3.2   Semantic Web

The content of the current World Wide Web is designed for human readers and not for computer programs to automatically parse or manipulate. Semantic Web (Berners-Lee, 2001) is an extension of the current web, aiming to enable computer programs to access the content of human readable web pages. Following the principle of separating technical and conceptual aspects, the semantic web moves to a conceptual level, adding information about the meaning of web content. Together with the commitment of a set of common concepts, i.e. a common *ontology* as harmonisation on a conceptual level, the semantic web enables the access of computer programs to heterogeneous information from different sources and supports interoperability.

4   Ontologies

When cooperating or even when interacting in social settings, people and organisations must communicate among themselves. However, due to different contexts and backgrounds, there can be different viewpoints, assumptions and needs regarding, essentially, the same domain or the same problem. They may use different jargons and terminologies that are sometimes confusing and overlapping, and concepts and evaluation methods that are mismatched or poorly defined. The consequence is the lack of a *shared understanding* that leads to a poor communication among people and organisations (Uschold and Gruninger, 1996).

The goal of an ontology is to reduce (or to eliminate) conceptual and terminological confusion. This is achieved by identifying and properly defining a set of relevant concepts that characterise a given application domain.

Ontology can be defined as a *formal, explicit specification of a shared conceptualisation* (Gruber 1993, Ding et al. 2002). A *conceptualisation* is an abstract model of some phenomenon in the world, which identifies the relevant concepts and relationships among concepts of that phenomenon. The conceptualisation has to be *explicit* since concepts and their constraints have to be explicitly defined. Furthermore, *formal* refers to the fact that the ontology should be machine-understandable. Finally, *shared* means that ontology captures consensual knowledge, accepted by a panel of experts in the given domain. Therefore, ontology contains a set of concepts (e.g. entities, attributes, processes), together with their definitions and their inter-relationships. The use of ontologies can facilitate:

Understanding: Ontologies allow the key concepts and terms relevant to a given domain to be identified and defined in an unambiguous way. They enable a deep understanding of the meaning of these concepts and their relationships.

Communication: One of the most important roles ontologies play in the communication is that they provide unambiguous definitions of terms. The presence of unambiguous definitions facilitates the use and exchange of data, information, and knowledge among people and organisations. Moreover, ontologies facilitate the integration of different user perspectives, while capturing key distinctions in a given perspective.

Cooperation: Ontologies allow cooperation among people at both levels: *internal cooperation* (within the same organisation) and *external cooperation* (among different organisations). Furthermore, they are an optimal support for *information integration* (from the same domain or from different, but inter-related, domains).

Nowadays, ontologies are mainly considered from the perspective of Semantic Web. For this reason, a lot of academic efforts have been put on how to represent ontologies. Instead, there is less research work, let alone applications, to address the issues of how to utilise ontologies in the areas like knowledge management, e-business and so on. It could be foreseen that in the next years ontologies will be studied and utilised more and more in the industrial environments.

5   Mediation

Mediators were first introduced as part of distributed information systems in the late eighties. Their role is to process data from possibly several data sources and to prepare them for the effective use by applications (Wiederhold, 1992). Typical examples of mediators' functions are, for example, information integration of data from multiple databases or semantic matching and translation services among data representations in multiple systems. In any case mediators have to deal with heterogeneity on both sides: at the data sources as well as at the site of target applications.

Together with mediators, the terms middleware and wrappers are frequently used. Their function can support or substitute the mediation process. However, their purposes are clearly distinguished:

- to support the communication between data sources and users, traditional *middleware* may be used. The distinction is that middleware connects and transports data, but a mediator also transforms the content (Kleinrock, 1994)

- in order to make the mediation task manageable, the so-called *wrappers* are placed between mediators and information sources. Wrappers shield mediators from some aspects of heterogeneity in information sources. Typically wrappers only resolve structural differences in data representation at data sources without touching the content. (Tork Roth and Schwarz, 1997).

Due to the development of the WWW with its increasing number of heterogeneous data sources the mediation problem is gaining importance. Information retrieval should be more efficient and leverage the requirements on human involvement in the process. This is the typical role for mediators, which have to be designed to deal with a large number of resources, and heterogeneity at all levels. To achieve this, technologies specifically designed for the Web, are adopted in mediators. Semantic Web provides solutions to enable explication of the knowledge in the Web, and ontologies are becoming the primary interface to the mediation process and simplify its wide use.

6   Harmonise: Putting the Puzzle Together

The three topics described in this paper can be considered as pieces of a "puzzle", which, if put together carefully, may create an image of a future harmonised electronic tourism environment. This new landscape may include separate data sources maintained by destinations or any other information sources, application servers providing value adding tourism services to the final user, and, finally, specific tourism mediators dedicated to the specific "translation" needs between these applications and data sources. In this context ontologies and ontology processing software will play a key role, offering to the participants a platform, where they can

64

harmonise their understandings of the tourism world. Obviously, there will never be one common language for everybody. Therefore, specific mediators can be foreseen providing translation services. These mediators look at information from a higher conceptual (semantic) level using this level of abstraction for the mapping purpose.

The aim is to allow information systems to cooperate without the need to modify their software or their data organisation. In Harmonise each organisation sees the world using its own concepts and data schema, which is called the "Local As View" (LAV) approach (Levy, 2000). Thus, even if there is a shared common ontology of the business sector, each organisation may keep its own view, given the mapping services of Harmonise.

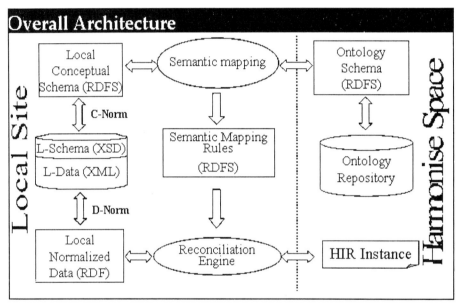

Fig. 1. The Harmonise Mediator

Figure 1 shows the overall architecture of the Harmonise mediator. The data model of a *source document* (i.e., the XML document) is first lifted to a *local conceptual schema* (*C-Normalisation*) and then *semantically mapped* to the terminology specified by the *shared ontology*. In the "lifting" process the granularity level of the local data models is also aligned to the representations used by the ontology. Note that the mapping process happens on a schema level. With this approach the transformation among local systems (standards) requires only linear number of "mappings", instead of an exponential one in the non-mediated scenario. The output of the mapping process is a set of *reconciliation rules,* which are used in order to transform the local data themselves and to code them according to the ontology content. Harmonise has chosen RDF(S) (Resource Description Framework) (W3C RDFS, 2000) as "language" for representing local conceptual schemata as well as the mediating ontology to enable the definition of mappings among their concepts. RDF(S) shows the needed expressiveness, simplicity as well as tool support. Additionally, it is the Semantic Web standard adopted by W3C.

Notice that the Harmonise approach to semantic interoperability is inherently different from other approaches that propose a unique interchange format, such as KIF (Knowledge Interchange Format). KIF is "neutral" with respect to the application domain, while the *Harmonise Interoperability Representation (HIR)* represents the exchange information by using ontology terms only.

Another approach in line of our proposal is PIF (Process Interchange Format). PIF is a translation language that has been conceived as a bridge among heterogeneous business process representations. It includes a core set of object types (e.g., activities, actors, etc.) that can be used to describe the basic components of any process, therefore allowing process descriptions to be automatically exchanged. However, the main difference is that PIF proposes a format, which resides on predefined (limited) process ontology, given with the respective standard. Conversely, HIR is based on a rich domain ontology, whose content is not linked with one specific method. It is initially built by the domain experts, and continuously evolves to keep pace with the changing reality.

7 Conclusions

In this paper we have presented a set of technologies, which are used to solve the interoperability problem in the tourist domain. We propose an ontology-mediated integration process based on a framework for data integration – instead of defining a new overall standard. Ongoing and future work concentrates on
1. setting up the "tourism harmonisation network" as the organisation responsible for ontology management as well the establishment of the Harmonise community;
2. continuing in the definition the tourism ontology with its rich concepts and their relationships as the basis of this ontology;
3. validating the Harmonise Mediator Prototype by connecting several organisations in the domain.

8 Acknowledgement

The Harmonise project is financed by the European Commission, under the 5th Framework RTD Programme, contract number IST-2000-29329. Hereby we would like to thank the partners of Harmonise project for their contribution to the realisation of this paper. These are as follows:

- T6 (Italy)
- ICEP (Portugal)
- LINK (Portugal)
- EC3 (Austria

References

ASCX12 (2002). American National Standards Institute Accredited Standards Committee X12 standards (ANSI ASC X12). ASC X12, www.x12.org.

Berners-Lee, T. et al. (2001). The Semantic Web. Scientific American, May 2001.

Ding, Y., Fensel, D., Klein, M., Omelayenko, B. (2002). The semantic web: yet another hip? Data & Knowledge Engineering 41 (2002) 205-207.

Gruber, T. R. (1993). A translation approach to portable ontology specifications. Knowledge Acquisition 5 (1993) 199-220.

HITIS (2002). Hospitality Industry Technology Integration Standards (HITIS). American Hotel & Motel Association, www.hitis.org.

Kleinrock, L. (1994). Realizing the Information Future: The Internet and Beyond; Computer Science and Telecommunications Board, National Research Council, National Academy Press, 1994.

Levy, A.Y. (2000). "Logic-Based Techniques in Data Integration." In Logic Based Artificial Intelligence, Jack Minker (ed.). Kluwer, 2000.

OTA (2001). Open Travel Alliance. www.opentravel.org.

RMSIG (2002). IFITT Reference Model Special Interest Group. www.rmsig.de.

TIN (1992). Die Touristische Informations-Norm (TIN) für den deutschen Fremdenverkehr - Empfehlung für Informations- und Reservierungssysteme. Bonn/München.

Tork Roth, M., Schwarz, P. M. (1997). Don't Scrap It, Wrap It! A Wrapper Architecture for Legacy Data Sources. Proeedings 23rd VLDB Conference, Athens, Greece, pp. 266-275, 1997

UN/EDIFACT (2002). United Nations Directories for Electronic Data Interchange for Administration, Commerce and Transport. United Nations Economic Commission for Europe, www.unece.org/trade/untdid.

Uschold, M., Gruninger M., (1996); Ontologies: Principles, Methods and Applications, The Knowledge Engineering Review, V.11, N.2, 1996.

W3C RDFS (2000). Resource Description Framework (RDF) Schema Specification 1.0. W3C Candidate Recommendation 27 March 2000. http://www.w3.org/TR/2000/CR-rdf-schema-20000327.

Wiederhold, G. (1992). Mediators in the Architecture of Future Information Systems. IEEE Computer, March 1992, pages 38-49.

XSLT-based EDI Framework for Small and Medium-sized Tourism Enterprises

Martin Blöchl, Wolfram Wöß

Institute for Applied Knowledge Processing (FAW)
Johannes Kepler University Linz, Austria
{mbloechl, wwoess}@faw.uni-linz.ac.at

Abstract

In the field of tourism electronic data interchange is of vital interest to almost every tourism enterprise. Small- and medium-sized tourism enterprises are virtually forced to interchange data with other competitors. The approach presented in this paper enables electronic data interchange based on the technologies XML and XSLT. These technologies are recommended by W3C, well supported and free of charge. The presented architecture enable easy configurable, flexible and cost-efficient EDI by transforming data stored within XML documents into a requested data format corresponding to a destination TIS. The transformation specification is defined within XSL style sheets; thus, necessary changes and adaptations do not affect the applications of participating TIS providers.

Keywords: tourism information systems (TIS); electronic data interchange (EDI); business-to-business (B2B); XML, extensible style sheet language transformations (XSLT)

1   Introduction

In recent years a large number of EDI (electronic data interchange) solutions have been developed and various efforts have led to the creation of a variety of standards for B2B applications. However, most approaches are limited to large-scale industry, especially in certain business sectors such as the automobile industry. For SMEs (small and medium-sized enterprises) there are several problems related to existing EDI solutions:

- Development, introduction and operation of EDI systems are time and cost intensive, which is a considerable technical and financial obstacle for SMEs.
- EDI systems have a limited accessibility regarding easy to use transactions between buyers and sellers. Due to the complexity of general (semi-) standards, as for instance the well known EDIFACT (Electronic Data Interchange For Administration, Commerce and Transport) standard, there are usually different sub-standards for different domains. Therefore, one-to-one solutions are often established.
- EDI systems are rigid and do not allow for intuitive interactions of flexible exchanges, whereas a key characteristic of SMEs is their flexibility.

Consequently, a new approach to EDI for an effective support of B2B E-commerce processes in tourism information systems (TIS) is required.

The paper is organized as follows.

After a short introduction to the relevant base technologies XML and XSLT in chapter 2, the approach to use an XSLT-based framework for EDI between tourism information system is presented in chapter 3. Chapter 4 shows an exemplary application of the EDI-framework. Chapter 5 concludes the paper and gives an outlook to further research.

2 Base technologies and standards

The presented approach uses exclusively standards approved by W3C (World Wide Web Consortium). Further information can be found on http://www.w3.org [October 3, 2003].

2.1 Extensible Markup Language (XML)

In the last few years XML has evolved into the universal language for EDI. XML, as technology, enables a user to transport data in a structured, platform independent and human-readable way. In contrast to HTML (hypertext markup language) XML is not per se thought to alter data for presentation purposes or to process data in any way (Harrold, 1999). XML is merely a structured storage container for data, while this data is transported to another application. The structure for the XML documents is defined in the respective document type definition (DTD). XML documents always represent a tree structure starting with a root element, which is unique in its XML document. This root element contains an arbitrary number of attributes or elements. Each of these child elements may itself contain an arbitrary number of attributes or elements, which allows to build up complex tree structures.

2.2 Extensible Style Sheet Language Transformations (XSLT)

XSLT is a subset of the extensible style sheet language (XSL). XSL formats data contained in XML documents for presentation purposes within the World Wide Web (WWW). XSLT is able to transform an XML document to another XML document containing the same or a subset of the data in a different structure (Kay, 2001). XSLT used in combination with XSL is mainly responsible for the transformation of data contained in a certain XML documents in order to present and process this data in Web applications. XSLT is also capable of minor operations on strings, numbers and Boolean values. The transformation operation is handled by an XSLT processor which transforms the XML document according to the rules specified in the applied XSL style sheet.

3 The XSLT-based EDI-Framework

The vision, not to implement an EDI mechanism from scratch, but to use existing and well-known technologies for the same concept, leaded to the development of the XSLT-based EDI framework (Blöchl and Wöß, 2002). One of the main problems with EDI from multiple sources is the high variety of heterogeneous XML documents. Especially in the field of tourism the supported data structures of a TIS (tourism information system) vary from almost every other TIS which is the consequence of the broad spectrum of functions a TIS is able to provide. To achieve a flexible and easily extensible EDI mechanism based on XML, XSLT is used as base technology in the presented concept. Figure 1 shows the process of a simplified EDI transaction.

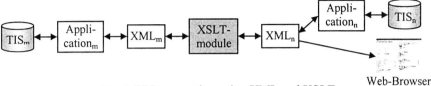

Fig. 1 EDI transaction using XML and XSLT

There are several general methods to transform incoming XML documents:

1. An incoming XML document is transformed by applying only a single XSL stylesheet which contains all transformation rules.
2. Necessary transformation rules are clustered in a number of different XSL stylesheets. A specific EDI application is used to determine which XSL stylesheet is required to perform the transformation.
3. An incoming XML document is transformed into a temporary XML document containing a reference to the XSL stylesheet, which is to transform the original XML document.

cf. 1): This alternative is suitable only for a limited number of XML documents. If there are too many different incoming XML documents the necessary XSL stylesheet used for the transformation becomes too complex. Hence, for this method there are considerable limitations concerning performance and stability.

cf. 2): This alternative bears significant disadvantages. If a new EDI scenario requires the development of a new or the adaptation of an existing XML document, the EDI application has to be updated as well. Adaptation of the software results in a time delay until the EDI application is ready to process the new XML documents. In addition, for this kind of adaptation a TIS provider needs a qualified team of software engineers.
A further problem is the stability and performance of the EDI application. Especially when using clusters of application servers there is a considerable problem, because if a new software release is required, it must be uploaded on every application server within the EDI cluster. Consequently after a software update, each application server needs to be restarted. In many cases this scenario causes additional problems and consumes time and money.

cf. 3): The presented approach adheres to this alternative. Two different transformations are sequentially processed on the original XML document. During the first transformation it is decided which XSL stylesheet is necessary to perform the second transformation for this particular XML document (see Figure 2). The decision is made in the XSL stylesheet by analyzing:

- element- and attributenames of the original XML document

- the structure of the original XML document, especially the exact sequence of required elements or attributes, specified by the used DTD
- the used DTD or the namespace extensions of the original XML document

The presented concept is shown in Figure 2.

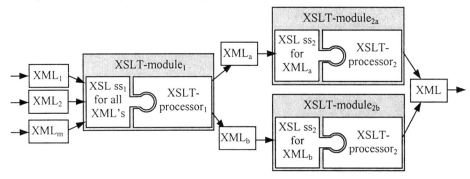

Fig. 2 The XSLT modules

In most cases analysing the used element- and attribute names of an XML document yields sufficient information to decide which XSL stylesheet is appropriate to process the second transformation step. For other cases, especially if a DTD was altered to cover the requirements of two or more TIS, the necessary information for this type of decision is extracted by analyzing the XML document structure. The third option, to analyze the used DTD or the used namespace extensions, is executed if the results of option one and two are not sufficient to determine which XSL stylesheet should be used for the transformation.

The XSLT module transforms different incoming XML documents into an XML document which corresponds to a predefined DTD. Data contained in this XML document is then either presented by a Web browser or stored within a database of the receiving TIS to be available for further tasks and processes.

The advantages of the XSLT-based concept are:
- Firstly, XSL stylesheets allow to easily extend an existing EDI-framework, thus, offering an important advantage for TIS providers. Secondly, due to an XSL stylesheet based determination of the XML document transform operations, it is possibly to react to new requirements just in time and without complex operations.
- If the XSL stylesheet which covers the transformation specification becomes too complex, it is possible to split it into a number of "smaller" XSL stylesheets, thus, splitting and reducing complexity and enhance performance.
- Generally, an EDI communication link is established by an XSL style sheet which defines the mapping between the source and the destination XML document. If a TIS provider alters its internal data structure and consequently the corresponding XML DTDs, it is sufficient to adapt the corresponding XSL style sheet by adding new transformation rules, removing obsolete rules, or modifying already existing transformation rules

in order to further support these new EDI requirements. The applications of the participating TIS providers are not affected by this customization process.

- The applied technologies are W3C recommended standards, free-of-charge and well supported. XML parser, XSLT processors and XSL editors are in many cases free available software packages.

- Due to an effortless maintenance of the EDI application and free available technologies the introduced concept is especially suited to satisfy the needs of small- and medium-sized TIS providers.

4 Example

This section gives a simplified example of transformations which are necessary to achieve the workflow illustrated in Figure 2.

```
<?xml version="1.0" encoding="UTF-8"?>
<!DOCTYPE room SYSTEM
  "http://www.url.com /dtd1.dtd">
<room>
 <name>3 bed room</name>
  <priceperday>
   <occupation>3</occupation>
   <board>HB</board>
   <price>120 EUR</price>
  </priceperday>
  <priceperday>
   <occupation>3</occupation>
   <board>FB</board>
   <price>132 EUR</price>
  </priceperday>
  <priceperday>
   <occupation>4</occupation>
   <board>BF</board>
   <price>108 EUR</price>
  </priceperday>
</room>
```

```
<?xml version="1.0" encoding="UTF-8"?>
<!DOCTYPE room SYSTEM
  "http://www.url.com/dtd/dtd1.dtd">
<?xml-stylesheet type="text/xml"
href="http://www.url.com/bsp12.xslt"?>
<room>
  <name> 3 bed room</name>
  <priceperday>
   <occupation>3</occupation>
   <board>HB</board>
   <price>120 EUR</price>
  </priceperday>
  <priceperday>
   <occupation>3</occupation>
   <board>FB</board>
   <price>132 EUR</price>
  </priceperday>
  <priceperday>
   <occupation>4</occupation>
   <board>BF</board>
   <price>108 EUR</price>
  </priceperday>
</room>
```

Fig. 3 The original XML doc. (XML$_1$)   **Fig. 4** The altered XML doc. (XML$_a$)

Figure 3 shows the original XML document which is planned to be transmitted to another TIS. This XML document was altered by using the XSL stylesheet in Figure 5. Figure 4 shows the resulting XML document, which also includes an additional processing instruction.

Firstly, the <xsl:output ...> element (Figure 5) specifies the output type "XML", the encoding and the version of the new XML document. In the next step the <xsl:for-

each ...> operation is executed and therefore each element of the XML document is processed. For each relevant element its name is assigned to a variable. Within the <xsl:choose> part the content of this variable is compared with hard coded strings, by using <xsl:when ...> operations. If the stored element name matches with one of the fixed strings, a collection of hand picked element names, then the corresponding DTD node and a new processing instruction for the second transformation are added to the output XML document. After the entire process the XML content of the original XML document is transformed into the new (output) XML document.

```
<?xml version="1.0" encoding="UTF-8"?>
<xsl:stylesheet version="1.0" xmlns:xsl="http://www.w3.org/1999/XSL/Transform">
  <xsl:output method="xml" version="1.0" encoding="UTF-8" indent="yes"/>
   <xsl:template match="/">
    <xsl:for-each select="*">
     <xsl:variable name="elemname" select="name()"/>
     <xsl:choose>
      <xsl:when test="$elemname = 'room'">
       <xsl:text disable-output-escaping="yes">
&lt;!DOCTYPE room SYSTEM "http://www.url.com/dtd/dtd1.dtd"&gt;
       </xsl:text>
       <xsl:text disable-output-escaping="yes">
&lt;?xml-stylesheet type="text/xml" href="http://www.url.com/dtd/bsp1_2.xslt"?&gt;
       </xsl:text>
      </xsl:when>
      <xsl:when test="$elemname = 'ROOM'">
       <xsl:text disable-output-escaping="yes">
&lt;!DOCTYPE ROOM SYSTEM "http://www.url.com/dtd/dtd2.dtd"&gt;
       </xsl:text>
       <xsl:text disable-output-escaping="yes">
&lt;?xml-stylesheet type="text/xml" href="http://www.url.com/dtd/bsp2_2.xslt"?&gt;
       </xsl:text>
      </xsl:when>
     </xsl:choose>
    </xsl:for-each>
   <xsl:copy-of select="."/>
   </xsl:template>
</xsl:stylesheet>
```

Fig 5 The first XSL stylesheet (XSL ss_1 for all XML's)

The XSL stylesheet presented in Figure 6 was added as processing instruction by the first XSL stylesheet (presented in Figure 5) to the altered XML document shown in Figure 4. This XSL stylesheet processes and transforms the altered XML document into the demanded final XML document (Figure 7).

Additionally, a number of string operations performed by this XSL stylesheet are necessary to achieve the correct results. For example, in the original XML document "price" and "currency" are stored within the <price> element. By applying substring() operations the values of the <price> element are extracted and two new strings are generated and stored as price and currency strings. The values stored within the

\<board\> elements of the original XML document are processed in an equivalent way. If the value of the \<board\> element of the original XML document matches with a fixed string a new value for the \<Board\> element in the result document is generated and inserted into the \<Board\> element.

```xml
<?xml version="1.0" encoding="UTF-8"?>
<xsl:stylesheet version="1.0" xmlns:xsl="http://www.w3.org/1999/XSL/Transform">
<xsl:output method="xml" version="1.0" encoding="UTF-8" indent="yes"
    doctype-system="http://www.url.com/dtd/dtd.dtd"/>
  <xsl:template match="room">
   <xsl:element name="Hotel">
    <xsl:element name="Room">
     <xsl:element name="Roomname">
      <xsl:value-of select="name"/>
     </xsl:element>
     <xsl:for-each select="priceperday">
      <xsl:variable name="pr_curr" select="price"/>
      <xsl:variable name="len" select="string-length($pr_curr)"/>
      <xsl:variable name="curr" select="substring($pr_curr, ($len)-3, $len)"/>
      <xsl:variable name="price" select="substring($pr_curr, 1, ($len)-4)"/>
      <xsl:element name="Prices">
       <xsl:element name="Occupation">
        <xsl:value-of select="occupation"/>
       </xsl:element>
       <xsl:element name="Board">
        <xsl:variable name="brd" select="board"/>
        <xsl:choose>
         <xsl:when test="$brd = 'BF'">
          <xsl:text>Breakfast</xsl:text>
         </xsl:when>
         <xsl:when test="$brd = 'HB'">
          <xsl:text>Half Board</xsl:text>
         </xsl:when>
         <xsl:when test="$brd = 'FB'">
          <xsl:text>Full Board</xsl:text>
         </xsl:when>
        </xsl:choose>
       </xsl:element>
       <xsl:element name="Price">
        <xsl:value-of select="$price"/>
       </xsl:element>
       <xsl:element name="Currency">
        <xsl:value-of select="$curr"/>
       </xsl:element>
      </xsl:element>
     </xsl:for-each>
    </xsl:element>
   </xsl:element>
  </xsl:template>
</xsl:stylesheet>
```

Fig. 6 The second XSL stylesheet (XSL ss_2 for XML_a)

```
<?xml version="1.0" encoding="UTF-8"?>
<!DOCTYPE Hotel SYSTEM
  "http://www.url.com/dtd/dtd.dtd">
<Hotel>
 <Room>
  <Roomname> 3 bed room</Roomname>
  <Prices>
   <Occupation>3</Occupation>
   <Board>Half Board</Board>
   <Price>120</Price>
   <Currency> EUR</Currency>
  </Prices>
  <Prices>
   <Occupation>3</Occupation>
   <Board>Full Board</Board>
   <Price>132</Price>
   <Currency> EUR</Currency>
  </Prices>
  <Prices>
   <Occupation>4</Occupation>
   <Board>Breakfast</Board>
   <Price>108</Price>
   <Currency> EUR</Currency>
  </Prices>
 </Room>
</Hotel>
```

```
<?xml version="1.0" encoding="UTF-8"?>
<!DOCTYPE ROOM SYSTEM
  "http://www.url.com/dtd/dtd2.dtd">
<ROOM>
 <ROOMNAME>3 bed room Deluxe
 </ROOMNAME>
 <CURRENCY>EUR</CURRENCY>
 <TAXRATE>20</TAXRATE>
 <PerPERSONPerDAY>
  <OCCUPATION>3</OCCUPATION>
  <BOARDPRICE>
   <BOARD>breakfast</BOARD>
   <PRICE>27.50</PRICE>
   <TAX>5.50</TAX>
  </BOARDPRICE>
  <BOARDPRICE>
   <BOARD>half board</BOARD>
   <PRICE>30</PRICE>
   <TAX>6</TAX>
  </BOARDPRICE>
  <BOARDPRICE>
   <BOARD>full board</BOARD>
   <PRICE>33.30</PRICE>
   <TAX>6.70</TAX>
  </BOARDPRICE>
 </PerPERSONPerDAY>
...
</ROOM>
```

Fig 7 The result XML document   **Fig 8** A corresponding XML document

The <xsl:element ...> operations generate the new elements and store the corresponding values within them. After the entire set of <xsl:element ...> operations is executed all XML elements of the resulting XML document are fully generated.

Figure 8 shows a further XML document that also could be used as input XML document similar to the XML document presented in Figure 3. The XSL stylesheet in Figure 5 already considers the structure of this XML document, and therefore it may be processed without an adaptation effort. Due to the lack of space it is not possible to present this particular transformation in detail.

5 Conclusion and further research

In this paper an approach for electronic data interchange based on XML and XSLT as flexible, powerful and standardized key technologies is introduced. The key characteristics of XML documents are the separation of the content from its presentation and the platform independence. The XSLT processor based concept has the advantage that new EDI requirements concerning the data interchange specification and data structures respectively, cause only low adaptation effort and requires neither a database administrator nor a software engineer. Hence, this concept offers a promising perspective for small- and medium-sized tourism information systems to participate in far reaching EDI processes.

Further research is required on the following topics:

- *Processing of other input types besides XML documents:* There are a number of TIS that do not use XML as EDI format, especially if such a TIS uses an aged quasi standardized EDI system. In most of these cases a structured text format is used as basis of the EDI process. To enable EDI also with this type of TIS and to correspond to their special needs, the introduced EDI framework has to be extended in order to support structured text as input type.
- *Security aspects concerning the XML based EDI-process:* During the last years several new security technologies and features have been developed in order to compensate exiting deficits concerning XML documents. Especially interchange of personal data or of sensitive information requires higher security standards than a common EDI process. In some cases, e.g. interchange of accounting transactions, there is a lively interest of both EDI partners to protect these data against potential misuse. Technologies like XML encryption, XML envelope or certain XML security packages are able to cover this special security requirements.

References

Blöchl M., & W. Wöß (2002). Flexible Data Interchange Based on XML and XSLT for Small- and Medium-Sized Tourism Enterprises. In K. W. Wöber, A. J. Frew, & M. Hitz (Eds.), *Enter 2002 – Information and Communication Technologies in Tourism 2002* (pp. 59-67). Innsbruck: Austria

Harrold E. R. (July 1999). *XML Bible 2nd Edition*, Hungry Minds Inc.

Kay M. (2001). *XSLT Programmer's Reference 2nd Edition*, Birmingham: Wrox Press Ltd

Heterogeneous Market Research Information in Tourism - Implementing Meta-Analytical Harmonization Procedures in a Visitor Survey Database

David M. Wagner
Karl W. Wöber
Institute for Tourism and Leisure Studies
Vienna University of Economics and Business Administration
(VUEBA)
{david.wagner, karl.woeber}@wu-wien.ac.at

Abstract

In the field of tourism various market research data from guest surveys is available. However, this valuable information cannot be compared as each entity uses different survey designs and therefore obtains heterogeneous data. Hence, extensive data is available but wise use of this data is impossible unless the information is harmonized in a comprehensive database. In this paper a comparable database is suggested that serves as a marketing information system and offers new opportunities for analyzing data from different sources. It discusses various comparability problems and shows how to harmonize heterogeneous data. Final result is a proposal of a quantitative model that integrates guest survey data from several European cities and allows access to a comprehensive database via the Internet.

Keywords: data heterogeneity; decision support system; comparability; data harmonization

1 Introduction

Comparability may defy precise definition, but it is an important and useful concept. By this we mean that data (estimates) for different entities can be legitimately (i.e. in a statistically valid way) aggregated, compared and interpreted in relation to each other. Comparability is a relative concept, as the best case scenario of absolute comparability is not attainable (Verma, 2002a). *Standardization* is a useful tool for ensuring that conditions for comparability are actually met. A more general concept is termed *harmonization*, which is taken "to encompass consistency, similarity, standardisation, etc., depending on the context" (Verma 2002b:192). In this paper, the concepts of standardization and harmonization are used as synonyms. The creation of highly standardized data sets is the main objective of this paper. Tourism marketing is becoming increasingly sophisticated as a result of greater importance attached to the reliability of information and the competent analysis of that information for the effective planning, monitoring and management of tourism entities (Bar-On, 1989). However, there are severe problems when someone wants to compare the results of market research initiatives. Clearly there is a need for a comparable international tourism database which contains data from different sources and harmonizes it in order to facilitate comparative analysis with competitors. By means of a harmonized data pool each user could enhance his learning opportunities and implement more sophisticated marketing strategies. As each entity lances its own research project, methods and design differ from case to case. Hence, there are a number of different guest surveys which all pursue the same objectives but try to reach them by using different strategies. In general there is little consistency between the different sample surveys in terms of survey techniques, definitions and presentation of analyses, so that comparability is greatly impaired (Devas, 1991). The first attempts to introduce a recognized uniform international system for tourism statistics were made by the World Tourism Organization (WTO). It published several manuals and reports "for providing guidance to national and local governmental statistical offices and the private industry in the implementation of WTO/UN Recommendations on Tourism Statistics" (World Tourism Organization 1995:i).

As a first step towards the long-term goal of comparable (city) tourism data European Cities Tourism (ECT) elaborated a proposal for a questionnaire design for guest surveys in European cities (FECTO, 1999). This common "blue-print" questionnaire serves as the point of departure for all national surveys. The questionnaire puts together a common set of questions which city tourism managers are recommended to use as the basis for tourism visitor surveys in order to achieve the level of consistency necessary for comparative analysis. The proposal's task can be seen as a certain stage of pre-harmonization, while this paper tries to post-harmonize both structural and content relations between the surveys. Since 1999, when ECT introduced the Eurocity survey, only a few cities have been able to participate in this project. Up to now, surveys in five European cities (Dublin, Edinburgh, Heidelberg, Tallinn and Vienna) were carried out and are examined in this paper. Each single survey was administered face-to-face by interviewers. In total 13,219 interviews are available where one interview consists of 180 variables on average. This number is going to rise as data from other international cities will join the database in the near future (e.g. Berlin, Amsterdam).

The study of the present questionnaires has shown that there exists a set of core questions that forms the basis of almost every guest survey. The main focus of data harmonization is on this set of questions as they provide the decision maker with the most important information needed. From a wide point of view, the questions are quite the same in terms of their content. The problem is that the characteristics of the questions (kind of survey, used wording, proposed answers in item lists) are different. That's why at first glance identical questions finally have different variables and values. This results in differences between guest surveys that can be separated into three main categories: technical, conceptual and semantic differences (Kotabe, 2001). Although the proposal with its set of core questions helps to improve the comparability and integration of statistics, each city adds or leaves out specific questions or changes some details concerning variables or values of the data set as it wants to adapt the questionnaire to its own needs. Therefore the issue of comparability still remains a problem to be solved. This leads to three main research questions to be dealt with in this paper:

1. Which types of data heterogeneity in guest surveys do exist?
2. How can heterogeneous data be compared and combined (showing an example of a method used for the purpose of comparison)?
3. How can these findings be used to generate a general model that can be applied to overcome some of the limitations associated with heterogeneity in (city) tourism guest surveys?
4.

The objective of this paper is to outline the database model and to introduce a decision support system offering access to a large number of information describing and analyzing visitor behavior in the field of international tourism. The present article is structured in three sections: First, the nature of heterogeneity in guest surveys is analyzed and different reasons for heterogeneity are listed. The second section points out some methods that can be used to standardize data from different sources and how these procedures can be applied in a harmonization model (illustrated with a case example). Finally, the last section describes how these procedures are evaluated in order to measure the overall success of the system.

The reasons for heterogeneous data can be divided into three categories while the boundaries between these concepts are blurred and open to discussion.

Non-textual differences: Technical aspects of heterogeneous data. Technical problems that can occur in comparing guest surveys are all deviations that are not tied to the content of the survey's questions, e.g. type of survey, sampling techniques and sample sizes (Häder et al. 2002), linguistic idiosyncrasies (Harkness, 2002), the order of questions (Schuman and Presser, 1996) or the use of multiple response questions. These differences are of technical origin as they are caused by the characteristics of data collection as a "tool" for research issues.

Textual differences: Conceptual aspects of heterogeneous data. Conceptual aspects are due to discrepancies in terms of the content of the question as they refer to the overlap in the definition of a given concept across the different surveys (Kotabe, 2001). They mainly result from different questions or variables respectively values used in guest surveys.

Textual differences: Semantic aspects of heterogeneous data. In contrast to conceptual differences the semantic aspect is a more subtle one. The differences concern the wording of the question, i.e. the meaning of a basically identical object. The same words can have different meanings in different sources, or different words can have the same meaning. Incompatible uses of the same word can be resolved by separating the words into different contexts and defining the mappings between words in these contexts (Tawakol and Singh, 1995).

2 Harmonization Model

2.1 Examples of methodological possibilities to overcome some of the heterogeneity problems

As a famous comparative researcher once put it over 30 years ago, "it is easier to explicate (the problems) than to suggest ways of dealing with them" (Verba 1969:54).

So the question is whether *sufficiently* comparable data can in fact be compiled in an international database. First of all it should be pointed out that data from guest surveys are harmonized at an aggregated (meta)level as the problem of missing data at the disaggregated level (i.e. within a particular item of the data set) is not target-oriented as this paper's objective is to harmonize data from different guest surveys. Furthermore, the problem of missing data has been dealt with extensively in the literature (Allison, 2001; Rubin, 1976). Thus, the analysis takes place at an aggregated and higher level. According to the different aspects of data heterogeneity in each particular question of the surveys under comparison, different meta-analytical harmonization procedures are used. Therefore it is called a "meta-analytical" approach. Main emphasis of the model is on the development of a system for supporting the production of harmonized aggregates from different data sources. General guidelines for the construction of the model are the following principles:
- The system should be able to cope with different types of data sources
- Production of statistical aggregates should be formalized as far as possible
- For aggregate production the model should be able to use data are not necessarily stored in one homogeneous database but come from heterogeneous sources
- The final model has to be implemented in the Tourism Marketing Information System TourMIS

The central idea of the harmonization model (Figure 1) is to use only standardized terminology inside the system and to translate actual data sets into this "language". In this case example, ECT's questionnaire is the language of standardized data (*reference frame*), its variables and values are the defined terms (*conceptual variables* resp. *conceptual values*) and the harmonization procedures are the dictionary to be used. So the main issue is to decide on corresponding variables and to compare their particular values. The connection between the conceptual variables in the reference frame and the actual observed *source variables* of the different data sets (*source frame*) is achieved by diverse harmonization procedures (*data source mappings*). These procedures may be different, but their target is the same: To translate actual values from different sources into the unifying language of the reference frame. This can be done by rules that define the semantic relationships between different concepts (Singh, 1998). Note that, although some information might be lost, the harmonization process allows inspection of the loss of information due to standardization. Generally, due to the different complexity of the harmonization problem, semantic discrepancies between datasets are easier to be harmonized than conceptual or technical aspects. Irrespective of the type of difference, in each specific stage of data harmonization the procedure that loses the fewest information in comparison with the original data is used. To provide for the best solution, a decision tree is implemented in a database model in order to make heterogeneous data comparable by choosing the most convenient harmonization procedure.

One methodological possibility to overcome the heterogeneity problem is the use of estimation methodologies. Additionally, missing time series (variables) can be calculated by using forecasting procedures or by applying general calculation rules. In the case of missing variables and/or values, substitution by average figures (estimates based on surrogate values or pattern-mixture models) might also be appropriate (Wöber, 1994).

Fig. 1. Overall architecture of harmonization model

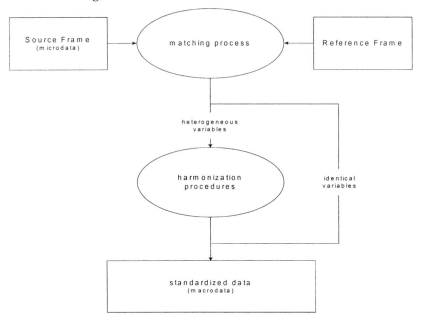

There is one main research question that can be split in many sub-hypotheses resulting from comparability problems between special variables and/or values: How can heterogeneous data from Eurocities be modified according to the Eurocity standard without losing more information than it is gained out from the inter-city comparisons? Figure 2 shows the very general concept of the matching process. Existing data from guest surveys is compared with the reference frame (ECT's proposal) and then split into one of three groups: there are either additional or missing variables that can not be compared with the core variables and therefore are not harmonized, or there are slightly different variables that vary on one or more of the aspects presented before. Only the last case poses problems in terms of standardizing data from different guest surveys. These difference may be due to missing values, additional values or semantic differences (Behling and Law, 2000). Heterogeneous data is harmonized and finally comparable with ECT's core variables. Finally, data following the European Cities Tourism standardization proposal is not affected by the harmonization procedures.

2.2 Case Example

As an example for a possible harmonization procedure the question concerning means of transportation used by the visitor was selected here. For this question the ECT survey proposes "On your journey TO Eurocity, what was your main form(s) of transport?" (FECTO 1999:46) and suggests items listed in the right column of Table

1. This multiple response question is compared with the same question from Vienna's guest survey. In Vienna, 8,869 interviews were held in 1999 while the city was not following the proposal at all.

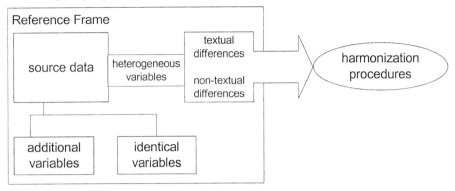

Fig. 2. Matching process of heterogeneous data

As Table 1 shows, Vienna's values are quite different from the proposal as some forms of transport are named differently (semantic aspect) while others are additional/missing variables (conceptual aspect).

In addition, contrary to the proposal, Vienna's variable of form of transport allows only one response (technical aspect). Thus, there are two problems (different items and the lack of a multiple response question) which lead to two stages in the process of harmonization, namely harmonization of textual and non-textual differences.

Table 1. Means of transportation used (different candidate values for a scale)

Source **Vienna** guest survey	Values **ECT** proposal
Private car/van	Private car/van
Hired car/van	Hired car/van
Train	Train
Plane	Plane
Charter flight	*Boat/ferry*
Bicycle	Bicycle
Motorbike	Motorbike
Bus	*Public bus/coach*
Caravan	*Private bus/coach*
	Walked/hitch-hiked
Other	Other (specify)

First the problem of semantic adaptation has to be solved (Table 2). The term "charter flight" can be seen as a synonym for airplane, while this latter expression can be found in the reference frame. "Caravan" is not included in the proposal, therefore it is added to the most similar term, in this case "private car/ van". The missing values "boat/ferry", "public bus/coach", "private bus/coach" and "walked/hitch-hiked" are added so that eventually Vienna's variable corresponds to the reference frame from a semantic point of view. After this step, the structure of the data is the same, but the harmonized numbers are still missing. Therefore, the next issue is to generate and calculate modified and/or missing values (Table 3). General calculation rules (equation 1) are applied for values with the same concept but different notions, e.g. the items "plane" (ECT) and "plane" and "charter flight" (Vienna).

Table 2. Example for scale formation

Vienna's original data	Semantically adapted data	Values ECT proposal
Private car/ van	Private car/ van	Private car/ van
Hired car/ van	Hired car/ van	Hired car/ van
Train	Train	Train
Plane	Plane	Plane
Charter flight	Plane	
	Boat/ ferry	Boat/ ferry
Bicycle	Bicycle	Bicycle
Motorbike	Motorbike	Motorbike
Bus	Public bus/ coach	Public bus/ coach
Bus	Private bus/ coach	Private bus/ coach
Caravan	Private car/ van	
	Walked/ hitch-hiked	Walked/ hitch-hiked
Other	Other	Other (specify)

The two separate values of Vienna's survey simply are summarized to ECT's more general value "plane":

$$34.1\% + 1.8\% = 35.9\% \tag{1}$$

This is a convenient strategy, especially when one value is actually very small ("charter flight" with 1.8%). The same approach is used for Vienna's "caravan", which is added to the more general concept of "private car/van".

In order to complete the other missing values, a harmonization procedure quite similar to pattern recognition is designed. Hence, the issue is to create a model that adapts Vienna's answers to the proposal, i.e. to modify missing and additional values of the variable "means of transportation". In this case a pattern-mixture model with Heidelberg (Table 4) as reference city is used. Heidelberg was identified as being the best available reference city for Vienna since this city allows multiple responses for the question under evaluation (Heidelberg's guest survey follows ECT's proposal in this particular item of the questionnaire) and has similar topographical characteristics as Vienna. Missing values ("boat/ferry" and "walked/hitch-hiked") are taken directly from Heidelberg's pattern. In this case the factor for multiple response is already included. Vienna's value "bus" is split into "public bus/coach" and "private bus/coach" using the pattern's code of 42.3% respectively 57.7%.

$$\text{Public bus} = 4.1\%; \text{key} = 4.1 / 9.7 = 0.423$$
$$\text{Private bus} = 5.6\%; \text{key} = 5.6 / 9.7 = 0.577$$
$$\text{Total} = 9.7\% \tag{2}$$

From a textual point of view, after this step Vienna's data corresponds to the proposal. The second level of the harmonization procedure deals with the technical aspect of heterogeneous data. As Heidelberg uses a multiple response question, the total percentage of items chosen by the respondents is 115.7% instead of 100% as in Vienna's guest survey. To consider this aspect each form of transport that already exists in Vienna's item list is multiplied by the factor of 15.7%. This approach can be adopted assuming that there are no preferences within the items of the multiple response question, i.e. the items of the question are equally distributed. Finally, the data of Vienna's guest survey concerning the main form of transportation corresponds with ECT's proposal and can be used for comparative analysis.

Table 3. Harmonization steps

Vienna's original data		Step 1: textually adapted data		Step 2: harmonized data	
Private car/van	21.1%	Private car/van	21.1%	Private car/van	25.2%
Hired car/van	2.3%	Hired car/van	2.3%	Hired car/van	2.7%
Train	26.9%	Train	26.9%	Train	31.1%
Plane	34.1%	Plane	35.9%	Plane	41.5%
Charter flight	1.8%				
		Boat/ferry		Boat/ferry	0.4%
Bicycle	1.2%	Bicycle	1.2%	Bicycle	1.4%
Motorbike	0.3%	Motorbike	0.3%	Motorbike	0.4%
Bus	10.6%	Bus	10.6%	Bus	
Caravan	0.7%	Caravan	0.7%	Caravan	
		Public bus/coach		Public bus/coach	5.2%
		Private bus/coach		Private bus/coach	7.1%
		Walked/hitch-hiked		Walked/hitch-hiked	0.7%
Other	1.2%	Other (specify)	1.2%	Other (specify)	1.4%

Table 4. Means of transportation of visitors to Heidelberg

Private car/van	53.6%	Motorbike	0.3%
Hired car/van	7.6%	Public bus/coach	4.1%
Train	29.7%	Private bus/coach	5.6%
Plane	11.5%	Walked/hitch-hiked	0.7%
Boat/ferry	0.4%	Other (specify)	1.1%
Bicycle	1.1%	total	115.7%

2.3   The Eurocity Database in TourMIS

European Cities Tourism has established a City Marketing Information System (CityMIS) which contains data from participating cities and which is updated continuously. The system is part of the Tourism Marketing Information System TourMIS (www.tourmis.info [September 15, 2002]) which is developed and maintained by the Institute for Tourism at VUEBA and the Austrian National Tourist Office. The system allows the rapid retrieval of data, offers procedures for basic statistical operations and the comfortable production of tables and graphs (Wöber, 1998). The presented harmonization model is implemented in TourMIS. The nature of heterogeneity that occurs when comparing city tourism guest surveys is disassembled in its generic components and categorized by applying the technical solutions to overcome these problems of heterogeneous data. Of course, the developed model has not only been designed for this special issue. It can also be used in an even more international context and is not only restricted to the field of tourism.

3 Outlook

3.1 Evaluation of the model

In order to measure the accurateness of the proposed concept and the overall success of the system, the paper discusses an evaluation strategy which will be conducted in the near future.

An experimental design will be used, consisting of one experimental and one control group of tourism managers, where the experimental group uses reports generated by the CityMIS system for responding to the case problem, while the control group will use original reports which were compiled for each individual city. The final result should demonstrate the accurateness and effectiveness of the model, where this effectiveness is defined in terms of achieving its objectives, specifically its role in increasing decision-making effectiveness (Silver, 1991). Hence, effectiveness should be evaluated on two criteria: Decision quality and decision-making efficiency. Decision support systems help decision makers in the process of generating information and help them avoid the potential mistakes in forming the judgement. Therefore, users of CityMIS should make better decisions. Concerning decision-making efficiency, members of the control group may have to turn to outside resources, leading to more time to make a decision. Therefore, users of DSS should make decisions faster. The perceptions of the managers on the impact of the DSS are measured using a self-reporting questionnaire. All hypotheses are one-tailed with an expectation of improvement in the decision parameter (e.g. speed, organization of thoughts, confidence about decision taken etc.). Persuasive evidence that the use of CityMIS improves the decision-making process is expected.

3.2 Conclusions

The multiple harmonization procedures presented here lead to a quantitative model integrating guest survey data from various cities and offering a way to perform on-line data comparisons. The model is conceptualized as a retrieval system for storage and access to key marketing variables by a user-friendly interface. Information can be accessed via the Internet and statistical operations can be computed for a subset of cities or for all cities in the database.

Advantages of this City Tourism Marketing Information System are:
- The increased sample size that will generate more precise estimates of relevant effects and facilitate subgroup analyses that were not possible before.
- Comparing findings of diverse studies can increase evidence of the generalizations of the studies, generate new hypotheses and direct attention to areas which need further research.
- Meaningful comparative analyses give tourist managers the opportunity to develop strategies taking into account the competitors' situation.

Although the model offers its users a wide variety of information and statistical operations one should bear in mind that its output is only the second best solution. "The only really good solution to the missing [comparable] data problem is not to have any. Statistical adjustments can never make up for sloppy research" (Allison 2001:2). In order to get access to an even more comprehensive database on (city) tourism, efforts should be made to agree on a mandatory survey-proposal to ensure that different sources deliver comparable data.

References

Allison, P.D. (2001). *Missing Data*. Sage University Papers Series on Quantitative Applications in the Social Sciences, 07-136. Thousand Oaks, CA, Sage.

Bar-On, R. (1989). *Travel and Tourism Data: A Comprehensive Research Handbook on World Travel*. London, Euromonitor.

Behling, O. and Law, K.S. (2000). *Translating Questionnaires and Other Research Instruments: Problems and Solutions*. Sage University Papers Series on Quantitative Applications in the Social Sciences, 07-133. Thousand Oaks, CA, Sage.

Devas, E. (1991). *The European Tourist – a Market Profile*, 5th edn. London, Tourism Planning and Research Associates.

FECTO (1999). *Proposal for a Questionnaire Design for Tourism Visitor Surveys in European Cities*, 1st edn. Vienna, ASART.

Häder, S., Gabler, S., Laaksonen, S., Lynn, P. and Purdon, S. (2002). *Sampling for the European Social Survey*. ESS Work-package 3.

Harkness, J.A. (2002) forthcoming. Questionnaire Translation. In: J.A. Harkness, F. van de Vijver and P.Ph. Mohler (eds.) *Cross-cultural Survey Methods*. New York, Wiley and Sons.

Kotabe, M. (2001). Using Secondary Data in International Business Research: Opportunities and Risks. In: B. Toyne, Z.L. Martinez and R.A. Menger (eds.) *International Business Scholarship: Mastering Intellectual, Institutional, and Research Design Challenges*. Wesport, Quorum: 158-179.

Rubin, D.B. (1976). Inference and Missing Data. *Biometrika* 63, 581-592.

Schuman, H. and Presser, S. (1996). *Questions & Answers in Attitude Surveys*. Thousand Oaks, CA, Sage.

Silver, M. (1991). Decisional Guidance for Computer-Based Decision Support. *MIS Quarterly* 15 (1): 105-122.

Singh, N. (1998). Unifying Heterogeneous Information Models. *Communications of the ACM* 41 (5): 37-44.

Tawakol, O. and Singh, N. (1995). A Name Space Context Graph for Multi-Context, Multi-Agent Systems. In: *Proceedings of the 1995 AAAI Fall Symposium Series*. Cambridge, MS.

Verba, S. (1969). The Uses of Survey Research in the Study of Comparative Politics: Issues and Strategies. In: S. Rokkan, S. Verba, J. Viet and E. Almasy (eds.) *Comparative Survey Analysis*. The Hague, Mouton.

Verma, V. (2002a). *Comparability in International Survey Statistics*. Paper presented at the International Conference on Improving Surveys 2002, Copenhagen.

Verma, V. (2002b). Comparability in Multi-country Survey Programmes. *Journal of Statistical Planning and Inference* 102: 189-210.

Wöber, K. (1994). Tourism marketing information system. *Annals of Tourism Research* 21: 396-399.

Wöber, K. (1998). TourMIS: An adaptive distributed marketing information system for strategic decision support in national, regional, or city tourist offices. *Pacific Tourism Review* 2: 273-286.

World Tourism Organization (1995). *Collection and Compilation of Tourism Statistics*. Madrid, World Tourism Organization.

Italian Tourism Virtual Communities: Empirical Research and Model Building

Magda Antonioli Corigliano[a],
Rodolfo Baggio [a]

[a] Master in Tourism and Economics
Bocconi University, Milan, Italy
{magda.antonioli, rodolfo.baggio}@uni-bocconi.it

Abstract

Virtual Communities are considered an effective mean to achieve good and stable relationships with customers, thus attaining higher levels of economic returns from a website. Italian tourism websites suffer from a general poor capacity in terms of offering good quality contents and services to their visitors. As a consequence, online Italian tourism market is one of the smallest in the world of Internet. This paper analyses the usage of community features by Italian online tourism operators and derives a viable model for a successful implementation of such features.

Keywords: virtual communities, tourism websites, Italy.

1   Introduction

Nowadays it has been well accepted that tourism on the Internet represents a sector of great importance; the digital travel segment, one of the most significant on the Web, is characterized by a steady growth, and the revenue forecasts are still on the rise. Even if the demography of the Italian internet tourism is in general conformity with that of the rest of the world, the e-commerce penetration is at a much lower level than the one of other countries. This consideration is particularly true concerning online tourism field, where the proportion of online transactions is one of the lowest in the world. It has been estimated that in 2002, electronic sales in our country will be only 0.25% of the whole tourism market share, while the European average could be around 2.8% and the American average (that is to say USA's and Canada's) may reach 10% (Antonioli and Baggio, 2002a).

Such a difference can be considered a consequence of the gap in terms of general e-commerce development, with worries concerning the security of data and payments and with the historical reservations of customers about mail order sales. But these explanations are, obviously, inadequate. Recent surveys (see, for example, http://www.cyberatlas.com or http://www.nua.com), on the reasons why cybernauts do not buy tourism and travel products online, say that they prefer the traditional travel agent, they do not trust web sites or they maintain they do not know "appropriate" sites. Basically, this means that the contents and the services offered online are not considered satisfactory by the users. The aforementioned has proven to be truthful for the Italian tourism web sites (Antonioli and Baggio, 2002b). The main findings are shown in the following figures.

Fig. 1 shows the distribution (in terms of percentage of sites which contains that particular type of content) of the main kinds of website contents: customer relationship services (CUST.REL), e-commerce functionalities (ECOMM), informational materials (INFO) and general interactive services (SERV).

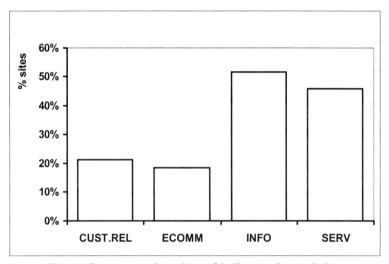

Fig. 1 Contents and services of Italian tourism websites

The user evaluations (given on a scale from 0 to 10) of the main characteristics of Italian tourist websites are shown in Fig. 2. The total average value is 6.58, a hardly sufficient grade.

For tourism intermediaries (tour operators and travel agencies) such a low quality represents a huge risk of disintermediation. Since traditional intermediaries do not offer valid web sites, the user is more tempted to turn straight to the main providers (hotels, airline companies, etc.) to fulfil his vacation needs.

For other tourism operators, the relative poverty in terms of contents and services generates directly a general distrust that can eventually affect also the economic performance of the "real world" (see, for example, Gaudin, 2002). And this situation becomes even more serious if the operator is a destination marketing organization because of the effects that these circumstances can have on the development of the whole destination. This is mainly true for public organisations that, in Italy, cannot directly market and may only attract visitors by offering interesting contents.

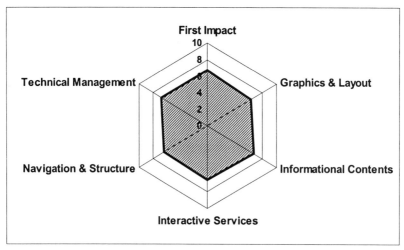

Fig. 2 Evaluations of Italian tourist websites

One possible help in building good quality websites and giving the users more interactive services is the exploitment of a virtual community.

Online travel communities are every day more utilized by tourists to satisfy their information and communication needs, to exchange ideas, impressions and suggestions. On the other hand, tourism organisations may appreciate the importance to use this "tool" to improve their capabilities and their relationships with the customers. This is necessary in order to develop products which could better respond to the market, thus increasing the possible economic returns. Creating a virtual community requires a thoughtful understanding of a range of elements: the overall concept, the necessary technologies supporting community building and the management requirements able to translate community systems into practical marketing tools.

The general hypothesis in assessing the value of a virtual community comes from the fact that, if an individual identifies himself with a community, then he will be probably accepting the models coming from that community. His behaviour as a consumer will be influenced in such a way that he will overcome much of the distrust which seems to be the main inhibitor to the spread of online commercial transactions. This business model is based on the belief that a community represents a set of loyal consumers that have commercial value, thus extending this value to the whole community. This work seeks to analyse the possibility to build efficient and effective virtual tourist communities, to assess the validity of this business model and to design a practical model able to reach, in short times, a self sustaining development in terms of users and contents. All these objectives are essential in order to give all the participants the feeling they have a place where it is possible to satisfy their travel needs and desires from an information and a commercial point of view.

2 A theoretical approach

A Virtual Community (VC) may be seen as a group of people that gathers electronically to discuss specific topics, ranging from hobbies to professional or academic matters. The participants are linked by a common interest. A virtual community does not have any geographic boundary and, usually, any person can be part of it.

Two main approaches can be identified in definitions found in the literature. The first one (Hof, 1997; Hagel and Armstrong, 1997) simply regards a VC as a set of subjects that communicates by using the tools provided by the *Computer Mediated Communication* (CMC), not giving much importance to the emotional aspects that arise from the relationships created, and emphasizing the utilitarian side of the connections deriving from the achievement of commercial objectives. The second one (Rheingold, 1993; Jones, 1997) mainly stresses the emotional and social aspects of the relationships among the subjects forming a VC.

All the definitions have some common points: the connection by means of telematic equipments, a common interest shared by the members of the community, the feeling they belong to the community, the idea that the latter can initiate a virtuous circle capable of creating a good amount of high quality content that may provide significant returns in terms of reputation and revenues for community organizers.
Virtual communities can be classified in four main categories (Hagel and Armstrong, 1997), based on the needs which people wish to satisfy with their participation: interest, relationship, fantasy, and transaction.

Interest aggregates people with a specific interest or expertise in a certain area; relationship gives participants with common experiences the possibility to share them and to develop significant personal interactions. Fantasy provides an opportunity for people to come together and to deal with matters of imagination and entertainment. The transaction communities are based on the exchange of information and/or products and services among the participants.

Similar classifications, even with some differences in the wording, have been given by several authors (Carver, 1999; Jones and Rafaeli, 2000).
It must be noted, however, that very seldom a VC has a simple connotation; more frequently they express hybrid characteristics making difficult, as a consequence, their classification.

Many virtual communities show an economy of increasing returns: they start providing very low revenues at the initial stage, then the revenues increase as the audience builds up and they have a sharp acceleration when the threshold called "critical mass" (Hagel and Armstrong, 1997) has been reached.

According to Subramaniam, Shaw and Gardner (1999) virtual communities show increasing returns due to marginal cost effect, learning curve effect and network externalities effect.

In the building of a VC Hagel and Armstrong (1997) suggest a four steps process:
1. *attract* members to join the community with a marketing strategy;
2. *promote* the participation by carrying the contents and providing facilities and services;
3. *build* a sense of belonging and loyalty to the group by enhancing the interactions;
4. *capture* the value generated in terms of unique content and revenue coming from targeted advertising or trading of information, products or services.

Obviously the process is a circular one and, by capitalizing the achieved activities and results, the cycle starts again.

The main advantages in creating and managing a virtual community may be summarized as follows (Kambil and Ginsburg, 1998:96): "First, it generates stocks of new, unique and proprietary content that is hard for competitors to replicate. Second, ... virtual communities can potentially create significant returns to scale in content production, reputation, and revenues for community organizers. As a site becomes more useful it can generate more subscribers who, in turn, can contribute new useful content, generating positive externality".

3   Methods

Aim of this work is to define a viable model of virtual community for a tourist website. As it has been seen above, one of the most critical problem for Italian tourism websites is the poverty in terms of contents and interactive services for the visitor.

To achieve this objective, a different classification of virtual community, this time based on the communication tools that are used, is needed. The tools generally utilized to provide community services are (Long and Baecker, 1997):

- *email*: the first minimal level of communication with interested customers;
- *newsletters*: sent regularly after a registration;
- *board*: electronic boards or guest books where visitors can publish online comments or specific requests;
- *forum*: one or more areas of active discussions among visitors with the participation of the community organizers;
- *chat*: interactive channels of simultaneous discussions (with textual and/or graphical interface).

These tools are considered a conventional value (see Table 1). By adding values corresponding to the features offered on the website, a *community index* can be calculated.

Table 1 Community tools values

Tool	Value
email	0
newsletter	1
board	2
forum	3
chat	4

It is possible to classify the level of community offered by a site by calculating the *community index* as shown in Table 2.

Table 2 Community levels

Level	Description	Community Index
0	No community	0
1	Minimal	$1 \div 3$
2	Medium	$4 \div 6$
3	Good	$7 \div 9$
4	Full	10

A random sample of 300 Italian websites has been chosen. The websites have been analysed to identify community services and to classify their types.

All the sites received a (e-mail) questionnaire asking for quantitative data on the following topics:

- average monthly number of visitors;
- average monthly number of HTML pages viewed;
- average monthly revenues;
- average monthly number of visitors making a commercial transaction.

A number of 85 websites have compiled the questionnaires that have been collected and analysed. A very small number of answers, not statistically significant, contained data regarding the economical side (revenue and buyers) of the survey, still considered as "reserved matters". Therefore the number of visitors and the pages viewed have been used as an index of website popularity and success.

4 The results arising from an empirical test

All the websites of the sample may be considered going thorough their mature stage (ages have a mean of 5.1 years and a median of 5). The distribution of the community tools used is shown in Fig. 3; only a small proportion of the sites offer their visitors advanced community tools.

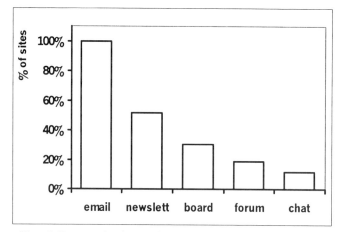

Fig. 3 Communication tools used in Italian tourist websites

The *community indexes* measured for the sites of the sample are shown in Fig. 4.

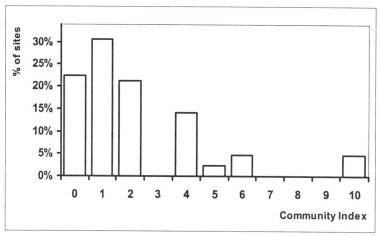

Fig. 4 Distribution of community indexes for Italian tourist websites

Comparing the popularity of the sites, measured by using the average number of pages (html) read online by the users and the average monthly number of unique visitors, with the community index, it is possible to show the importance of community tools to increase the audience of a website and, as a consequence, the higher value of a virtual community for the online tourism sector.

A linear regression on the data shows only a slight positive trend, but with a very low significance ($R^2 = 0.02$). This is mainly due to the high concentration of very low usage of interactive services by the Italian sites. Of greater interest are the distributions obtained plotting the popularity data with the community indexes (Fig. 5 and Fig. 6).

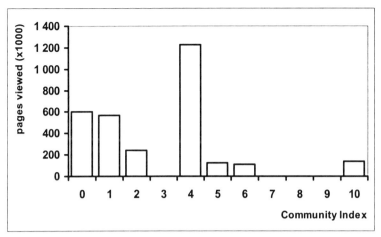

Fig. 5 Pages viewed and community index

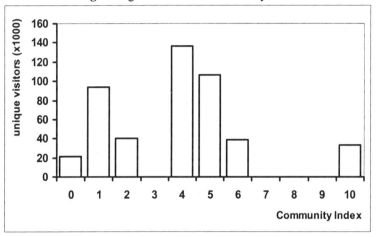

Fig. 6 Unique visitors and community index

It is rather clear that the highest popularity (in terms of the mean values of pages viewed and unique visitors) belongs to sites with medium levels of community features, usually consisting of newsletters and discussion forums. Implementing and running a virtual community is a growth factor for the audience of a tourist website, able to insure the production of a significant amount of high quality content and increase the level of economic returns that may be expected from a commercial website. The high loyalty of the members a virtual community can eventually bring to the building of a good and stable customer base.

Further analyzing sites with the highest popularity, it is possible to derive a list of elements that have shown to be very successful in building and managing a virtual community. Apart from the obvious richness and completeness in terms of informational contents and tools to keep visitors informed (newsletters) about the different areas of the website, these sites usually show:

- multiple levels of interactivity: well organized multithematic forums dedicated to different possible destinations or different types of vacations, including areas where it is possible to meet potential fellow travellers by requesting or offering company for a trip;
- some kind of "human" guide proposing and commenting trips and destinations and keeping a good contact with the site visitors via online (forums or chatboards) or offline (e-mail) channels;
- proper use of different types of media, offering services through various channels (web, e-mail, sms, etc.), including the possibility to have a direct contact via green telephone numbers;
- large areas where the visitors are the main contributors to the content of the website through stories, observations and comments about real travel experiences or new travel proposals.

These elements may be rightly considered "best practices" according to the definition given by O'Dell and Grayson being a "practice, know-how, or experience that has proven to be valuable or effective within one organization and that may have applicability to other organizations" (O'Dell and Grayson, 1998: 13).

5   Conclusions

Virtual communities are a very important aspect of the Internet world, both from a social and an economical point of view. It has been suggested that virtual communities may play a central role in marketing tourist destinations and services over the Internet (Dellaert, 1999) and may become an important tool to overcome the mentioned poverty of Italian tourism websites in terms of contents and services (Antonioli and Baggio, 2002a, 2002b).

As a matter of fact, communities' participants provide valuable information to other tourists about travel destinations or tourism service providers.

The analysis of Italian tourist websites shows a relatively low presence of community tools: less than 20% offer a discussion forum and only 12% has some chat facilities.

An empirical test performed on a sample of Italian tourism online websites clearly shows that the sites that use these tools, even at a moderate level, have higher amounts of visitors and pages viewed. Sites with a community index value between 4 and 6 (i.e. using three community tools) exhibit 40% more pages viewed and 140% more unique visitors than sites with minimal or no use of community features.

It is therefore possible to suggest that the adoption of interactive features encourages the building of a community. In particular, a discussion forum and a regular newsletter prove to be very useful in fostering the creation of noteworthy informational contents, which are able to increase the audience of a tourist website and to raise the time spent by the users in browsing the site's pages.

The content generated will then help to improve the frequency rate of returns, the average time spent by visitors in browsing the site and the loyalty of the members to the community may assure more reliable and committed customers.

References

Antonioli Corigliano, M. & Baggio, R. (2002a). Italian Tourism on the Internet - New Business Models. *Proceedings of the International Conference "LEISURE FUTURES"*, 11 – 13 April 2002, Innsbruck, Austria.

Antonioli Corigliano, M. & Baggio, R. (2002b). *Internet & Turismo*. Milano: EGEA.

Carver, C. (1999). Building a Virtual Community for a Tele-Learning Environment. *IEEE Communication Magazine*, March.

Dellaert, B.G.C. (1999). The Tourist as Value Creator on the Internet. In D. Buhalis & W. Schertler (Eds.), *Information and Communication Technologies in Touris* (pp. 66-76). Wien: Springer-Verlag.

Gaudin, S., (2002). Is Your Site Getting The 'Internet Death Penalty'? *Datamation*, May 30.

Hagel, J. & Armstrong, A.G. (1997). *Net Gain - Expanding markets through virtual communities*. Boston, MA: Harvard Business School Press.

Hof, R.D. (1997). Internet communities. *BusinessWeek European Edition*, 5 May 1997, pp. 38-47.

Jones, Q. (1997). Virtual-Communities, Virtual Settlement & Cyber Archaelogy: A Theoretical Outline. *Journal of Computer Mediated Communication*, No.3, 3 December 1997, online: http://www.ascusc.org/jcmc/vol3/issue3/jones.html, last access: 3 August 2002

Jones, Q. & Rafaeli, S. (2000). Time to Split, Virtually: 'Discourse Architecture' and 'Community Building' as means to Creating Vibrant Virtual Metropolises. *International Journal of Electronic Commerce & Business Media*, Vol.10, No.4.

Kambil, A. & Ginsburg, M. (1998). Public access Web information systems: lessons from the Internet EDGAR project. *Communications of the Association for Computing Machinery (CACM)*, Vol. 41, No. 7, July 1998, pp. 91-97.

Long, B. and Baecker, R. (1997). A Taxonomy of Internet Communication Tools. *Proceedings of WebNet'97*, AACE, Toronto, Canada: October 1997.

O'Dell, C. & Grayson, C. J. (1998). *If Only We Knew What We Know: The Transfer of Internal Knowledge and Best Practices*. New York: The Free Press.

Rheingold, H. (1993). *The Virtual Community: Homesteading on the Electronic Frontier*. Reading, MA: Addison-Wesley.

Subramaniam, C., Shaw, M. and Gardner, D. (1999). Product Marketing on the Internet. In M. Shaw, R. Blanning, T. Strader & A. Whinston (Eds.), *Handbook on Electronic Commerce*. Wien: Springer-Verlag.

Entrepreneurial Networks in Italian eTourism

Elisa Molinaroli
Dimitrios Buhalis

Centre for eTourism Research (CeTR)
School of Management, University of Surrey, Guildford,
GU2 7XH, UK
e_molinaroli@hotmail.com and D.Buhalis@surrey.ac.uk

Abstract

Information technology and tourism are both driven by a consistent change in structures, players and products in order to adapt to fast changing consumer behaviour. The key objective of this study is to explore to what extent the Italian tourism sector is ready to adopt a common and more flexible technological infrastructure that can facilitate the development of entrepreneurial networks at a local level. Destinations can increasingly be seen as "clusters" or aggregation of businesses, consisting of mainly Small Medium-sized Tourism Enterprises (SMTEs) that through new web technology tools and interoperability can assist each other to provide a "complete" experience to the customer by delivering complimentary products.

Results of extensive web research, email questionnaires and in-depth interviews suggest that while SMTEs lack inter-organisational links, which are fundamental to increased co-operation with other operators, larger enterprises have already developed common platforms mainly for business-to-business activities. The implications of this are that in an increasingly competitive e-marketplace, destinations, like Italy, that lack a Destination Management System (DMS) will only be able to maintain their attractiveness if they manage to enhance interconnectivity among suppliers at the destination, interactivity with customers, and Destination Marketing Organisations (DMOs).

Keywords: entrepreneurial networks, infrastructure, technology, destination, SMTEs.

1 Introduction

This research focuses on eTourism or electronic tourism. Advances in technology and increased interactivity and connectivity among tourism enterprises can offer a viable alternative to destination systems to interconnect consumers with suppliers. Firstly it examines deals with e-commerce in the tourism sector and examines how the Internet and new technological tools are contributing to the growth of the distribution of tourism information and the sales of tourism products online in Italy. Secondly it explores how the Internet is changing the structure of the tourism industry in Italy enabling all players to participate in the e-marketplace.

This study also identifies the main reasons that have so far prevented the adoption of a destination management system in the Italian tourism market. It also examines how the e-marketplace is developing, and to what extent individual organisations, which include SMTEs and larger organisations that are not supported by a DMS or an infrastructure able to aggregate them can successfully participate in the online market. The study also examines how advances in web technology can transform potential weaknesses of the Italian tourism sector into opportunities. The concept of networked organisations and entrepreneurial networks in the tourism sector is examined as a workable alternative to DMS, whose implementation in Italy seems not to be a viable solution.

2   Entrepreneurial networks in Tourism

However, in the near future, the destination and its range of attractions and activities are likely to remain the core product, the focus of tourists' motivation and experiences (Buhalis and Laws, 2001). This requires emphasising local differences and sources of unique value (Jackson and Murphy, 2002). Increasingly, the tourist service economy will depend on the capacity of companies to, collaborate, adapt to market changes and provide personalised information and services (CeltTa, 2001). By moving towards "complementarity" and interaction, each activity can gain from the other. Destinations will be able to increase value by enabling revenue increases (Amitt & Zott, 2001) and constant innovation and maintenance of their attractiveness.

Entrepreneurial networks have been defined as networks of economic "actors" and entrepreneurial activities aimed at creating new resources or combining existing resources in new ways to develop new products and service new customers (Hitt, Ireland, R.D., Camp, S.M. and Sexton, 2001). In the tourism industry intra-industry networks can be developed among clusters of firms that are more alike than others, i.e. similarity of products and services (Gulati, Nohria, & Zaheer, 2000). More importantly vertical networks of complimentary services, such as hospitality with transportation and leisure organisations are critical for the creation and delivery of the tourism product. The co-ordination of all these activities and services is of paramount importance for the performance of the destination and for customer satisfaction. Franchising, consortia, code-sharing airlines are already examples of co-operation among several actors of the tourism value chain (Buhalis & Laws, 2001). Within the accommodation sector for example, organisational networks could tie together hotels, bed and breakfast, campsites, which would constitute "strategic blocks". This will facilitate e-commerce activities throughout the network and increase the quality of services, time to market, market enlargement through multi-channel distribution, product customisation and cost reduction (European Commission DG Enterprise, 2001).

3   Technology tools for tourism entrepreneurial networks

The Internet is changing the structure of the industry enabling all players to participate in the e-marketplace. This underlines the far-reaching implications of the Internet and new e-tools on the structure and management of tourism enterprises. The technological changes in the sector not only condition the evolution of customer behaviour and demands, but they also generate important changes in the production and planning strategies of tourism enterprises (Horrillo, 2001).

Tourist experiences happen in relatively small geographical areas or destination "clusters". Each micro-area supports tourist experiences and satisfies customer needs and hence value is created. It is evident that the needs of the customer cover an entire performance bundle, which consists of the services supplied by many different (local) industry operators, mainly SMTEs (Osti &Pechlaner, 2001).

Destination regions can create their competitiveness by extending the integration of tools for electronic intermediaries, Destination Marketing Organisations (DMOs), end-consumers and tourism suppliers. This is achieved through speed, globalisation, differentiation, personalisation, enhanced products, market reach, innovation and collaboration across the entire tourism value chain at the destination (Barnes and Hunt, 2001). Therefore the needs for co-ordination of entrepreneurial networks in tourism can increase their competitiveness. An "organic" technological infrastructure that enables customers to link directly to suppliers and increase collaboration among providers of tourism products and services could be developed in the Italian tourism sector.

Italy as a country has never implemented a Destination Management System (DMS) and lacks a common technological infrastructure able to bridge supply and demand at the destination level. Decentralisation, lack of a strong decision making power at a central level and fragmentation of the tourism industry constitute the main barriers to the adoption of such systems in Italy. These factors are driving the trend towards generic platforms, based on · open standards, independent from applications (F.E.T.I.S.H., 2002). The move is towards "distributed networks" and a more fluid, agile and innovative form of collaborative commerce empowered and enabled by new collaborative web tools and technologies, such as interoperability, standards, and XML (Pollock, 2001).

4   Methodology

This exploratory study is divided into two sections. The aim of the first phase of the research was to examine the Italian tourism sector, both offline and online, by identifying all the major players that have a web presence and are involved with domestic and inbound tourism. The focus is on how potential tourists, using the Internet to look for information on the destination products and services and who wish to make a reservation are able to reach individual suppliers, both SMTEs and larger enterprises. This is of particular interest since a DMS or an infrastructure does not support tourism enterprises, which is able to aggregate them.

Extensive web research identified providers of tourism products and services at the destination who have developed e-commerce activities with end-consumers. These were then clustered according to the 6As model (Buhalis, 1997), which identifies the six most important categories in the tourism value chain: Attractions, Activities, Accessibility, Amenities, Ancillary Services and Available Services. Further research was conducted through structured questionnaires emailed to 185 SMTEs, who have developed basic e-commerce activities. This aimed to examine how the technology infrastructures available to SMTEs belonging to the main sectors of the tourism value chain were effective in reaching the end-consumer through collaboration with other partners/operators.

Twelve out of 185 SMTEs responded to the emailed questionnaire, giving an overall response rate of 6.5%. As expected, two thirds of respondents (9 or 75%) belonged to the accommodation sector. This was not only because the sample size was greater than those of other categories, but also because of a higher level of IT adoption, since trade organisations in Italy have launched initiatives to stimulate the use of email and Internet within the accommodation sector (Trademark, 2002). However, one third of respondents (3 or 25%) belonged to other sectors, namely activities (2) and accessibility (1).

In the second phase, in-depth qualitative interviews were carried out with members of the e-commerce and e-business teams of five large tourism companies, namely Costa Crociere, Alitalia, Welcome Travel group, eDreams and Turisanda. The aim was to explore to what extent tourism enterprises are ready to adopt a more open and integrated infrastructure. This would permit the development of entrepreneurial networks among economic actors at a local level and would increase co-operation across all stakeholders. Thus interviews discussed the facilitation of e-commerce and e-business activities among suppliers, customers, destination organisations and all potential partners in the context of the Italian tourism value chain.

Quantitative data from the questionnaires served primarily for descriptive purposes where counts and percentages (frequencies) for individual variables were used to obtain pie and bar charts. Interviews were analysed by using content analysis.

5   Findings

This section provides the results of the analysis of data collected through the entire range of research instruments including web research, email questionnaires and interviews. Results from the web research suggest that the Italian online travel sector is very active and key players come from a wide range of sectors, both inside and outside the travel industry. The main stakeholders in the Italian tourism sector have achieved an online presence and there are linkages amongst them.

Domestic and international tourists, researching travel information on the Internet or trying to make a reservation online are able to reach suppliers at the destination, in particular SMTEs. The research demonstrated that the Italian tourism sector is less centralised and integrated than a DMS-based destination. It is dominated by a vast number of large, small and medium sized enterprises most of whom have already developed an online presence. However, due to the low propensity of Italians to buy online, both larger enterprises and SMTEs have developed only basic e-commerce activities and still rely heavily on intermediaries to reach their end consumers. The role of new e-intermediaries such as: the main generic portals, that offer tourism products via the web by exploiting their commercial and brand image potential; travel portals; as well as regional and local tourist portals supported by the private sector is becoming quite important for Italian tourism.

Findings from the questionnaires suggest that SMTEs are visible on the Internet and regularly interact with consumers. The Internet represents a powerful channel for the promotion and provision of information for both larger enterprises and SMTEs in the Italian e-marketplace. As is evident from the web research and questionnaires, e-commerce activities are mainly limited to the accommodation sector, which has developed portals, that give visibility to different types of products and services in highly developed tourist destinations. SMTEs are still making very limited use of Internet technology to increase the level of co-operation with other partners/operators. lack inter-organisational links, thus preventing increased co-operation among other operators. The telephone is still the main means of communication. Results show that the use of email is growing, while the extranet and intranets are present only in one case. Email is used as a tool for communicating with other suppliers and operators, e.g. restaurants, car rentals, and intermediaries. This suggests that small operators want to provide a better service to customers and gradually adopt IT tools. These results seem to confirm research carried out by ISTAT that in 2001 on the use of IT among small and medium enterprises of all sectors.

Interviews demonstrated the Internet and web-based tools have penetrated all aspects of the business process of large enterprises. It is evident that large Italian enterprises have realised that data interoperability exchange is a fundamental requirement to allow faster and more effective market and management dynamics (Harmonise, 2002). Larger enterprises have the technology infrastructure in place and are working towards the development of common platforms, mainly for B2B activities. From interviews with experts in the field, it emerged that travel intermediaries seek to increase collaboration with smaller operators at the destination. This will enable them to offer more customised products and assist travellers arriving at the destination or those wishing to experience a particular type of niche tourism.

Tourism organisations in Italy are looking at changing trends in technology. According to the various operators, technologies can undoubtedly offer a quicker and more effective communication tool between consumers and suppliers as well as with agencies. Each sector is currently working towards the integration and standardisation of information, which will interact with any device. Alitalia is very proactive towards new technologies, as it believes in UMTS, Wifi and wireless solutions for targeting business travellers. Alitalia has completed a technological platform, which can be accessed via the Internet, mobile phones or other devices. In all other sectors, it is evident that operators are very sceptical about how soon Italian travellers, in particular, will make use of new wireless devices such as mobile phones, PDA and digital television to interact with tourism products and purchase products and services. This demonstrates that although most services target international incoming travellers, technological developments are paced according to the IT adaptation at the national level.

6 Analysis

In light of the fact that Italy lacks a common infrastructure, namely a Destination Management System able to bridge supply and demand, this research sees the creation of "entrepreneurial networks" among tourism enterprises at a local level as a fundamental instrument. They are expected to support the Italian tourism sector to disseminate important information; attain more visibility on both domestic and especially international markets; linking the local product to the real and potential demand; improve management of all activities in general (Jacucci, La Micela & Roberti, 2002).

Little research has so far been conducted on the Italian tourism sector to try to identify a viable solution to DMS implementation to co-ordinate and manage the fragmented supply at the destination and improve communication and collaboration in tourism to create value for consumers.

Results from web research, questionnaires and interviews, indicate that in the Italian travel market, representatives of all players involved in the sector value chain are, already represented to a certain extent on the Internet. However, neither larger enterprises, nor the majority of SMTEs participate in sophisticated and highly interactive e-commerce activities. This is mainly due to the nature of domestic travellers, which account for most of the tourism presence in the Italy. The low propensity of Italians to buying online is mainly due to cultural and technological factors as well as due to the high concentration of travel agencies across the country.

On one hand the strong dependence of online firms on the offline channels for booking confirmations and payments seems to benefit mainly domestic tourists. On the other hand it represents a barrier to potential incoming tourists trying to reserve tourist products over the Internet. For the majority of self-contained SMTEs, the low interactivity with customers online means less opportunities to attract incoming tourists, who may wish to pre-book a series of services and activities online, directly with suppliers at the destination. As emerged from the study, the majority of foreign arrivals in Italy originate from the US, Germany, UK and France. These are countries with a high Internet usage and developed e-commerce activities. Also, the self-packaging segment in Italy has increased from 62% in 1998 to 64.3% in 2002 (Trademark Italia, 2002) along with the growth of domestic short breaks and holidays.

There is an opportunity for providers of tourism products and services to transform even domestic short breaks into "complete travel experiences" by supporting collaboration among suppliers of complementary services. This can develop a "supply community" or a network of inter-dependent suppliers who share the objective to satisfy the diverse and complex needs of customers (Pollock, 2002) and to add value to their travel experience.

However, this research demonstrates that only large Italian tourism enterprises have started to implement web technology to support e-business activities and develop a more integrated infrastructure that permits greater flexibility, quick updating of information and the possibility to increase electronic communication with other partners. Although SMTEs managers are aware of the advantages that increased co-operation could bring to their competitiveness and to the attractiveness of the destination, issues related to the under-developed technological infrastructure that favours B2B communication need to be addressed.

7   Conclusions and recommendations

Based on the analysis a wide range of recommendations can be offered for each particular sector examined. These recommendations may support the growth of the online Italian tourism market and assist its competitiveness.

SMTEs
improved communication and exchange of information would facilitate e-commerce opportunities among businesses belonging to different sectors but within the same destination in order to put together more complex, customized packages. This will allow consumers to purchase online through the different devices including the Internet. However issues relating to payment security, privacy, ease of use have to be resolved first. Relationships could also be developed with the main search engines, in particular the Italian ones to achieve visibility, as 80% of domestic tourists visit Italian sites. Partnerships could also be developed with local DMOs to provide institutionalized content and ensure the quality of information provided online by individual suppliers.

Hotel consortia: whereas hotels used to produce their own facilities and services in-house, hotel consortia could develop relationships with private providers of complementary services. This could enable them to share information and availability in real time with complimentary services and to develop customised services based on resources and services being available locally. Suppliers will be able to meet fast changing consumers' requirements and perhaps improve customer satisfaction. This will facilitate e-commerce activities throughout the network and increase the quality of services, time to market and market enlargement (European Commission DG Enterprise, 2001). The hospitality sector could be at the centre of a series of activities that make up the travel experience in a certain destination.

Intermediation: larger enterprises are re-organising their products offering following the trend towards a "rediscovery" of the Italian resources. Research shows that the majority of Italians going on short breaks tend not to book through a travel agency. However, no benchmarking so far has been conducted to compare the cost of the same holiday by contacting private destination portals and travel agencies. The intermediation sector should therefore focus on providing a complete experience and adding value to the whole experience, taking advantage of the long-standing relationships with local suppliers. The lack of computerised communication with local operators as well as with DMOs needs to be addressed either through a comprehensive DMS or through Extranets.

Public destination representation: the Italian State Tourist Office (Ente Nazionale Industrie Turistiche –ENIT) should increase the level of co-operation with the private sector in marketing and promotion. They have a natural responsibility for the content made available online by individual actors. This can be through integrating, classifying and standardising information coming from disparate sources. It becomes increasingly apparent that DMOs will need to take further responsibility if destinations are to remain competitive in the long term.

This study sought to provide a general understanding of how the Italian online tourism sector has developed. Further research could specifically address all main private portals developed at a local level (regional, town, area) to explore the current level of interconnectivity among organisations and identify their level of success. In addition, by segmenting customers and identify emerging trends in tourism demand, niche markets could be developed and integrated with existing offer to innovate and maintain competitiveness of each cluster.

References

Amit, R. and Zott, C. (2001), Value Creation in E-Business, *Strategic Management Journal*, 22 (6-7), 493-520.

Barnes, S. and Hunt, B. (2001), *E-Commerce & V-Business*. Oxford: Butterworth-Heinemann.

Buhalis, D. (1997), Information Technology as a Strategic Tool for Economic, Social, Cultural and Environmental Benefits Enhancement of Tourism at Destination Regions, *Progress in Tourism and Hospitality Research*, 3 (1), 71-93.

Buhalis, D. and Laws, E. (2001*), Tourism Distribution Channels, Practices, Issues and Transformation*. London: Continuum.

CeltTa (2001), *Study on electronic commerce in the value chain of the tourism sector*, http://www.sociedaddigital.org/Espanol/Paginas_Secundarias/Comercio_Electro nico/Evolucion_Sectores_Turismo.htm [accessed 22 April, 2002].

European Commission DG Enterprise (2001), *Managing the Impact and the Use of "Information and Communication Technologies Based Services on the Tourism Sector*, Report of the Working E Party, Brussels.

Federated European Tourism Information Service Harmonization (F.E.T.I.S.H) (2002*), Jini Network Technology Link Online Travel and Tourism Service*, available at: http://fetish.t-6.it/project.html [accessed 4 February, 2002].

Gulati, R, Nohria, N. and Zaheer, A. (2000), Strategic Networks, *Strategic Management Journal*, 21 (1), 203-215.

Harmonise (2002), *Harmonise Project*, available at: http://www.harmonise.org/mission/mission_001.html [accessed 9 April, 2002].

Hitt, M., Ireland, R.D., Camp, S.M. and Sexton, D.L. (2001), Strategic Entrepreneurship: Entrepreneurial Strategies For Wealth Creation, *Strategic Management Journal*, 22 (6/7), 479-491.

Horrillo, A. (2001), Resources and Capabilities of Tourist Firms in the Knowledge Economy: Towards a Framework. In P. Sheldon, K. Wöber and D. Fesenmaier (Eds.), *Proceedings of the International Conference on Information and Communication Technologies in Tourism*, ENTER'01, (pp. 33-41). Wien: Springer Verlag.

Jackson, J. and Murphy, P. (2002), Tourism destinations as clusters: Analytical experiences from the New World, *Tourism and Hospitality Research*, 4(1), 36-52.

Jacucci, G., La Micela, A. and Roberti P. (2002), From Individual Tourist Organisations to a Single Virtual Tourism Organisation for Destination Management. In K. Wöber, A. Frew and M. Hitz (Eds.), *Proceedings of the International Conference on Information and Communication Technologies in Tourism*, ENTER'02, (pp 87-96). Wien: Springer-Verlag.

Osti, L. and Pechlaner, H. (2001), Reengineering the Role of Culture in Tourism's Value Chain and the Challenges for Destination Management Systems – The Case of Tyrol. In P. Sheldon, K. Wöber and D. Fesenmaier (Eds.), *Proceedings of the International Conference on Information and Communication Technologies in Tourism*, ENTER'01, (pp 294-301). Wien: Springer-Verlag.

Pollock, A. (2001), *Shifting Sands: The Tourism Ecosystem in Transformation*, available at: http://www.eyefortravel.com/papers.asp [accessed 12 December, 2001.]

Pollock, A. (2002), *Customers: The G-force that will Pull Web Services into the Frame.* DestiCorp Limited.

Trademark Italia (2002), *Dove vanno gli italiani in vacanza*, (in Italian), available at: http://www.trademarkitalia.com/Dove%20vanno%20in%20vacanza..2002.pdf [accessed 20 June, 2002].

Networking for Small Island Destinations –
The Case of Elba

Harald Pechlaner[a,b],
Valeria Tallinucci [a]
Dagmar Abfalter[b]
Hubert Rienzner[a]

[a] Department of Tourism Management
EURAC research, Italy
harald.pechlaner@eurac.edu

[b] Department of General and Tourism Management
University of Innsbruck, Austria
dagmar.abfalter@uibk.ac.at

Abstract

There is a tradition in tourism literature dealing with the problems and prospects of small and medium-sized enterprises associated with the applications of IT solutions. This article focuses on the special case of small islands in the Mediterranean, addressing the peculiar problems of seclusion and isolation. An empirical inquiry among the tourist entrepreneurs of the Isle of Elba deals with the respondents' attitude and behavior towards IT- and e-commerce solutions but also cooperation, networking and employee qualification. The results are used to develop a business model that is based on the Internet as a local network platform and capable to manage a small island tourist destination.

Keywords: small Mediterranean island tourism, destination management system, management of networks, tourist SMEs

1   Theoretical Background

The problems and prospects of small and medium-sized enterprises (SMEs) associated with the application of IT solutions have become an important issue within tourism literature (Evans et al. 2000). In this respect, also questions concerning the behavior of these SMEs' executives and employees have generated interest among academic researchers (Pechlaner et al. 2002; Sigala et al. 2001; Baker et al. 1999). They agree that general management competence, on the one hand, is essential for the successful implementation of IT strategies within these companies, but that, on the other hand, specific capabilities for the management of IT systems are also increasingly important in order to enable enterprises to cope with market requirements. The studies mentioned above, part of them have been empirically verified, are mostly referring to tourist destinations in general or to special cases (e.g. alpine destinations) and emphasize the necessity of setting up destination management systems in the sense of virtual destination networks. However, there is virtually no literature dealing with the special case of small islands with regard to the introduction of destination management systems. Nevertheless, island tourism has been the object of scientific debate for several years and enjoys increasing popularity (Lockhart & Drakakis-Smith 1997; Lockhart 1993; Keane et al. 1992). In these studies, the characteristics of island tourism with special consideration of economic and social aspects come to the fore, to an increasing extent accompanied by questions of tourist management. It is particularly remarkable that these studies are focusing on a certain type of island tourism, namely that of nation states. So, although in most cases the conviction prevails that island tourism is characterized by separateness and difference (Wong 1993; Butler 1993), these studies mostly deal with island tourism of considerable size. Small islands have become the object of scientific research only recently. At the center of debate are planning and development issues (Henderson 2001), on the one hand, and questions concerning marketing and market segmentation (Mykletun et al. 2001), on the other hand. Empirical studies investigate small islands that, from a tourist point of view, can be put under the heading of traditional destinations, but increasingly also small islands as peripheral and/or rural areas. The Mediterranean region is an ideal stage to examine the problems and prospects of small islands. Altogether, this area accommodates about 250 million tourists per year, a significant part of them on its approximately 4000 islands and islets (Lanquar 1999). The common ground of small islands in the Mediterranean, on the one hand, is their seclusion or "insularity" as a result of their small physical extension, which also limits their development potentials and allows for socio-economic individualities that cannot be compared to those of the mainland. At the other hand, it is their isolation that results in a particular sense of togetherness of the island's inhabitants which seems significantly more apparent than on the mainland (Inguanez 1999). Such an environment is especially suitable for the study of systemic connections, in this specific case with the example of the tourist industry. An essential basis of these system-theoretical interrelations (Boulding 1969) – applied to economic systems – is the growing understanding of those involved in the system that under certain conditions an enterprise may lack the resources it would need in order to operate competitively (Becattini 2000). This insight is further intensified by special socio-cultural conditions, arising from the Mediterranean islands' seclusion/insularity and isolation (Tallinucci 2001). The island of Elba is part of the national park *Tuscany Archipelago* in Italy. 250 entrepreneurs are forming the tourist network that has been the object of empirical analysis.

2   Method of the Study

The objective of this study was to develop a business model based on the Internet as a local network platform capable to manage a small island tourism destination of the

Mediterranean area. Therefore, an empirical mail inquiry among tourist entrepreneurs at the Island of Elba has been conducted in order to analyze the needs and the behavior of local tourist operators concerning the introduction of a destination management system that is centrally coordinated by a destination company. At the center of interest are their attitudes towards cooperation, the use of IT and e-commerce development but also their competences, capabilities for networking and qualifying employees as well as professionalism. This empirical study has been realized by the University of Innsbruck together with the European Academy Bolzano (EURAC research) and TIScover, one of Europe's leading providers of destination management systems, and took place in September 2002. The questionnaire has been sent via e-mail to all of Elba's 250 tourist entrepreneurs and was composed of both closed questions and open reflection sentences, completed with personal interviews. This approach allows for an integration of quantitative and qualitative analysis and should generate reliable results together with a deep understanding of the small island entrepreneurs' attitude towards the development of a network system and a destination company as well as the opportunity and the threats of the use of IT.

3 Results of the Empirical Study

The self-administered questionnaire used in this study included questions to assess

(1) the use of IT and e-commerce
(2) needs, attitudes and behavior of tourist entrepreneurs
(3) competences and qualifications of employees.

57 fully completed and utilizable questionnaires were returned. These 57 respondents represent 80 tourist enterprises of Elba, since some of the respondents own more than one enterprise. These 80 enterprises represent approximately one third of all tourism enterprises in Elba. The following sections present the key findings of this study.

3.1 The Use of IT and E-Commerce

Only a small part of the entrepreneurs in Elba (5.5%) stated that they neither have their own website nor are listed on a tourism portal. 88.8% have their own homepage and 68.5% dispose of a site within a tourism portal. Table 1 shows the top 6 services offered on the individual website of the enterprises included in the study. Most of the homepages (54.5%) include a presentation of the island and one out of five (21.8%) has a link to other tourism enterprises and/or tourism portals in Elba. These results suggest that there is a strong tendency towards networking and cooperation. This aspect will be further discussed later on in this paper.

Table 1. Services Offered (Top 6)

Services offered	
Presentation of the offer	100%
Form for reservation request	65.5%
Short presentation of Elba	54.5%
Last-minute booking	23.6%
Links or banners to other enterprises	21.8%
Links or banners to tourism portals	21.8%

55 valid cases

Investment in internet activities is growing. Table 2 shows that the number of enterprises willing to spend more than 1,500 per year has doubled from 2001 to 2002. At the same time, the percentage of enterprises spending less than 250 per year has decreased from 25% (22.5% + 2.5%) to 10.4%.

Table 2. Investment in Internet Activities

Year		Investments				
	0	Less than 250	251- 500	501 – 1,500	1,501- 2,500	More than 2,500
2000	5.6%	16.7%	36.1%	27.8%	5.6%	8.3%
2001	2.5%	22.5%	30%	27.5%	10%	7.5%
2002	-	10.4%	18.8%	33.3%	20.8%	16.7%

Regarding the motivations for setting-up their website, the results (Table 3) revealed that the internet is mainly considered a distribution channel. 94.5% of the respondents declared the possibility to achieve visibility on a wider market as a reason for their internet presence, 87.3% stated that their presence on the Internet is a necessity, and 76.4% appreciated the speed of communication within this medium and therefore put their offers online.

Table 3. Motivations (Top 6)

Services offered	
Allows to be present on a wider market	94.5%
Online presence is a necessity	87.3%
Allows fast communication	76.4%
Required by our guests	43.6%
To simplify e-booking	32.7%
To offer last-minute bookings	29.1%

55 valid cases

As regards the usefulness of the internet in communicating with customers (Table 4), one third (31.5%) defined it "indispensable", 51.9 "good", 7.4% "average" and 9.3% "low". The possibility to acquire new customers through the Internet has been emphasized by 87%, the reduction of response times by 83.3% and the enhancement of customer satisfaction by 57.4%. Only 7.4% of the respondents mentioned the possibilities offered by e-commerce (including online payment) as a particular benefit.

Table 4. Usefulness of the Internet

Indispensable	31.5%
Good	51.9%
Fair	7.4%
Low	9.3%
	54 valid cases

Respondents have also been asked to estimate the number of new customers acquired through the Internet. The results show a clear growth: 12.7% in 2000, 15% in 2001 and 19.1% (preliminary results) in 2002.

These results reveal that Elba's entrepreneurs are generally interested in using the Internet. With regard to e-commerce, the respondents still showed some doubts. Three out of four (73.6%) stated that the need for personal contact remains and consequently limits the application of e-commerce. For 56.6% of the respondents the Internet is not safe enough for financial transactions and for 43.4% traditional means of communication are sufficient to satisfy the customers' expectations.

3.2 Needs, Attitudes and Behavior of Tourist Entrepreneurs

A second set of questions referred to the needs and the behavior of tourist entrepreneurs in Elba. To start with, it was of research interest to know who produced the companies' websites. Surprisingly, one out of three respondents (31.9%) stated that he/she had produced the site personally. 7.2% engaged a friend/relative to set up the page, 30.4% outsourced the production to a web designer and another 30.4% to a multimedia agency specialized in the tourist industry.

Since the number of "self-made" websites had been surprisingly high, the researchers tried to further explore the reasons for this behavior through personal interviews. The results are discussed further on in this paper.

Another question referred to the means of communication with customers. More than half of the respondents (54.7%) defined the telephone as the principal means of communication. The second most important means indicated by 40.4% was e-mail, followed by fax and mail. The respondents were also asked to rank the most important promotion channels. As shown in Table 5, 38.1% ranked catalogs of tourism intermediaries (travel agencies, tour operators) first, for 32.7% their own website was the most powerful promotion channel.

Altogether (including rank 2 and 3), the website has become the most important promotion channel.

Table 5. Most Important Promotion Channels

	Most important channel	Second most important channel	Third most important channel
Internet banner	13.3%	10%	3.3%
Advert in specialized Internet portal	22.6%	9.7%	22.6%
Catalog of tourism intermediaries	38.1%	19%	19%
Brochure	28%	18%	20%
Tourist guide	12.1%	9.1%	15.2%
Advert in specialized magazine	8.1%	13.5%	8.1%
Own website	32.7%	30.6%	24.5%

57 valid cases

Relationships with individual customers are mainly kept on an individual level. According to table 6, 80% of the entrepreneurs prefer direct personal relations to their existing customers.

Table 6. Who is in Charge of CRM (Customer Relationship Management)

	Rank 1	Rank 2	Rank 3
Directly	80.0%	3.6%	3.6%
Travel agency / tour operator with Internet portal	32.3%	19.4%	41.9%
Local travel agency	10.8%	64.9%	21.6%
Local tourism organization	28%	16%	56%

3.3   Competences and Qualifications of Employees

The third part of the questionnaire consisted of a set of questions to explore the competences and qualifications of employees. 68.6% of the respondents expect computer knowledge (Word, Excel and knowledge of booking systems) from their employees. 23.5% do not request any computer skills. While 11.9% of the respondents do not offer any training in this field, 17.6% offer computer courses to their employees during the employment and 4.5% regularly offer computer courses to their employees. 45.8% of the entrepreneurs are ready to invest up to   500 (per employee) for training and 18.8% even up to   1,000. Every third employer surveyed (35.4%), though, is not ready to pay anything to increase its employees' computer skills.

Table 7 shows the employees' acceptance of the Internet. 46.3% rated the acceptance as "high", 37% as "average" and 14.8% as "low".

Table 7. Employees' Acceptance of the Internet

Low	Average	High
14.8%	37%	46.3%

54 valid cases

47.3% of the respondents declared that their employees are answering incoming e-mail messages twice a day, while 41.8% sent answers to e-mail requests only once a day. In 7.3% of the enterprises e-mail answers are sent 3 times a week.

4   Interviews Results

More detailed investigations concerning the behavior of local tourist operators were employed to identify the reasons for some rather surprising answers. A qualitative analysis was carried out using the method of personal interviews addressed to 25 tourist enterprises on Elba Island. The key elements investigated were:

(1) Specific behavior of SMEs on a small Mediterranean tourist island;
(2) "Self-made" websites;
(3) doubts about the use of e-commerce.

Elba is a small Mediterranean tourist island characterized by the factors of insularity and isolation which represent elements of tourist attraction but at the same time restrict the local population's opportunities of communication. For the islanders, the Internet represents the opening of a window to the world, allowing for direct access to the guests and communicating the personal philosophy of Elba's mostly family-run tourist enterprises. This explains the extremely high rate of self-made websites (31.9%) and consequently also the remarkable investments in IT compared to the low interest in the employees' qualification. Most of the respondents appendant to three-star tourist structures prefer the direct contact with their guests (80%) in order to increase customer loyalty, and to run their business personally, creating an individual brand to link with the destination image of Elba.

The positive attitude towards the Internet is increased by the "generation change" in the management/ownership of the SMEs, introducing innovation and professionalism. Local structures show a significant interest in e-commerce and the support of a destination company capable to manage innovative promotional and payment services. At the same time better security of transactions is required in order to offer clear and easy procedures to order and delete e-payments.

5   Conclusions and Implications

The systemic approach on the research of tourist destinations is based on the General Systems Theory which analyzed the system as an autonomous organism composed by a mix of interacting elements reacting on a cause-effect relationship (Von Bertalanffy 1969). This approach was first applied to the industrial districts (Becattini 2000) and analyzed the building and management of relationships among Italian SMEs. The development of the concept of destinations, defined as areas consisting of all services and offers a tourist consumes during his/her stay, as well as the growing importance of the Internet suggest an adaptation of the systemic approach to the development and management of a tourist destination.

Italy has known three different trends of destination development: (1) the *tourist districts,* which can be defined as a further development of the Italian industrial districts focusing on the creation of knowledge-based destination networks, (2) the *product-oriented destination development approach,* which interprets destinations as products without considering regional limitations and (3) the *local tourist systems,* where destinations are focusing on local points of attraction.

Tourist districts have emerged from a direct application of the industrial district theory to the tourist sector (Lorenzoni et al. 1988). The *product-oriented destination development approach* was created in the Italian region of Emilia Romagna in order to experiment the cooperation opportunity between the public regional organization and private entrepreneurs. Finally, the *local tourist system* (Manente 2001) is defined in the new Italian normative "legge quadro 135/2001 art. 5" as "a homogeneous or integrated tourist context, which could be extended to different regions and is characterized by an integrated offer of cultural, natural and recreational products, of typical goods or by the presence of single or associated tourist enterprises". It has been applied in the SLOT model (local tourist offer system) which focuses on the offer of a composed tourist product furnished by a networked organization of local tourist SMEs (Rispoli & Tamma 1995).

Small islands are characterized by factors such as insularity and isolation, which led to the development of a nearly self-sufficient economy of subsistence based on the exploitation of the local natural, marine and mineral resources. After the 1950s, tourism emerged as an opportunity to develop a wealthy economy that could guarantee the survival of the island population. Nevertheless, the need for a professional management of the Mediterranean islands' offer and a sustainable tourism development in balance with the preservation of the natural and cultural environment has only been identified within the last few years. Tourist businesses have to cope with scarce resources such as space, water and electricity, but also building elements and food, with difficult access by ferry or plane, with a lack of infrastructures and furthermore with the specific need for preservation of the natural and cultural environment. Consequently, missing professionalism and coordination of destination management can cause a severe loss of opportunities.

Following these theoretical considerations regarding the local tourist offer systems, a small island tourist offer system is defined as a theoretical model of development, comprising the structures and points of attraction which characterize the tourist destination, with a strong focus, however, on the quality of hospitality and the lasting development of a small island's typical tourist offer. The system component is understood as an integrated model of strategic leadership of the network of the small island's tourist operators. The offer model of a small island is typically an integration model, based on long-term cooperation between tourist enterprises at the local level and therefore able to generate a unique and competitive offer. In contrast, the fragmentation model is characterized by an incomprehensive offer and low marketability. These basic conditions are frequently underdeveloped in small islands, due to the fact that isolation leads to a distinct common insight into the necessity of joint action, as has also been demonstrated by the essential results of the empirical analysis. Furthermore, also the dependence model, where the use of local resources and the access to the market depend on the enterprises acting outside of the destination, is hardly found on small islands such as Elba, as their limited size does not allow for the generation of substantial economies of scale (Rispoli & Tamma 1995).

Some indications about the strategic problems of tourist enterprises can be derived from this model. Every organization is involved in the entire product, through the production of specific products. The operation of the organization, however, is influenced by the reactions of the system to which it consciously or unconsciously belongs. The possibility to activate resources and to acquire lasting competitive advantages can differ enormously for individual organizations. The integral way of looking at things makes clear that development and success of a tourist destination's offer systems depend to a large extent on its ability to integrate resources and activities. At the same time, it becomes obvious that coordination and control of the system take place by the system itself.

It can be recorded that the system of a small island is characterized by special peculiarities, which renders a direct comparison with other island destinations difficult. The empirical study. it made it possible to show that among the entrepreneurs of the island of Elba a distinctive interpretation of quality with regard to hospitality is predominating; this, in turn, has consequences for the entrepreneurs' attitude and behavior towards the Internet. Furthermore, this leads to the assumption that the seclusion (insularity) and isolation of the insular system reveal socio-cultural peculiarities, which, on the other hand, facilitate the establishment of destination management systems. The differentiating clues with regard to a small island's entrepreneurs' cooperative behavior on the isle of Elba are an important indication for this hypothesis.

References

Baker, M., Sussmann, S. and M. Meisters (1999). The Productivity Paradox and the Hospitality Industry. *Information and Communication Technologies in Tourism 1999.* D. Buhalis and W. Schertler (eds.). Springer, Wien: 300-309.

Becattini, G. (2000*). Il distretto industriale.* Rosenberg & Sellier, Torino.

Boulding, K.E. (1969). General System Theory – The Selection of Science. *Modern Systems Research for the Behavioral Scientist.* W. Buckley (ed.). Aldine Publishing Company, Chicago: 3-10.

Butler, R. (1993). Tourism Development in Small Islands. *The Development Process in Small Island States.* Lockhart, D.G., Drakakis-Smith, D. and J. Schembri (eds.). Routledge, London: 113-216.

Evans, G., Bohrer, J. and G. Richards (2000). Small is Beautiful? ICT and Tourism SMEs: A Coparative European Survey. *Information Technology & Tourism* 3(3/4): 139-153.

Henderson, J.C. (2001). Developing and managing small islands as tourist attractions. *Tourism and Hospitality Research* 3(2): 120-131.

Inguanez, J. (1999). Turismo e comunicazione culturale. *Strategie di comunità nel turismo mediterraneo.* Guidicini, P. and A. Savelli (eds.). Franco Angeli, Milan: 287-297.

Keane, M.J., Brophy, P. and M.P. Cuddy (1992). Strategic Management of Island Tourism: The Aran Islands. *Tourism Management* 13(4): 406-414.

Lanquar, R. (1999). Turismo e ambiente: la posta in gioco nel Mediterraneo. *Strategie di comunità nel turismo mediterraneo.* Guidicini, P. and A. Savelli (eds.). Franco Angeli, Milan: 179-189.

Lockhart, D.G. and D. Drakakis-Smith (eds.) (1997). *Island Tourism: Trends and Prospects.* Pinter, New York.

Lockhart, D.G., Drakakis-Smith, D. and J. Schembri (eds.) (1993). *The Development Process in Small Island States.* Routledge, London.

Lorenzoni G. and O. Ornati (1988). Constellations of Firms and New Ventures. *Journal of Business Venturing* 3(1): 41-57.

Manente M. (2001). Sistemi turistici locali. *Rivista del Turismo* 3(5/6): 23-27.

Mykletun, R.J., Crotts, J.C. and A. Mykletun (2001). Positioning an island destination in the peripheral area of the Baltics: a flexible approach to market segmentation. *Tourism Management* 22(3): 493-500.

Pechlaner, H., Rienzner, H., Matzler, K. and L. Osti (2002). Response Attitudes and Behavior of Hotel Industry to Electronic Info Requests. *Information and Communication Technologies in Tourism 2002.* K.W. Wöber, A.J. Frew and M. Hitz (eds.). Springer, Wien: 177-186.

Rispoli M. and M. Tamma (1995). SLOT – Sistema Locale d'Offerta Turistica. *Risposte strategiche alla complessità.* Rispoli M. e M. Tamma M. (eds.). Giappichelli, Torino.

Sigala, M., Airey, D., Jones, P. and A. Kockwood (2001). Investigating the Effect of Multimedia Technologies on the Employment Patterns in Small and Medium Tourism and Hospitality Enterprises in the UK. *Information and Communication Technologies in Tourism 2001.* P.J. Sheldon, K.W. Wöber and D.R. Fesenmaier (eds.). Springer, Wien: 201-214.

Tallinucci, V. (2001). SOTIM – Il sistema d'offerta turistica insulare minore. *Turistica* 10(3): 129-135.

Von Bertalanffy L. (1969). General System Theory. *Modern Systems Research for the Behavioral Scientist.* W. Buckley (ed.). Aldine Publishing Company, Chicago: 18-19.

Wong, P.P. (ed.) (1993). *Tourism vs. Environment: The Case for Coastal Areas.* Kluwer Academic Publishers, The Netherlands.

Location-based Mobile Tourist Services - First User Experiences

Barbara Schmidt-Belz [a]
Heimo Laamanen [b]
Stefan Poslad [c]
Alexander Zipf [d]

[a]Fraunhofer FIT, Sankt Augustin, Germany
Barbara.Schmidt-Belz@fit.fraunhofer.de

[b]Sonera, Helsinki, Finland
heimo.laamanen@sonera.com

[c]Queen Mary University of London, UK
stefan.poslad@elec.qmul.ac.uk

[d]European Media Lab, Heidelberg, Germany
Alexander.Zipf@eml.villa-bosch.de

Abstract

The vision of nomadic users having seamless, worldwide access to a range of tourist services seems within reach, within only a few years from now. While much of the underlying technology is already available, there are challenges with respect to usability that need intelligent solutions. CRUMPET has realized a personalized, location-aware tourism service, implemented as a multi-agent system with a concept of service mediation and interaction facilitation. The system has been validated in terms of technical and user-perceived qualities at four European trial sites. This paper reports the findings of the user validation and draws conclusions concerning mobile tourism services.

Keywords: Mobile Tourism Service; Location-Based Services; User Validation; Software Agents.

1   Introduction

Emerging new technologies such as handheld mobile devices with wireless connections to the Internet open up new prospects for eCommerce and eTourism. The vision of a broad range of services for tourists being available, from everywhere and at every time, becomes realistic for the near future. Location-based and personalized services are considered key features of such services. A promising technical approach is software agent technology, which has additional advantages with respect to scalability, service mediation, and the management of seamless mobility. The provision of content and services, adaptable to, and tailored for mobile users, may soon become another important channel of destination marketing for cities and regions.

But what do users think about such a mobile tourism service? Do they feel the service has added benefits, compared to traditional media and Web-based services? What would be the crucial applications and qualities that "make the big difference"? A range of usability issues concerning mobile services is being discussed in the science community; are there already viable, good solutions? And, last but not least, would there be a future market for such systems?

Once prototypical realisations are available, users can validate the implemented approaches and assess concepts and realization details from their point of view. Such first user experiences are a valuable guidance for further improvements, design decisions and market strategies for the new technology.

This paper is structured as follows: Section 2 gives an overview over the CRUMPET system, the features of the prototypes and how the trials have been performed. In Section 3 we report the outcome of the user validation, focussing on location-based and personalized services. Section 4 contains the more general results of our survey when looking for the added value of mobile tourism service. Finally, in Section 5, we draw our major conclusions and outline future work.

2 CRUMPET system and Trials

The goal of the European IST project CRUMPET was the "Creation of User-friendly Mobile Services Personalised for Tourism". CRUMPET has two main objectives:

- To implement and trial tourism-related value-added services for nomadic users across mobile and fixed networks

- To evaluate agent technology in terms of user-acceptability, performance and best-practice as a suitable approach for fast creation of robust, scalable, seamlessly accessible nomadic services

Figure 1 illustrates the functional architecture of the CRUMPET system. It has in essence a three-tiers structure, with the mobile clients on the one hand, the local services on the other hand, and a multi-agent system between both, which implements the value-added integrated services. For more details and a discussion of the design rationale of this system architecture we refer to (Poslad et al. 2001) and (Schmidt-Belz et al. 2002). In this paper here we rather focus on the user's point of view on this system.

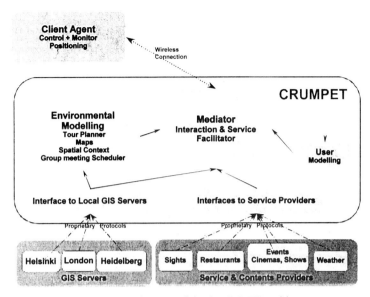

Figure 1: Functional architecture of the CRUMPET multi-agent system

The system offers a simple user interface and handling of services. The main functionality is:

- Recommendation of services, e.g. tourist attractions (based on personal interests and the vicinity to the current location)

- Interactive maps (overview maps of the area, highlighting the current position of the user; maps highlighting sites of interest and tours; maps can be panned and zoomed)

- Information about tourist attractions (short text, more detailed information, maps, directions and pictures).

- Proactive tips, giving an unobtrusive tip when the user gets near a site that might interest him or her.

The client device is a handheld computer (e.g. iPAQ), the user location is determined by GPS sensor data. Modern handheld computers offer a screen size and resolution that is adequate to display maps and simple HTML pages. The system has not been realized for extremely limited displays such as offered by WAP enabled mobile phones. The project has developed a functional prototype, available with local content at four trial sites: Heidelberg/G, Helsinki/Fi, London/UK, and Aveiro/Pt. The trials allow validating the system and the approach, both with respect to technical achievements and user assessment. The trial sites each have a special focus, which allowed comparing variants, such as differences in local contents available.

Figure 2 List of recommended sites

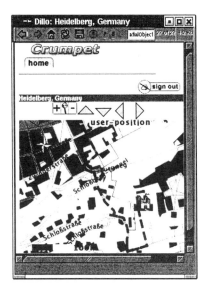

Figure 3 Map with highlighted tour

An essential part of the trial is a user validation of concepts and solutions. The ultimate question that needs to be investigated on basis of the trials is: "Will this type of system be a success?" This overall question has several aspects on a more concrete level, which have been addressed directly in the user validation:

- Does the system meet users' needs, i.e. does it offer the required functionality?
- Is the system considered useful and usable?
- Has it benefits and added values compared to other systems or media available in this application area, i.e. tourism related services?
- What would be "killer applications" for a mobile tourism service?

The method chosen for this user validation is a field experiment, where users have to perform some typical tourist tasks, while using the service given by the CRUMPET

system. The test persons answered a questionnaire, which has been developed to clarify the above-mentioned questions. The questionnaire ensures comparable validation results throughout all trial sites. We have also applied the standardized questionnaire SUMI (Kirakowski 2002) to measure usability. We also observed the test users during the experiment, which gave us deeper insight in their experiences. This combination of quantitative and qualitative methods is appropriate especially for validating and exploring a highly innovative technology.

The selection of test persons was oriented at the prospective market of such a system. Basically, every tourist should be able to use CRUMPET. On the other hand, we assume that the CRUMPET user owns and uses a PDA for everyday tasks (such as personal planner and for taking notes) as well as when travelling. In this respect CRUMPET differs from a kiosk solution or a museum guide. The future CRUMPET user is more likely to be a mobile knowledge worker at a higher level of computer literacy. More than 75 persons have taken part in the CRUMPET trials. The familiarity with related systems (such as WWW, WAP) has been documented for each test person but not made a criterion in selecting participants.

In the end, we wanted to know whether the users see the added benefits in a CRUMPET system and whether they would be prepared to pay something for this service. The rather high percentage of people (>75%) who clearly see the added value of a system like CRUMPET is very encouraging. It is also good to see how many of these (64%) are in principle willing to pay for the service. Considering that the many services in the WWW are usually for free, this was not a save bet. Open questions and a few interviews gave a first idea of which modes of payment would be acceptable for users. There seems a clear preference for pre-paid or subscription modes, but this would need further investigation as most users may not be aware of all the alternatives.

Figure 4

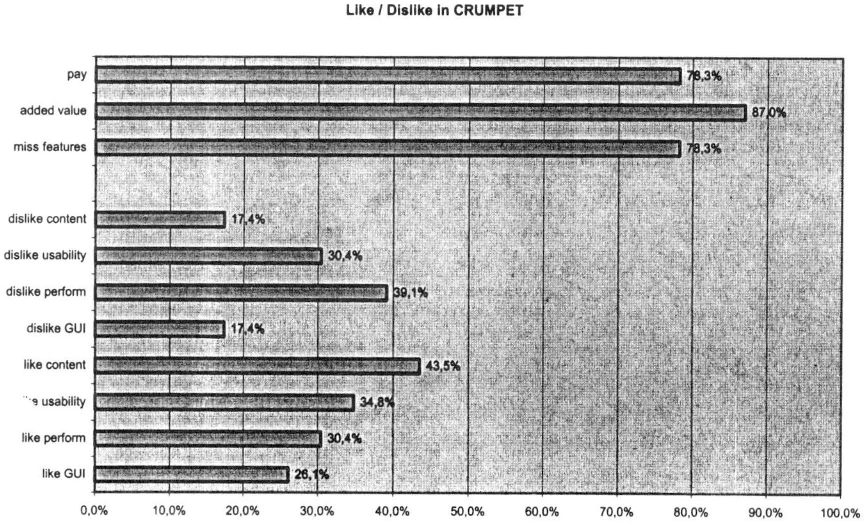

Like / Dislike in CRUMPET

The correlations of these two parameters to other demographic data were interesting: there seems a tendency that male users would more easily see the added value of the system; there seems a tendency that people who travel more are more ready to pay for such a service. And the most important correlation is to "purpose of travelling": people who travel often for business purposes are more likely to see the added values of a CRUMPET and to be willing to pay for it. So "business traveller" is another important characteristic of the target group for a mobile tourism support.

3   Location-Based Personalized Services

In general, location-based services are considered crucial for the success of mobile applications (Oertel et al. 2002). It is also widely assumed that the mobile services should be personalized, see for instance (Short 2000). By location-based services (LBS) a broad scope of value-added features is understood that are based on the system's awareness of the current user location. The user acceptance of location-based services has recently been investigated in more detail (Kölmel and Wirsing 2002). In CRUMPET, the user location serves to facilitate the user request of services and to add functionality to maps (Zipf 2002). Personalization adds consideration of the user's personal "taste" in the available service types; the user interest can be automatically learned. The pro-active mode of service is another option to use positioning and personalization for a value-added tourism service. The ranking of mobile services as discussed in the next section, confirms the general importance of personalized LBS services.

The central role of maps in a tourist guide is obviously confirmed by our experiences. Maps can meanwhile be rendered in a good quality on a modern PDA. Navigation support on very small screens (such as WAP phones) is often given by maps that show only a schematic picture of a route. In our experiments, however, we observed that users, i.e. pedestrian tourists, asked for many details they wanted to find in a map. Depending on the task at hand, they looked for a match between aspects of their physical environment and the map representation on the screen. There is certainly a trade-off between avoiding cognitive overload, giving task-specific information, and adapting to personal preferences, which needs further and more detailed investigation.

We also found evidence that a few people would need textual tour descriptions, as they were unable to interpret a map. If textual directions are provided, in addition to maps, this could also be used as an audio guide, which is more adequate in some situations when the user cannot look at the screen.

The interaction with maps in order to pan or zoom, also in order to include and display specific objects, needs to be as simple and straight forward as possible. We observed many users who intuitively tried to directly manipulate the map in order to pan and zoom. We also had the impression, that some would prefer scrolling to panning (i.e. a larger map is transmitted, the user then scrolls to see the area of interest. We found diverging opinions about the orientation of maps, i.e. should the map be always oriented to the north or should it be turned to match the user's direction. The latter would require sensor data (e.g. electronic compass) and might confuse some, when they turn constantly to have a look around. This also needs a more detailed investigation.

4 In search of the crucial application

Information needs differ for the planning vs. the travelling situation. For instance, Hotels are usually looked up (and booked) before travelling (~ 90%). Sights to visit at a destination are often looked up before travelling but equally while on tour (both ~70%). The interest in events that might be available at a destination during the intended period of time is also rather high (both ~60%) .

The most interesting finding, however, is that transportation ranks very high both when planning a tour and while travelling (~75%, ~60%). Transportation was also considered an important aspect of a destination ("Very important or "important" for 78%). This does probably not mean that people are interested in transportation but rather that transportation is very often a problem.

The sources from which people get this information are various: most people in our sample (95%) use the WWW for planning. This was to be expected of highly computer-literate people. While on tour, the WWW is currently not very important (~25%), the reason probably being that mobile access to the Web is still limited.

Figure 5

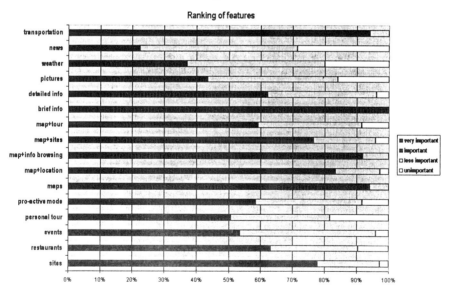

We found it remarkable, however, that the same people in addition to using the WWW also buy books (60%) and use maps on paper (~55%). We conclude that for the near future various sources of information for travelling people will continue to co-exist.

Finally we asked for the importance of several features that would be supplied by a mobile tourism service such as CRUMPET.

Information about transportation again ranks very high, and opinions do not diverge much in these points. From interviews we learned that people hope for transportation information that are frequently updated, reliable and personalized. From a content provider's point of view this is rather a challenge, unfortunately. Maps also rank very high, and even more so when enhanced by highlighting the current position of the user, a tour, or sites of interest.

All in all, location-based services and transportation information would certainly be crucial applications for a mobile tourism support, which have the potential to most clearly demonstrate the added value of mobile support compared to existing media.

We have also tested these preferences for correlation with gender and age. Overall, correlations of gender or age with other variables were usually rather low. In other words, within our rather homogeneous sample, these variables played no significant role in determining the other variables.

5   Conclusions and Future Work

It is by personal experience only that users get a feeling for innovative technologies and become more explicit and confident in what their needs and requirements are. The CRUMPET system has been acknowledged by users for its simplicity of use and for its focus on location-based services. It was very encouraging that a high percentage of the test users saw the added benefit of the system compared to other available information sources, and that a rather high percentage of users would also be willing to pay for such services. The paying modes acceptable for users need further investigation.

But there is still a way to go before a CRUMPET system could become a marketable product. The most important improvements of CRUMPET in future implementations would be to include more service types, especially such as related to transportation, events and restaurants.

For mobile tourism services in general the importance of added value by location-awareness has been confirmed, also the importance of providing interactive maps. Essential applications would be content about local transportation (especially when personalized and reliably updated), background information about local sites (in a choice of granularity) and advanced LBS services.

Several usability issues still need further investigation, which is already subject to several ongoing research projects in HCI, see (Schmidt-Belz and Cheverst 2002).

Examples of usability-related issues are the adequate use of personalisation data, the adaptation to individual preferences in visualization of tours and directions given, and more advanced interaction modes.

Acknowledgements

This work has been undertaken in the context of the EU-funded project CRUMPET (IST-1999-20147), and we wish to thank all project partners: Queen Mary University of London (UK), Emorphia (UK), European Media Lab (D), Fraunhofer Institute for Applied Information Technology (D), PTIN (PT), Sonera (FI), and University of Helsinki (FI). We namely want to thank Hidir Aras, Mikko Laukkanen, Alastair Duncan, Leonid Titkov, Rossen Rashev, and Michael Charalambous for their merits to the trials.

References

Kirakowski, J. (2002). "SUMI - Software Usability Measurment Inventory."

Kölmel, B., and Wirsing, M. (2002). "Nutzererwartungen an Location Based Servcies - Ergebnisse einer empirischen Analyse." Geoinformation mobil, J. Strobl, ed., Huethig Verlag, Heidelberg.

Oertel, B., Steinmüller, K., and Kuom, M. "Mobile Multimedia Services for Tourism." *ENTER 2002*, Innsbruck.

Poslad, S., Laamanen, H., Malaka, R., Nick, A., Buckle, P., and Zipf, A. "CRUMPET: Creation of User-friendly Mobile Services Personalised for Tourism." *3G 2001*, London.

Schmidt-Belz, B., and Cheverst, K. (2002). "HCI in Mobile Tourism Support." GMD.

Schmidt-Belz, B., Makelainen, M., Nick, A., and Poslad, S. "Intelligent Brokering of Tourism Services for Mobile Users." *ENTER 2002*, Innsbruck.

Short, M. (2000). "My generation. Third generation wireless mobile communication." *Electronics and communication engineering Journal*, 12(3), 119-122.

Zipf, A. "User-Adaptive Maps for Location-Based Services (LBS) for Tourism." *ENTER 2002*, Innsbruck, Austria, 329-338.

Services on the Move :
Towards P2P-Enabled Semantic Web Services

Alexander Maedche[a]
Steffen Staab[b]

[a] FZI Research Center for Information Technologies at the University
of Karlsruhe
D-76131 Karlsruhe, Germany
maedche@fzi.de

[b] Institute AIFB, University of Karlsruhe, D-76128 Karlsruhe,
Germany
& Ontoprise GmbH, D-76131 Karlsruhe, Germany
staab@aifb.uni-karlsruhe.de

Abstract

Service-driven architectures promise a new paradigm providing an extremely flexible approach for building complex information systems. However, at the current moment, service architectures go little way beyond standardized remote procedure calls and textual directories to locate and describe a service provider based on human intervention. In this paper we consider three important dimensions for building next-generation service-driven systems building on enabling technologies such as Web services, Peer-2-Peer and Semantic Web.

Keywords: Semantic Web, Web Services, Peer-2-Peer

1 Introduction

Recently, so-called service-driven architectures, including a services-driven middle tier that mediates back-end resources with multiple channels (PCs, PDAs, wireless phones, etc.) to the end user have been proposed (see http://www.sun.com/executives/sunjournal/v4n1/Feature3.html). The vision of service-driven architectures promises a new paradigm providing an extremely flexible approach for building complex information systems. In the future the overall vision of fully-enabled services within such architectures promises great benefits for tourism information systems, because they will *(i)* ease participation of SMEs to larger networks, *(ii)* facilitate virtual organizations, and *(iii)* allow the user to create bundled products tailored to his needs.

So far, however, the definition of these service architectures go little way beyond standardized remote procedure calls and textual directories to locate and describe a service provider requiring human intervention. In this paper, we argue that we need to consider three dimensions to fully enable service-driven architectures that accomplish the above mentioned objectives (cf. Figure 1):

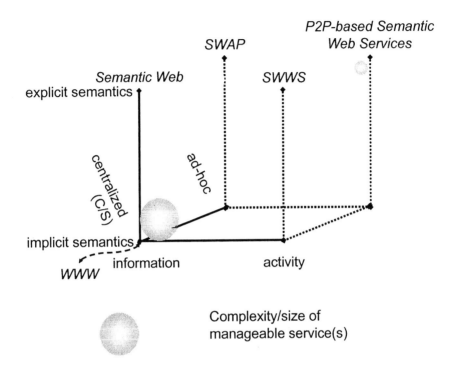

Figure 1: Towards P2P-based Semantic Web Services

1. **Information vs. activity**: This is the difference between a static HTML page or an information repository like the initial WWW on the one hand and a service that provides some complex activity (e.g., a room booking) on the other hand.

2. **Centralized vs. ad-hoc**: This dichotomy describes the difference between a client/server paradigm, where clients call dedicated servers (even if roles may switch), and a system that is assembled ad-hoc. Note that current Web Service standards assume one (or few) central directories (e.g., UDDI) to locate other services.

3. **Implicit vs. explicit semantic descriptions**: This dimension distinguishes between semantics that are only implicitly available, e.g. by convention (cf. EDI-FACT), and semantics that are explicitly specified, e.g. by an ontology.

Each of these three dimensions facilitates the automatic identification, location and invocation of services for tourism information systems increasing system flexibility. Together, they will create a new paradigm of building systems.

However, we here also argue that automation is limited because each of these dimensions adds new complexity to the overall task that may be accomplished by a service. Therefore, there will be a trade-off: ad-hoc configured, activity-oriented, services with explicit semantics will be created, offering new chances to the tourism industry (e.g. dynamic product configuration by customers). However, they will not replace conventional approaches that will still be able to manage tasks that need a higher degree of complexity or reliability (e.g. enterprise resource planning).

2   Enabling Technologies

In this section, we briefly survey the underlying technologies necessary to achieve ad-hoc configurable, activity-oriented, semantic-descriptions based services.

Web Services.   Web Services can be defined as software objects that can be assembled over the Internet using standard protocols to perform functions or execute business processes. The key to Web Services is on-the-fly software creation through the use of loosely coupled, reusable software components. This has fundamental implications in both technical and business terms. Software can be delivered and paid for as fluid streams of services as opposed to packaged products. They also facilitate interoperability between systems to accomplish business tasks. Businesses can be released from the burden of complex, slow and expensive software integration and focus instead on the value of their offerings and mission critical tasks. Then, the internet will become a global common platform where organizations and individuals communicate among each other to carry out various commercial activities and to provide value-added services.

Peer-to-Peer (P2P).   The essence of Peer-to-Peer computing is that nodes in the network directly exploit resources present at other nodes of the network without intervention of any central server. For this purpose P2P platforms like JXTA (see http://www.jxta.org) provide a layering over communication networks that abstract from lower level transport protocols. They provide a namespace mechanism that allows for direct P2P communication even if that means to cross the boundaries of different networks or firewalls. New nodes may join the network and be integrated ad-hoc. On the basis of such communication services higher-level application services like indexing, search, file exchange or querying may be built. Every peer may fully participate in the services offered by the network. However, there exist possibilities to structure the overall network into *peer groups* facilitating the creation of interest groups and groups with dedicated rights for accessing critical information.

Semantic Web. The Semantic Web, a term coined by the inventor of the WWW, Tim Berners-Lee, describes the next generation of the Web, which does not only provide information as text and graphics understandable to the human reader, but also gives a semantic description interpretable by machines. By defining the semantics of terms, the Semantic Web will include, will build on, but will also significantly extend the current XML revolution in order to achieve better interoperability between tourism information systems (cf. (Maedche & Staab, 2002)). Therefore, we consider the Semantic Web as a prime enabler of future advanced Web-enabled applications and services such as intelligent Web services, next generation knowledge management solutions, and collaborative e-business in dynamic value constellations. Ontologies are universally recognised as an essential technology to achieve the Semantic Web. Ontologies provide both human-understandable and machine-processable semantic mechanisms needed to let enterprises and application systems collaborate in a smart way. An ontology is a conceptual information model that describes "the things that exist" in a domain (hence the name): concepts, relations, facts and rules, in a consensual and formal way. An ontology thus acts as a standardised reference model, providing a stable baseline for shared understanding of some domain that can be communicated between people and inter-organisational application systems.

3 Towards Services on the Move

This section describes our vision towards ad-hoc configurable, activity-oriented, semantic-descriptions based services. The vision builds on technologies currently being developed in four recent EU-IST projects Ontologging, WonderWeb, SWAP and SWWS.

3.1 KAON — Semantic Web Infrastructure

In (Maedche & Staab, 2002) we have argued for the wide spread adoption of Semantic Web ideas in information systems. However, most current systems still take a static view of the Semantic Web and support only a simplistic kind of "read and query" modus. In the EU IST projects WonderWeb and OntoLogging we have been developing the KAON ontology and Semantic Web infrastructure (Bozsak et al., 2002; Motik et al. 2002,a; Motik et al. 2002b; we refer the interested reader to http://kaon.semanticweb.org, where the KAON open-source software is available for download). KAON targets semantics-driven business applications including a comprehensive infrastructure allowing easy ontology management and application.

The main focus of KAON is on integrating traditional technologies for ontology management and application with ones used in typical business applications, such as relational databases. It is based on an ontology model introduced in (Motik et al., 2002a), derived as a minor extension of RDF(S), with some proprietary extensions, such as inverse, symmetric and transitive relations, cardinalities, modularization, meta-modeling and representation of lexical information. The infrastructure and its associated server does not only take into account the interactions as required by modifications of atomic logical statements, but rather provides mechanisms for dealing with complex behavioral requirements, e.g. transactions and evolution mechanisms. Thus, it provides the basis for the building blocks that follow.

3.2 SWWS — Semantic Web enabled Web Services

SWWS (see http://swws.semanticweb.org) is about bringing web services and fully enabled E-commerce to reality. Everybody must be able to trade and negotiate with everybody else. However, such an open and flexible E-commerce has to deal with many obstacles before it becomes reality (cf. (Fensel et al. 2002a)):

- Current web service technology around UDDI (Universal Description, Discovery and Integration of business for the web; http://www.uddi.org), WSDL (Web Service Description Language: http://www.w3.org/TR/wsdl), and SOAP (Simple Object Access Protocol; http://www.w3.org/TR/SOAP/) does not yet provide a mature technology as it leaves the semantics of data, business logics and message exchange sequences undefined. SWWS combines ontology technology with workflow approaches in order to support automated discovery, invocation and composition of web services.

- Means for scalable mediation between different and heterogeneous services have to be developed. The mediation framework will substantially rely on the semantics-driven descriptions of data and business logics. This framework will also include means for configuration, composition and negotiation of Web Services.

SWWS Case Study. The vision of Semantic Web Enabled Web Services is introduced by providing the following concrete B2B case study: A newly hired employee requires a laptop for his workplace. In order to buy one he defines the characteristics of the laptop like processor speed and disk size. The purchasing process makes sure that the new employee's manager authorizes the purchase. Based on the configuration of the laptop, alternative hardware vendor's offerings have to be collected and a price comparison has to be done. Once the cheapest laptop is determined, a service contract for three years has to be found and purchased together with the laptop. The service contract might be from the laptop vendor or might be from an independent insurance company. It might be that a same or better-equipped laptop together with the service contract is cheaper than if both are bought separately. Once the cheapest combination is found, the laptop purchase order is issued. Finding alternative laptops and service contracts as well as the final purchase should be automatically supported without human user involvement. The mediation related problems that have to be solved within this use case are:

- Business process execution: The whole process must be modeled and executed. The set of tasks may include human ones as well as ones performed by machines.

- Transmissions over networks must be secured according to the trading requirements of the partners and the B2B protocols involved.

- Possible laptop vendors as well as service contract vendors have to be discovered and messages have to be exchanged with them. The search should be reduced to 10 service providers.

- Different document type formats have to be understood and transformed into each other to make the prices and products comparable.

- Alternative purchase approaches will be necessary if a separate company buys the service contract. The reason is that in this case the purchases, the laptop and the service contract, must both succeed simultaneously or they must not be accomplished at all.

SWWS technology will offer the mechanism to realize the scenario introduced above. Various aspects can be identified as being necessary to achieve such an intelligent and automatic web service discovery, selection, mediation and composition into complex services. Among others, the focal aspects of the SWWS framework are:

- Semantics: First, elements of document types must be populated with correct values so that they are semantically correct and are interpreted correctly by the service requesters and providers. This requires that a vocabulary is defined that enumerates or describes valid element values, for example, a list of product names or products that can be ordered from a manufacturer. Further examples are unit of measures as well as country codes. Ontologies provide a means for defining the "concepts" and therefore the semantics of the data to be exchanged. If ontologies are available then document types refer to the ontology concepts. This ensures consistency of the textual representation of the concepts exchanged and allows the same interpretation of the concepts by all trading partners involved. Finally, the intent of an exchanged document must be defined. For example, if a purchase order is sent, it is not necessarily clear if this means that a purchase order needs to be created, deleted or updated. The intent needs to make semantically clear how to interpret the sent document.

- Data & Process Mediation: It is not expected that there will be one global and consistent definition of a vocabulary. Actually, there will be many vocabularies leading to data heterogeneity. One of the most important paradigms of SWWS is to support strong mediation capabilities. Thus, mechanisms for mapping heterogeneous vocabularies will be provided. Additionally, not only data that is exchanged may be heterogeneous, also processes may not match to each other.

3.3 SWAP — Semantic Web and Peer-to-Peer

SWAP (see http://swap.semanticweb.org) is about demonstrating that the power of Peer-to-Peer computing and the Semantic Web can actually be combined to support decentralized, ad-hoc environments where participants can maintain individual views of the world, while sharing knowledge in ways such that administration efforts are low, but knowledge sharing and finding is easy (cf. (Fensel et al., 2002b)):

- Key to the success of combining Peer-to-Peer solutions with Semantic Web technologies is the use of Emergent Semantics. Emergent semantics builds on light-weight and/or heavy-weight ontologies that different individuals, departments, or organizations have created.

- SWAP develops intelligent tools and graphical user interfaces that allow for the use of precise definitions such that knowledge may be nicely structured and easily re-found. The exchange of knowledge in a virtual tourism organization will be enabled without without overhead through central administration.

SWAP Case Study. The SWAP case study involves the connection of SMEs from the tourism industries in the Balearic Islands for the purposes of tourism quality management and sustainable development. Tourism is a highly competitive industry, and the European tourism sector can no longer compete on the basis of cost alone. Quality is therefore a key element for the competitiveness of the tourism industry. It is also important for the sustainable development of the industry and for creating and improving jobs (Lladóo et al., 2002). An integrated approach to quality management is necessary because so many different elements affect the tourist's perception of a destination (such as transport, accommodation, information, attractions, the environment etc.). Integrated Quality Management (IQM) needs to take into account tourist businesses, tourists' interests, the local population and the environment, and to have a positive impact on all of them. The agreed-upon definition of quality in tourism is in line with that of the World Tourism Organisation, emphasising that quality is the perception by the tourist of the extent to which his expectations are met by his experience of the product. Quality is not to be equated to luxury, and must not be exclusive, but must be available to all tourists, including those with special needs. Tourism providers in a location such as the Balearic island are in a cooperation situation. A tourism enterprise there is competing for tourists and therefore aims at outperforming its peers. However, it is not sufficient to be better than the immediate competitor. The quality criterion as defined above require that the natural environment, transport, cultural events and other factors are given special consideration. This cannot be achieved by any enterprise there alone. Rather, all enterprises need to cooperate in order to achieve a high-quality impression by their tourist clients.

The SWAP technology will offer a natural way to exchange information on tourism quality management and sustainable development as it allows for easy, minimal invasive knowledge exchange between tourism enterprises. For instance, a hotel manager may want to exchange information on the measures of sustainable development that his company pursues (for instance, water preserving measures). Common water preservation means will not only help to bring down water costs of one's own enterprise, but it may reduce water costs in general and it will improve the general impression that a tourist receives of the Balearic islands.

In SWAP a hotel manager will simply select appropriate folders (and subfolders) as well as databases (or views on them) for information exchange with selected peer groups (e.g. to competitors on the Balearic islands, but not to ones in Turkey or *vice versa*). He himself may similarly query other peers in his group for related information taking advantage of the precision resulting out of semantic technologies and peer group selection.

4   The Vision: P2P-based Semantic Web Services

P2P-based Semantic Web Services combine the three enabling technologies in order to support scenarios like the following:

A customer might plan an itinerary in the southwest of Germany visiting restaurants with at least 15 of 20 gourmet points (http: //www.gaultmillau.de/gmd/index.php3) and some classical concerts.

Because such a specific wish may hardly be pre-configured, it will be necessary to locate and integrate several services on the fly. Thereby, providers want to facilitate the use of information to their customers and describe it via semantics that can be processed by machines and understood by the customer. The customer may — ad-hoc — integrate constraints coming from different services (a restaurant may be closed on Mondays, hotels may have restricted number of non-smoking rooms, events are pre-scheduled, other constraints may come from a route planning service) in order to answer his request. In order to facilitate assembling the services from virtual networks (e.g. restaurants know about hotels and event organizers in their vicinity and automatically forward requests) based on P2P query routing. Services, such as hotel reservations, may then be invoked (semi-) automatically.

5 Conclusion

Service-driven architectures promise a new paradigm providing an extremely flexible approach for building complex information systems. However, at the current moment, service architectures go little way beyond standardized remote procedure calls and textual directories to locate and describe a service provider based on human intervention. In this paper we have considered three important dimensions for enabling the development of fully enabled service-driven systems building on enabling technologies such as Web services, Peer-2-Peer and Semantic Web. The vision of fully enabled service-driven systems is pursued within the four different EU-IST funded projects Ontologging, WonderWeb, SWAP and SWWS that have been shortly introduced in this paper.

Acknowledgements. Research for the paper has been partially funded by the EU IST projects SWAP (EU IST-2001-34103), SWWS (EU IST-2002-37134) and WonderWeb (EU IST-IST-2001-33052) and Ontologging (EU IST-2000-28293). We thank our colleagues at Institute AIFB, University of Karlsruhe, Research Group Knowledge Management at the FZI Research Center for Information Technology and at Ontoprise GmbH as well as our partners in the above mentioned projects for many fruitful discussions on the topics of this paper.

References

E. Bozsak, M. Ehrig, S. Handschuh, A. Hotho, A. Maedche, B. Motik, D. Oberle, C. Schmitz, S. Staab, L. Stojanovic, N. Stojanovic, R. Studer, G. Stumme, Y. Sure, J. Tane, R. Volz, and V. Zacharias. Kaon - towards a large scale semantic web. In *Proceedings of EC-Web 2002*, LNCS, pages 304–313. Springer, 2002.

D. Fensel, C. Bussler, and A. Maedche. Semantic web enabled web services. In *Proc. of the International Semantic Web Conference 2002*, LNCS, pages 1–2. Springer, 2002a.

D. Fensel, S. Staab, R. Studer, and F. van Harmelen. Peer-2-peer enabled semantic web for modern knowledge management. In N.J. Davies, D. Fensel, and F. van Harmelen, editors, *Towards the Semantic Web – Ontology-based Knowledge Management*. Wiley, London, UK, 2002b.

E. Lladóo, I. Salamanca, and B. Llodrá. D7.1 first user environment definition. Technical report, Fundación IBIT, www.ibit.org, 2002. SWAP Deliverable D7.1.

A. Maedche and S. Staab. Applying semantic web technologies for tourism information systems. In K. Woeber, A. Frew, and M. Hitz, editors, *Proceedings of the 9th International Conference for Information and Communication Technologies in*

Tourism, ENTER 2002, Innsbruck, Austria, 23 - 25th January 2002. Springer Verlag, 2002.

B. Motik, A. Maedche, and R. Volz. A conceptual modeling approach for building semantics-driven enterprise applications. In *Proceedings of the First International Conference on Ontologies, Databases and Application of Semantics (ODBASE-2002)*, LNCS. Springer, 2002a.

B. Motik, D. Oberle, S. Staab, R. Studer, and R. Volz. KAON Server architecture. Technical Report 421, Institute AIFB, University of Karlsruhe, 2002b. http: //wonderweb.semanticweb.org/internal/deliverables/D5-V2.0.pdf.

Travelers and Location-information in the Mobile Environment – Consumer Attitudes and a Prototype of a Service for Early Adopters of Mobile Internet

Ingvar Tjostheim
Bjorn Nordlund
Joachim Lous
Kristin Skeide Fuglerud

Norwegian Computing Center

{ingvar, bjorn, joachim, kristin.fuglerud}@nr.no

Abstract

Survey results show that travelers with e-commerce experience are among the early users of mobile Internet services. The members of this segment seem to be relatively willing to share personal information on the Internet. It can be hypothesized that in most case the individual does not experience misuse of this information. In the mobile environment, location based service are now starting to appear. The user of the mobile phone will then be aware of the fact that others will know his/her location. Hence, the question of how users can make privacy decisions in this new mobile environment is important. In the paper, a prototype of a new service is described. This service or application gives the user the control over to whom information about his or her location is revealed.

Keywords: consumer behavior, privacy enhancing technology, personal information, travelers, location-based services, mobile phones.

1   Introduction

The main function of the mobile phone is to make phone call. However, some recent changes can be easily observed and SMS-messaging is just one example. The telecom industry has shifted the focus from providing voice telephony (on the GSM networks) to providing new services. The 2.5G and 3G networks are currently being built and advanced m-phones are introduced to the market. Travellers are per definition mobile, and mobile Internet is expected to become popular among travelers. According to Koivumäki (2002) travel companies have been among the forefront of companies introducing mobile services. Mobile phones with the possibility of accessing Internet, WAP-phones, have been in the market place for several years. The market for m-services such as location-aware services is expected to grow substantially (O'Grady & O'Hare 2002). Still, WAP-services are only used by very few. Two studies in Norway showed that the first owners of WAP-phone had many similarities with the frequent web-users, a relatively high percentage of them had e-commerce experience, and the WAP-phones were most popular among managers (Tjostheim & Heier 2001;

Tjostheim & Boge 2001). The purpose of this paper is to study attitudes of current and potential users of advanced mobile phones and to present a prototype of a service that gives the user the control over to whom information about his or her location is revealed. And based on the surveys among Internet-users and mobile-phone users, the following two hypotheses are tested:

H1 Individuals with e-commerce experience are more willing to share personal information online than the general Internet-user.

H2 The travelers with e-commerce experience will be among the early adopters of mobile Internet (MI) services.

2 Literature and theoretical foundation

Two research disciplines are particularly important for this study, computer science and consumer behavior a sub-field in marketing science. Some researchers are have start to use the term "Privacy Enhancing Technology" (Borking & Raab, 2001). According to Westin (1967) an individual's information privacy is the right to determine when, how, and to what extent information about a person is communicated to others. The control or lack of control of access that others have to ones personal information is the key issue. Rust et. al. (2002) define privacy as the degree to which personal information is not known by others. According to Goodwin (1991) consumers' information concerns the two dimensions of environmental control and social use of information. Hoffman et. al. (1999) distinguish between environmental control and control over secondary use of the information. In this study the focus is on the first of these two dimensions. Privacy is recognized as very important in relation to the Internet market place and mobile commerce. *"How must consumer privacy policies be designed? What m-commerce activities should be placed in the set that mobile devices users would need to opt in to, and what m-commerce activities should be placed in the set that they would need to opt out of?"* (Balasubramanian, 2002, p.359). This author also addresses the issue of the costs of lost privacy related to what is called "consumer location-based initiatives". Rust et. al (2002) argue that technology is advancing over time, and when technology advances, the cost of obtaining and processing personal information declines. The need for examining the ongoing development in wireless technology is also emphasized. If for instance a tourist guide is made available on a mobile device, it will become a location-based service if the position of the users is integrated or can be integrated in the service. O'Grady & O'Hare (2002) write that it is widely projected that the market for location-aware services will increase substantially in the coming years. The authors conclude their article by underlining the importance of user privacy; *"Location-aware services as a general concept raise a number of ethical and legal issues. These include security, user privacy, etc. Obviously such issues have implications for everybody, including tourists, and the successful resolution of these issues is a prerequisite for the widespread deployment of electronic tourist guides."*

3 Method

This study is based on the analysis of results from two surveys, the first a large postal survey to members of a national panel the winter 2000/2001. The second, a travel survey targeted to members of the Internet-population with travel experience. A pre-requisite to participate in this survey was "have been on a holiday abroad the last 5 years". The travel survey was a web-based questionnaire sent by e-mail March 2002.

Table 1. Key figures

	National survey 2000/01	Travel survey 2002
Internet users	3094 (58%)	458 (100%)
Daily Internet user	1572 (12%)	367 (80%)
Has e-commerce experience	1544 (29%)	334 (73%)
Has an advanced m-phones	485 (9%)	-

The respondents in the travel survey were more frequent Internet users than the respondents in the national survey. This was an expected result since the travel survey was targeted at Internet-users only. Hence, the comparison of respondents in the two surveys is limited to individuals with e-commerce experience.

Table 2. The profile of the respondents with e-commerce experience.

	National survey 2000/01	Travel survey 2002
Men - women	52% - 48%	68% - 32%
15 – 29 years old	24%	22%
30 – 39 years old	32%	29%
40 – 49 years old	24%	30%
50 – 59 years old	16%	14%
60 years or old	5%	5%
Student	12%	-
Primary school	13%	20%
Technical/professional school	28%	23%
College- or university degree	48%	57%
Income less than 25000 EURO	22%	16%
" between 25 – 50000 EURO	56%	51%
" more than 50000 EURO	15%	28%
Un-answered	7%	5%
N=	1544	334

The main difference between the two sub-groups is related to gender and income. A significantly higher percentage of the respondents in the travel survey were men and a high percentage belonged to the highest income-segment. There is approximately one year between the two surveys. A very high number of the mobile phones that were sold in 2002 were WAP-phones compared to the winter 2000/2001. Hence, the question regarding WAP-phones was substituted with a question about the use of WAP-services in 2002. As a consequence, the results in the two surveys cannot be compared directly. Several authors have studies attitudes and the use of personal information on the Internet and it is documented that willingness to share personal information varies (Ackerman et. al. 1999). In the national survey the respondents were asked respond to a set of statements in order to reveal their attitudes to this topic.

Table 4. Attitudes to advertisement and willingness to share personal information on the Internet. *To what extent do you agree to the following;*

National survey Winter 2000/01	Internet-users **without** e-commerce experience	Internet-users **with** e-commerce experience	**Travelers** – has booked travel services online more than 3 times
It is OK to receive advertisements from a company in e-mails even though I haven't approved it in advance			
Strongly disagree	64%	64%	66%
Partly disagree	16%	18%	12%
Partly agree	11%	10%	14%
Strongly agree	5%	6%	8%
Impossible to answer	3%	2%	0
*For goods & services of interest to me, I am positive **to register** and receive advertisement*			
Strongly disagree	**47%**	29%	24%
Partly disagree	15%	16%	10%
Partly agree	25%	30%	**39%**
Strongly agree	12%	23%	26%
Impossible to answer	2%	1%	-
*I think it is OK **to register** in order to get access to info on the Internet if the info is interesting*			
Strongly disagree	**21%**	12%	9%
Partly disagree	13%	10%	5%
Partly agree	37%	35%	35%
Strongly agree	27%	**42%**	**49%**
Impossible to answer	2%	1%	1%
*I am positive to become a member/**registered** customer on the Internet of a supplier of a brand that I like and use.*			
Strongly disagree	**47%**	26%	18%
Partly disagree	16%	16%	12%
Partly agree	23%	**33%**	33%
Strongly agree	10%	23%	**34%**
Impossible to answer	4%	2%	-
N=	1525	1350	188

Ads that are not approved in advance are not well received. However, when the statements are paraphrased, the result is significantly different particularly for the last two segments. Indirectly, the answers reveal a willingness to share personal information. The travelers are the most positive, but on certain conditions. In the situations described in last three statements, the individual has to some extent control – it is he or she who decides to register information about them selves.

Table 3. Internet-behavior and mobile phones

	National survey		Travel survey	
	Has a MI phone	Is planning to buy a MI-phone	Are using SMS, but not WAP-services	Are using WAP-services
E-commerce experience	13%	9%	47%	12%
Without e-com. experience	10%	5%	45%	2%

The results from the national survey show that a slightly higher percentage of the respondents with e-commerce experience had bought or were planning to buy advanced m-phones compared to the other segment. More interestingly the traveler survey shows that a sub-segment of the travelers have started to use MI services. A sub-segment of 12% is not a large segment, but it is quite visible. It is still rare to find studies that document use of MI services in the marketplace.

Table 5. Personal information and search vs. buying behavior

*What kind of information are you willing to give in order **to get information / to buy** about products and services on the Internet?*

National survey Winter 2000/01	Internet-users without e-commerce experience	Internet-users with e-commerce experience	Travelers - has booked travel services online more than 3 times
... in order to get information about products and services			
E-mail address	76%	89%	95%
Name and address	50%	65%	68%
Shopping behavior information	26%	32%	31%
Info about gender, age, income	26%	34%	34%
Official id-number	3%	3%	6%
Credit card number	1%	3%	5%
...in order to buy products or services			
E-mail address	58%	78%	88%
Name and address	55%	89%	87%
Shopping behavior information	21%	26%	23%
Info about gender, age, income	20%	32%	33%
Official id-number	3%	7%	11%
Credit card number	5%	23%	43%
N=	1548	1356	188

The Internet-users are more negative to reveal information about themselves than the members of two segments. There are more similarities than differences between the last two segments. However, the travelers are more open that the Internet-users with e-commerce experience and they are particularly more willing to use credit cards online. To summarize, the findings in the surveys can be interpreted as a support to the H1 – individuals with e-commerce experience are more willing to share personal information online than the general Internet-user. The findings also support H2 - the travelers with e-commerce experience are, according to this study, among the early adopters of mobile Internet (MI) services. Eleven percent of the members in this segment had experience with MI services. The travelers with e-commerce experience seem to be quite willing to reveal personal information in an electronic environment.

3.1 Discussion

Several studies show that individuals are concerned about personal privacy when using electronic services (Spreitzer & Theimer 1993; Ackerman, et. al. 1999; Cranor et al. 1999; Dahlberg & Sanneblad 2000; Tjostheim & Boge 2001). The two surveys presented in this paper only address a few issues with regard to personal information and privacy. It can be argued that it is of great importance that the service providers respect the individuals' need of privacy, and that the personal information is protected against misuse. There are two major factors that affect what level of privacy individual desires according to Laukka (2000). These are the potential harm and the expected rewards from disclosure of the personal information. But there is a wide variety of what individuals consider negative or harming and what individuals find rewarding. However, it is neither an acceptable nor a realistic solution to stop offering services that also can be misused. The challenge is to develop mechanisms, which enable both service providers and customers to benefit from the evolving communication technologies. The question then is whether there is a convenient means of mediating what is of interest for a person without having to reveal too much personal information. This will be of particular importance when it comes to MI services, both because messages on the mobile phone may seem more intrusive than e-mail on a PC, and because a mobile phone is regarded as more personal. As use of electronic services becomes more and more integrated both in work and leisure it will not be possible or acceptable for the user to evaluate privacy matters in every single case. One possible way of meeting these challenges is to develop a system for defining, administering and enforcing privacy policies (Lau et al. 1999). In this case, the individual will be able to define who will be given access to his/her personal information, as well as when and for what purpose.

In order to study the feasibility of this concept, a prototype for defining privacy policies have been developed. There are many categories of personal information, and in order to limit the prototype a choice have been made to focus on location information. Location information is a relevant and interesting choice of several reasons. First, it is information that is based on the position of a mobile phone. Secondly, a variety of new business opportunities based on location information is foreseen, including m-commerce targeting consumers, (O'Grady & O'Hare 2002). Third, and equally important, our hypothesis is that information about the location of an individual may be highly sensitive, and that the individual would want to restrict the access to this information. There are several ways to implement a privacy management system. The prototype makes use of some of the ideas presented in the article: "Concepts for Personal Location Privacy Policies" (Snekkenes, 2001).

4. The prototype – controlling access to location information

Information about a person's location is personal information, and a key question is "who should have access to what location information under which circumstances?" In this section a prototype is presented. This prototype is a system that provides a single place for a user of a mobile phone to define his or her privacy policy regarding location information. The overall architectural model is primarily based on Snekkenes (2001), which describes a possible sequence of interactions for a request for a service based on location data. The model is outlined in **figure 1** and illustrates a scenario where a person includes the following entities:

* **Personal Location Privacy Policy:** Statement of what can be released to whom and when. Each located object will have an associated policy.
* **Policy custodian directory:** Public directory of Policy Custodians.

* Policy custodian: Where the policy is stored and possibly also enforced. The set of permitted operations may include read, write, modify, and query etc. depending on the identity of the requesting entity.

* Location provider: Entity providing the location data. Any release of data should be subject to the policy.

* Service provider: Entity that is combining location data with other data to produce some service. An example could be a company using a person's location to generate a graphical map indicating where the user is.

* Service consumer: Entity to which service is presented for consumption. An example can be a taxi driver wanting to know where to pick up a passenger.

* Service initiator: Entity that would like and/or accept that the service is produced. An example can be a person wanting a taxi. This person accepts that the taxi driver gets location information about himself.

* Service requestor: Entity that makes a request to the service provider for the service to be produced. The same example can be the used - a person wanting a taxi, thus requests the service provider to give the taxi location data.

* Located object: The entity whose location data will be required to deliver the service.

* Owner of policy and located object: The entity that owns the located object.

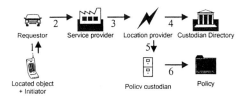

Figure 1. Service request: Sequence of interactions

The prototype is an implementation of the policy custodian in **figure 1,** with interaction interfaces 5 and 6 in the described model. The policy owner can access the custodian via different clients to define and manage their policy (interaction-step 6 in the figure). The policy is exposed to the location provider (interaction-step 5 in the figure). For simplicity the chain of service provider, service initiator and service requestor are handled as one entity, *the observer* in the prototype. The prototype can be further developed to handle all the entities. A typical user walkthrough can be described as follows:

- A new user chooses a custodian to trust, and registers with him/her as a user.
- S/he downloads the windows client and installs it (a web client is planned).
- On the first login, a default policy is created. It has some typical example elements and rules already present
- The user can then edit the policy. Rules are of the form *"for a certain **located object** in a certain **situation**, an **observer** sees only what is allowed by **visibility**"*
- Users (or their automation tools) can switch easily between different *situations* to modify their visibility in the field, without relating to the full details of the policy.

4.1 Architecture

The general design follows a classic multi-tier model: a database for persistence, an object layer implementing the fundamental logic, higher-level functions and external SOAP API, and finally various types of clients for presenting and manipulating the data (figure 2).

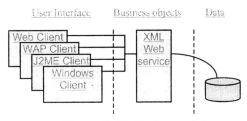

Figure 2. Multi-tier model

4.2 Policy model

The core concept is using rules to define allowable visibility levels depending on a set of circumstances. Finding out what information an observer is allowed corresponds to *visibility=f(infoOwner, infoType, observerRole, ownerSituation)* where all values are enumerations, some of them wholly or partly user-defined. A set of rules defines which combinations will result in which visibility level. A rule has no other function than to couple together named entities.

* **Visibility:** Visibility is a description of how much information is disclosed. Besides all and none one can degrade the precision of location data.

* **InfoOwner:** The policy owner must be identified to determine which rules apply.

* **InfoType:** The user may have different restrictions on different information. So far InfoType only selects which of the user's locatable objects is being queried.

* **ObserverRole:** The information requester must match one of a predefined set of Roles. Roles can be user defined, system-wide, or provided by an external service. Roles are hierarchical, so rules can function as exceptions to more general rules.

* **OwnerSituation:** What 'context' or 'situation' is the user in. This is defined by selection from a user-editable list of named alternatives. This facility can be used to implement time- and location dependencies.

4.3 The Custodian Service

The policy custodian runs the custodian service. This is implemented as a web service. The location providers must communicate with the custodian service trough the web service to retrieve policies tailing which information they can give away to service providers. The main methods available for the location providers are lookup-functions, returning a visibility to enforce based on provided parameters and the information owner's context (known already by the custodian and not disclosed).

4.4 The Clients

The admin interface is where the user defines the policy. The admin interface can potentially be implemented on any platform as long as it supports the web-services standard. A windows platform was chosen for the purpose of developing the initial client. The policy is defined to be dynamic based on the users' situation. When the users' situation changes, the rule for what observers should see also changes. The user needs to manually update his/her situation from a set of predefined alternatives. This is done trough the users mobile phone. Different software agents could also update the situation automatically, based on information about the owner.

4.5 Discussion

The prototype will be used as a tool in the investigation of user-preferences, as well as reactions and attitudes with regard to location privacy policies. The effectiveness of a privacy policy will depend on the possibilities for enforcing the policy. Cryptographic mechanisms and protocols can probably be utilized to develop systems for automatic enforcement of important parts of a policy. There is also probably a need for legal enforcement of the privacy policy. There is a substantial challenge in making the relation between the user interface and the requirements for a privacy policy consistent and at the same time making it understandable for the users. It is not trivial to avoid ambiguities in the policies and hence risking an unintentional privacy policy or policies that is not enforceable.

5. Concluding remarks

The survey results showed that people with e-commerce experience in Norway are more willing to share personal information online than the general Internet users. Moreover, the travelers that have booked travel services online are even more willing to share personal information than the average person with e-commerce experience. These results can be interpreted as a supported to two hypotheses. Moreover, according the surveys, there seems to be a clear tendency that people are willing to share personal information in exchange for information or services of interest to them. A consequence of sharing personal information is potential misuse of this information. To address this problem a technical solution, for dealing with the question of who should be given access to a person's location information, was proposed. This solution does not prevent misuse of personal information. We only suggest a solution where a person can set a policy for how personal information should be handled. To prevent misuse of the information there will be a need for national laws and regulations. Travel and tourism is in its nature international. Hence, systems that will function across borders are needed. It is highly desirable to develop a cross national and nonproprietary solution that operates independent of the separate network operators.

References

Ackerman, Mark S., Cranor, L. F. & Reagle, J. (1999) Privacy in E-commerce: Examining User Scenarios and Privacy Preferences, *Proceeding of the ACM Conference on Electronic Commerce,* November 3 – 5, pp. 1–8.

Ang, Peng Hwa (2001) The Role of Self-Regulation of Privacy and the Internet. *Journal of Interactive Advertising*, Volume 1, Number 2, Spring.

Balasubramanian, S., Peterson, R. A. & Jarvenpaa, S. L. (2002) Exploring the implications of M-commerce for Markets and Marketing, *Journal of the Academy of Marketing Science*, 30 (4): 348-361.

Bellman, S., E. J. Johnson, & Lohse, G. L. (2001). To Opt-In or Opt-Out? It Depends on the Question. *Communications of the ACM* 44 (2): 25-27

Borking, J. J. & Raab, C. D. (2001) Laws, PETs and Other Technologies for Privacy Protection. *The Journal of Information, Law and Technology.* Issue 1, published on 28 February 2001, http://elj.warwick.ac.uk/jilt/

Dahlberg, P. & Sanneblad, J. (2000). The use of Bluetooth enabled PDA's: some preliminary use experiences. *IRIS 23*, Unversity of Trollhättan, Uddevalla

Goodwin, Cathy (1991) Privacy: Recognition of a Consumer Right, *Journal of Public Policy & Marketing* 12 (Spring): 106-119.

Hoffman, Donna L., Novak, Thomas P. & Peralta, Marcos A. (1999) Information Privacy in the Marketspace: Implication for the Commercial Uses of Anonymity on the Web, *The Information Society* 15 (2): 129-139.

Koivumäki, T (2002). Consumer Attitudes and Mobile Travel Portal. *Electronic Markets*, Vol. 12, Number 1: 47-57.

Lau, T., Etzioni, O. & Weld, D. S. (1999). "Privacy Interfaces for Information Management." *Communications of the ACM* 42(10): 89-94.

Laukka, M. (2000). Criteria for Privacy Supporting Systems. Nordsec: *the Fifth Nordic Workshop on Secure IT systems*, Reykjavik, Iceland

O'Grady, Michael J. & O'Hare, Gregory M. P. (2002) Accessing Cultural Tourist Information Via A Context-Sensitive Tourist Guide, *Information Technology & Tourism*, Vol. 5; 35-47.

Rust, Roland T., P. K. Kannan & Na Peng (2002) The Customer Economics of Internet Privacy, *Journal of the Academy of Marketing Science*, 30 (4): 455-464.

Snekkenes, E. (2001) Concepts for Personal Location Privacy Policies. *Proceedings of the ACM Conference on Electronic Commerce (EC'01)*, 14-17 October 2001 Tampa, Florida, USA. ACM Press. 2001. pp. 48-57.

Spreitzer, M. and M. Theimer (1993). Providing location information in a ubiquitous computing environment. *14th ACM Symposium on Operating System Principles*, ACM SIGOPS.

Tjostheim, I. & Boge, K. (2001) Mobile Commerce - Who Are The Potential Customers? COTIM 2001, *Conference on Telecommunications and Information Markets*, Karlsruhe, Germany, July 18 - 20, 2001, ISBN: 0-965440-2-6

Tjostheim, I. & Heier, S. (2001) The characteristics of WAP-phone users - travel habits, Internet usage and demographics. *Proceedings, Information and Communication Technologies in Tourism 2001*, Sheldon, P., Wöber, K. W. & Fesenmaier, D. R. (eds.) SpringerWienNewYork, pp. 130-138.

Westin, Alan, (1967) *Privacy and Freedom.* New York, Atheneum.

Use of the Internet by 'silver surfers' for Travel and Tourism Information Use and Decision-making: A Conceptual Review and Discussion of Findings

Alice Graeupl, and
Scott McCabe

School of Tourism and Hospitality Management
University of Derby, UK
{A.Graeupl, S.McCabe}@derby.ac.uk

Abstract

Information search is a recognised aspect of tourism decision-making process theory. However, there remains relatively little empirical research that demonstrates the complexity of information search in relation to the temporal sequence of decision-making. Further the literature fails to problematise the relationships between age, knowledge acquisition and information needs. This paper discusses these conceptual issues in relation to a small pilot study of 'silver surfers' (older members of society – the so-called 'grey market' - who actively use the Internet). It suggests tentatively that information needed and used is different, not connected to an immediate purchase and proposes further study into this crucial aspect of consumer theory and information use.

Keywords: Tourism; Information Search; Internet; Grey Market

1   Introduction

The importance of information in the tourism decision-making process has been well documented in tourism literature. Furthermore it has been argued that the Internet can respond effectively in a particular way to the information preferences of the tourism market. However, the actual information-orientated search part of the decision-making process (i.e; not an immediate purchase intention) has received little attention from research so far. Although several papers (e.g. Beirne and Curry, 1999) touch upon the information-orientated search as a part of the broader decision-making process, the majority of papers concentrate on the Internet as a new distribution channel. It is argued that although selling via the Internet is on the increase (Cyber Atlas Statistics, 2002), the primary reason for the use of the Internet is still information retrieval in order to gain knowledge, the ultimate goal being to make an informed decision. Although the 'grey market' is one of the fastest growing segments online (Greenfield, 1999 and ISP -Planet, 2000), research has focused on the youth market. Information provided on the Internet should appeal to different market segments and therefore different market groups should be targeted with specialised information, knowledge and service. This paper addresses the information-orientated search as part of decision-making processes from the perspective of the 'grey market'.

2 Theoretical background and issues

As the world's largest industry (Page, Brunt *et al*, 2001), it was never more important for the tourism industry to adapt itself and its products to changing patterns of consumption. Consumers are changing, they are becoming more experienced at buying holidays and the industry is beginning to develop more targeted strategies to smaller groups or clusters of the market to create competitive advantages. Tailor made itineraries mean that travel companies need to have a better understanding of their consumers and the lifestage factor has become of supreme importance (Mintel, 2000). It has long been contended that individuals display different behavioural patterns at different life-stages (e.g. Rappoport and Rappoport, 1975). Different lifestyles, family and job situations as well as different market factors all contribute to these changing patterns. According to Mintel's report on U.K holidaymakers (the paper focuses mainly on the advanced economies of Europe), the empty nester and post-families life stages show similar holiday preferences. Having time, money and freedom they tend to go on several holidays a year, often mixing short breaks in the United Kingdom and/or Europe with longer long-haul holidays. Both of these groups are an important segment for the industry as they have a higher level of personal income available (Mintel, 2000).

The literature on consumer behaviour and in particular the explanatory frameworks for understanding buying behaviour recognises stage in the lifecycle as important factors (see for example East, 1997). Since buying behaviour patterns change over time, it is possible and common sense to suggest that the types of information sought about products and the types of products themselves will also change with time. This temporal dimension to information search in consumer behaviour is an overlooked issue. It is further possible to hypothesise that as people grow older, their knowledge of certain things in life becomes richer and so information needs may change. They may be, or more likely, perceive to be, more susceptible to 'risk' and therefore information needs may be of a different quality to those of other life-stage groups in society. In the context of tourism information search and decision-making, one may speculate that perhaps more detailed information on transportation; accommodation and (relevant to life-stage) facilities may be (sought or) required, together with issues such as security and health, etc.
So how does the Internet relate to different information needs of different life-stage groups? Due to the Internet and other Information Technologies (e.g. digital television), consumers are becoming more aware of the holiday options available to them. As a result, many tourism suppliers have responded by developing their websites to provide information on destinations, flights and accommodation. Increased access as well as consumer sophistication has added to an increase in consumers booking their holidays independently, however not necessarily on the

Internet. This has led to a slight decrease in travel agent's bookings as well as lower commissions from transportation, accommodation and package providers (Mintel, 2000). Given that the traditional role of the travel agent was to provide guided and directed information and choice options to consumers, and also that the increase in experience of travel, booking holidays, Hotels, etc is driving a move to independent booking, the types of information required by consumers should change concomitantly.

Today's society is an 'Information Society', consumers are constantly surrounded and dominated by information (Shenk, 1997). Generally seen as business orientated, a decision-making process is influenced by our lives in an 'information-dependent' society and therefore should be seen as a process through which knowledge is acquired over a long period of time. In leisure and tourism decision-making, an immediate buying decision is rather uncommon as a lot of different types of information e.g. on destinations, means of transportation, accommodation and available packages as well as time and price issues need to be considered. Society, culture and consumption patterns play an important role in the tourist information search and decision-making process. Due to the relatively increasing importance life stage issues, the aspect of tourist information search and the decision-making processes of distinct market (life-stage) groups are neglected by research. Additionally, there are only few available studies that relate the grey market to the Internet [e.g. Trocchia and Janda (2000), Williams and Nicholas (1998)] in fields other than tourism. Furthermore, no studies could be found that investigated the types of information that are used by specific 'grey market' market segments.

It has been suggested in several newspaper articles (Salzburger Nachrichten, August 2001 and February 2002) that the 'grey market' receives more and more attention from advertising companies and that it has been identified as one of the most important 'online' market segments. Interestingly enough, the grey market gets a lot of attention in statistical reports (e.g. ISP and Greenfield) but when it comes to academic research there is little to be found about it, especially in relation to the Internet and especially in relation to leisure travel decision-making processes and experiences on the Internet. Furthermore, as the grey market incorporates people from 45 onwards, they tend to be better off financially, often without children and have more free time (especially when they are retired), they seem the obvious emerging market segment for tourism related research (Mintel, 2000).

Increased accessibility and availability of tourism websites and increased consumer sophistication has added to consumers booking their holidays independently or directly with suppliers but not exclusively on the Internet (Mintel, 2000). However, Mintel's report also indicates that one of the fundamental driving forces for the tourism industry in the next few years will continue to be different requirements of the main lifestage groups. Consumers will be more experienced and confident in using and booking over the Internet, therefore driving changing patterns of purchase with new technologies and therefore holiday requirements will continue to be more sophisticated. Since information search is recognised as a fundamental part of the decision-making process, and as experiences of independent booking develop, the need for independently accessed travel and tourism destination information will also grow. It is proposed that the Internet can be considered a major channel in this changing pattern of information needs. However, it is suggested that information search behaviour between life-stage groups will not be homogenous. Not only do people's information needs change with an increase in independent booking and a decrease in the role of travel agent, but also older groups in society may have different access (to IT) issues as well as information needs.

Further to these complex issues, most of the decision-making process theory models [e.g. Horner and Swarbrooke (1996), Gilbert (1991), Wahab, Crompton and Rothfield (1976), Schmöll (1993), Mathieson and Wall (1982) as well as Lumsdon (1997)] are based on conceptual work and there is little empirical evidence available within the information search aspects of the process. Most of the best-known models are at least 15 years old and they do not acknowledge recent developments as for example the impact of the Internet on changing patterns of consumption. Additionally, the decision-making process is usually seen as a process that concludes in a decision to buy something almost always in a direct and immediate relationship to the information search process.

However, it is proposed that the individual is engaged in a perpetual process of knowledge acquisition, a process of learning which has an impact on touristic decision-making in a more organic and problematic way. Given these theoretical problems and the challenges of the emerging changes in the nature of touristic consumption patterns by life-stage, this study aims to explore the issues through a discussion of pilot-study data undertaken as part of an ongoing investigation into information search practices by older members of society.

3   Methods and preliminary results

An emergent, inductive study design (Silverman: 2001) to the issues of information search needs amongst the 'grey market' was taken. Remaining cognizant of the extant theory in consumer behaviour, yet wanting to test question styles and approaches to targeting and selection of participants. This discussion is based around a small-scale email survey study, which is part of a wider research project. Results are preliminary and are not meant to be conclusive in any way. The main part is a consumer questionnaire aimed at the 'grey market', in this case persons over 45 years of age and active Internet users. A loosely structured questionnaire was sent by e-mail – e-surveying (Litvin and Kar, 2001). The questions asked were divided in three main sections – information search, decision-making process and the Internet as an information source. Although the respond rate was low, several inclinations seemed to show.

Respondents were asked to answer the questions according to the last leisure holiday they took. This was considered an easier approach than asking the set questions on a more general basis. It is acknowledged that the information search and decision-making process is unique and never done the same way twice (Fodness and Murray, 1998). In general the results were as expected. Although only 42 percent of the respondents visited their destination for the first time, 67 percent did in fact search for information. People who did not search for information were – with only one exception – repeat visitors to their chosen destination. One respondent stated that he preferred the adventurous approach by not searching for information before his holiday. Interestingly, the most used information source was the Internet, closely followed by travel guidebooks (e.g. Lonely Planet) and brochures.

Table 1. Most used information sources

Information source	Number
Internet	64
Travel guidebooks	49
Brochures	47
TV programmes	11
Other	29

When asked to rank the used information sources, the Internet, travel guidebooks and friends were awarded equal first position. Respondents also stated that they used the Internet most – the main reason being its easy accessibility – followed by travel guidebooks. At this early stage, it seems that the Internet's main competitor are travel guidebooks. Although it was expected that the Internet is an important information source, it was rather a surprise that it led all categories. Rather disappointing was the response to the question asking participants when they started their information search. Answers ranged from six months to one week before the intended holiday was taking place. Most respondents failed to answer this question and therefore results are not useful. The section on decision-making offered the most surprising results. The final decision was usually made by the partner, which is rather interesting if considered that 92 percent of the respondents were male.

Table 2. Main decision-maker

Decision-maker	Number
Partner/spouse	49
Family incl. children	28
Mutual decision	18
Friends	9
Missing	16

One of the surprising results was that 87 percent of the respondents arranged their holidays independently and only 13 percent booked a package holiday. Furthermore 50 percent even booked their holiday arrangements online e.g. with Expedia or directly with the hotel.

The reasons why respondents decided to use the Internet as an information source were mainly the range of information available and convenience, a rather predictable result. Seventy-five percent of respondents started their search with a search engine, the remaining 25 percent relied on a known website. Respondents were asked if they believed that one could make an informed decision with the Internet as the only source of information. The result was quite close with 49 percent thinking that it is possible and 38 percent that it is not (13 percent missing). The main reason in favour of the Internet was that 'you can find everything that is necessary' and the main reason against the Internet was that 'you always need personal advice'. The most disappointing findings came from the question where respondents were asked to rate tourism related websites. Most people did not answer the question and those who did tended to tick the 'box in the middle' so that the result cannot be seen as valuable. The intended result was to get an insight into people's perception of tourism-related websites and if they cater for their needs.

The final two questions asked about the consumer – provider relationship more specific if respondents considered an online relationship the same as an offline one. Forty-two percent thought that the relationship was different, as 'there remains a need for personal contact' and 'providers take the relationship more seriously'. However, 29 percent (29 percent missing) considered the relationship to be the same because 'an offline relationship is usually anonymous at first too and more likely to be followed up by a phone call rather than a second personal visit'. Interestingly, 50 percent of the respondents prefer an online relationship as opposed to 50 percent who prefer an offline relationship. Sample profile: Respondents were all Austrians, 92 percent male and 8 percent female, all between 45 and 65 years old. The majority is employed and married.

4 Discussion

At the beginning of this Millennium it is important for the tourism industry to become or to remain consumer focused. As stated earlier, consumers are becoming more sophisticated, more experienced and they are demanding more tailor-made products. As the industry is responding to these growing demands it its vital not to forget that different market segments have a need for different information and ultimately for different products. The Internet has created an easy marketing and selling place for the industry, however it must not be forgotten that not all information provided is suitable for all target markets. Although the Internet was one of the most used information tool within this study, the majority backed up their information search with travel guidebooks and/or asking friends and acquaintances for their opinions. Even though the Internet is easy accessible, convenient and has a wide range of information available, the 'silver surfer' respondents within this study are not all convinced that an informed decision can be made with the Internet as the only used information tool.

This paper has taken a different approach to the tourist information search, decision-making process, 'grey market' and Internet theory. It builds on a more culture-related approach, employing theory of society and culture rather than business literature to help understand the problematic relationship between users of the Internet by age group, differing information needs and the information provided. As already mentioned, the tourism decision-making process is conceptualised as a knowledge acquiring process rather than a structured business-like procedure, where the latter is usually followed by an immediate buying decision. It is proposed that culture and society have a major influence on the structure of the decision-making process (and particularly on the information search). One example from this study is that friends were mentioned as the most important information source in several cases.

The Internet, however, could respond even more effectively to the need of this particular market segment. Transportation, accommodation and package providers could establish a new link to their websites that holds a wider range of information that is more detailed and specifically focuses on the needs of the 'grey market'. Although it is noted that specific website information oriented to the needs of these groups is available, in general, the information needs, search processes and decision-making processes are not clearly understood. This ongoing study hopes to map out the uses of information of 'silver surfers' in much more fine detail with an expectation that a better understanding of these processes will enable organisations to deliver targeted communications in a coherent and profitable way. As the research has shown, they are not reluctant to book online, however in some cases more detailed information would be of advantage. Another issue to be raised is that of security and risk. Although older people usually possess a credit card or debit card, they might be hesitant to use it on the World Wide Web, as they might not be so confident about online security.

It was suggested in this paper that when consumer start looking for information on tourism related products and services they often do not know what they are looking for. Frequently, the decision 'where to go' and 'what to do' will be made during the search. That is one of the reasons why the 'silver surfer' respondents of this study largely started their information search with the help of a search engine. However, the research has not investigated which element of the holiday is researched first. The authors suggest that the search for a suitable holiday destination would probably be the first choice followed by transportation and accommodation availability. This importance of information for the 'grey' market segment proves that the proposed suggestion that the decision-making process is usually not followed up by an immediate buying decision but more of a knowledge acquiring nature is true in this particular scenario.

5 Conclusions

As this paper is part of a wider research project, all findings to date are preliminary. However, certain conceptual frameworks have been proved right. The results of the 'silver surfer' questionnaires have shown some inclination towards statistical trends. The 'grey market' uses the Internet for tourism related decision-making, mainly for information search but they are not reluctant to book online. However, just a fragment of questionnaire respondents rely on the Internet fully, most of the 'silver surfers' prefer to back up their information search with other information sources. Although neglected by tourism literature up until now, the 'grey market' is a significant online market segment, even more interesting are its information needs as well as ways and experiences of dealing with knowledge acquisition and decision making.

References

Baldinger, I. *et al* (2001) Silver Surfer haben Zeit – und Geld. *Salzburger Nachrichten.*

Baldinger, I. *et al* (2002) Series – Die fidelen Alten, *Salzburger Nachrichten.*

Beirne, E. and Curry, P. (1999) The Impact of the Internet on the Information Search Process and Tourism Decision Making. In Buhalis, D. and Schertler, W. (eds) (1999) *Information and Communication Technologies in Tourism 1999.* New York, Springer Verlag Wien.

Cyber Atlas Statistics (2002) Cyber Atlas Newsletter <www.cyberatlas.com>

East, R. (1997) *Consumer behaviour: Advances and applications in Marketing.* Herts, Prentice Hall.

Fodness, D. and Murray, B. (1998) A Typology of Tourist Information Search Strategies. *Journal of Travel Research,* 37, 108-119.

Gilbert, D.C. (1991) An examination of the consumer behaviour process related to tourism. In Cooper, C. (ed.) (1991) *Progress in Tourism, Recreation and Hospitality Management,* Vol.3, London, Belhaven Press.

Greefield (1999) *Survey says silver surfers will spend,* <http://www.channelseven.com>

Horner, S. and Swarbrooke, J. (1996) *Marketing Tourism, Hospitality and Leisure in Europe,* London: International Thomson Business Press.

ISP-Planet (2000) *Seniors, Baby Boomers becoming hot ISP market segment,* <http://www.isp-planet.com>

Leisure Intelligence/Mintel (2000) *Holidays by Lifestage.* London.

Litvin, S.W. and Kar, G.H. (2001) E-Surveying for Tourism Research: Legitimate tool or a researcher's fantasy? *Journal of Travel Research,* 39, 308-314

Lumsdon, L. (1997) *Tourism Marketing.* London, International Thomson Business Press.

Mathieson, A. and Wall, G. (1982) *Tourism: economic, physical and social impacts.* London, Longman.

Page, S.J., Brunt, P., Busby, G. and Connell, J. (2001). *Tourism: A Modern Synthesis.* London, Thompson Learning.

Rappoport, R. and Rappoport, R.N.(1975) *Leisure and the Family Life Cycle.* London, Routledge.

Schmöll (1977) *The Schmöll model of the travel decision process.* In Swarbrooke, J. and Horner, S. (1999) *Consumer Behaviour in Tourism.* Oxford, Butterworth-Heinemann.

Silverman, D. (2000) *Doing Qualitative Research: A Practical Handbook.* London, Sage.

Trocchia, P. J. and Janda, S. (2000) A phenomenological investigation of Internet usage among older individuals. *Journal of Consumer Marketing,* 17(7), 605-616.

Wahab, Crompton and Rothfield (1976) *The Wahab, Crompton, and Rothfield model of consumer behaviour in tourism.* In SWARBROOKE, J. and HORNER, S. (1999) *Consumer Behaviour in Tourism.* Oxford, Butterworth-Heinemann.

Williams, P. and Nicholas, D. (1998) Not an age thing! 'Greynetters' in the newsroom defy stereotype. *New Library World,* 99(4), 143-148.

A Natural Language Query Interface for Tourism Information

Michael Dittenbach [a],
Dieter Merkl [a,b], and
Helmut Berger [a]

[a] Electronic Commerce Competence Center – EC3
Donau-City-Straße 1, A-1220 Wien, Austria

[b] Institut für Softwaretechnik, Technische Universität Wien
Favoritenstraße 9-11/188, A-1040 Wien, Austria
{michael.dittenbach, helmut.berger, dieter.merkl}@ec3.at

Abstract

With the increasing amount of information available on the Internet one of the most challenging tasks is to provide search interfaces that are easy to use without having to learn a specific syntax. Hence, we present a query interface exploiting the intuitiveness of natural language for the largest Austrian web-based tourism platform Tiscover. Furthermore, we will describe the results and our insights from analyzing the natural language queries collected during a field trial in which the interface was promoted via the Tiscover homepage. This analysis shows how users formulate queries when their imagination is not limited by conventional search interfaces with structured forms consisting of check boxes, radio buttons and special-purpose text fields. The results of this field test are thus valuable indicators into which direction the web-based tourism information system should be extended to better serve the customers.

Keywords: Tourism information system; Natural language processing; User behavior study.

1   Introduction

The development and availability of efficient and appropriate search functions are still a challenge in the field of database and information systems. Consider, for example, the context of tourism information systems where intuitive search functionality plays a crucial role for the economic success. The reason, obviously, is that users, i.e. tourists, are often computer illiterate, such that formal query languages like SQL in database systems or Boolean logic in retrieval systems are an enormous barrier. This is further impaired by the development of search engines that have subtle differences in their functionality that are not apparently visible (Shneiderman et al., 1998). When faced with a natural language interface, however, these users are able to express their information needs as they are used to when interacting with a human travel agent.

The two design goals for the interface were, first, to provide multilingual access, and second, to develop a generic framework independent of a particular application domain. To achieve these goals, we strictly separated natural language query analysis from the domain-specific processing logic. The analysis process identifies relevant parts of the query using language-dependent ontologies describing the concepts of the application domain. The domain-specific processing logic defines how these relevant parts are related to each other and builds the appropriate database query. Further features of our system are that additional languages can be added conveniently to the information system and the interaction with the system can be designed with respect to the differing capabilities of various client devices such as web browsers or PDAs.

2 Natural Language Query Processing

This section contains a brief description of the various steps performed during natural language query processing of the demonstrator system. The software architecture is designed according to a pipeline structure as shown on the right hand side of Figure 1.

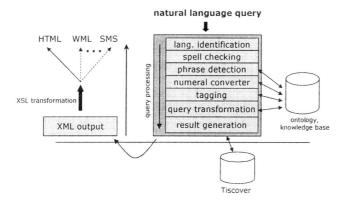

Fig. 1. Software Architecture

2.1 Language Identification

To identify the language of a query, we use an *n-gram*-based text classification approach (Cavnar and Trenkle, 1994) where each language is represented by a class. An *n-gram* is an *n*-character slice of a longer character string. As an example, for $n = 3$, the *tri-grams* of the string *"language"* are: *{_la, lan, ang, ngu, gua, uag, age, ge_}*. Dealing with multiple words in a string, the blank character is usually replaced by an underscore *"_"* and is also taken into account for the construction of an *n-gram* document representation.

This language classification approach using *n-grams* requires sample texts for each language to build statistical models, i.e. *n-gram* frequency profiles, of the languages. We used various tourism-related texts, e.g. hotel descriptions and holiday package descriptions, as well as news articles both in English and German. The *n-grams*, with *n* ranging from 1 to 5, of these sample texts were analyzed and sorted in descending order according to their frequency, separately for each language. These sorted histograms are the *n-gram* frequency profiles for a given language.

In the English text, *{the, and, of, in}* and the suffix *{ion}* are the most frequent *tri-grams*. Contrarily, in the German texts, the most frequent *tri-grams* are endings like *{en, er, ie, ch}* and words like *{der, ich, ein}*. To determine the language of a query, the *n-gram* profile, $n = 1 \ldots 5$, of the query string is built as described above. The distance between two *n-gram* profiles is computed by a rank-order statistic. For each *n-gram* occurring in the query, the difference between the rank of the *n-gram* in the query profile and its rank in a language profile is calculated. For example, the *trigram {the}* might be at rank five in a hypothetical query but is at rank two in the English language profile. Hence, the difference in this example is three. These differences are computed analogously for every available language. The sum of these differences is the distance between the query and the language in question. Such a distance is computed for all languages, and the language with the profile having the smallest distance to the query is selected as the identified language, in other words, the most probable language of the query. If the smallest distance is still above a certain threshold, it can be assumed that the language of the query is not identifiable with a sufficient accuracy. In such a case the user will be asked to rephrase her or his query.

2.2 Error Correction

To improve the retrieval performance, potential orthographic errors and misspellings have to be considered in our web-based interface. After identifying the language we use a spell-checking module to determine the correctness of the query terms. The efficiency of the spell checking process improves during the runtime of the system by learning from previous queries. The spell checker uses the metaphone algorithm (Philips, 1990) to transform the words into their soundalikes. Because this algorithm has originally been developed for the English language, the rule set defining the mapping of words to the phonetic code has to be adapted for other languages. In addition to the base dictionary of the spell checker, domain-dependent words and proper names like names of cities, regions or states, have to be added to the dictionary.

For every misspelled term of the query, a list of potentially correct words is returned. First, the misspelled word is mapped to its metaphone equivalent, then the words in the dictionary, whose metaphone translations have at most an edit distance (Levenshtein, 1966) of two, are added to the list of suggested words. The suggestions are ranked according to the mean of:

- the edit distance between the misspelled word and the suggested word, and
- the edit distance between the misspelled word's metaphone and the suggested word's metaphone.

The smaller this value is for a suggestion, the more likely it is to be the correct substitution from the orthographic or phonetic point of view. However, this ranking does not take domain-specific knowledge into account. Because of this deficiency, correctly spelled words in queries are stored and their respective number of occurrences is counted. The words in the suggestion list for a misspelled query term are looked up in this repository and the suggested word having the highest number of occurrences is chosen as the replacement of the erroneous original query term. In case of two or more words having the same number of occurrences the word that is ranked first is selected. If the query term is not present in the repository up to this moment, it is replaced by the first suggestion, i.e. the word being phonetically or orthographically closest. Therefore, suggested words that are very similar to the misspelled word, yet make no sense in the context of the application domain, might be rejected as replacements. Consequently, the word correction process is improved by dynamic adaptation to past knowledge. Another important issue in interpreting the natural language query is to detect terms consisting of multiple words. Proper names like *"St. Anton am Arlberg"* or substantives like *"swimming pool"* have to be treated as one element of the query. Regular expressions are used to identify such cases.

2.3 SQL Mapping

With the underlying relational database management system PostgreSQL, the natural language query has to be transformed into a SQL statement to retrieve the requested information. The knowledge base of the domain is split into three parts. First, we have an ontology specifying the concepts that are relevant in the application domain and describing linguistic relationships like synonymy. Second, a lightweight grammar describes how certain concepts may be modified by prepositions, adverbial or adjectival structures that are also specified in the ontology. Finally, the third part of the knowledge base describes parameterized SQL fragments that are used to build a single SQL statement representing the natural language query.

The query terms are tagged with class information, i.e. the relevant concepts of the domain (e.g. *"hotel"* as a type of accommodation or *"sauna"* as a facility provided by a hotel), numerals or modifying terms like *"not"*, *"at least"*, *"close to"* or *"in"*. If none of the classes specified in the ontology can be applied, the database tables containing proper names have to be searched. If a substantive is found in one of these tables, it is tagged with the respective table's name, such that *"Tyrol"* will be marked as a federal state.

In the next step, this class information is used by the grammar to select the appropriate SQL fragments. To illustrate this processing step, consider the following SQL fragment as the condition for an accommodation being located in or not in a particular city, where @OP is a placeholder for an operator and @PARAM for the city name.

```
SELECT entity."EID" FROM entity
WHERE entity."CID" = city."CID" AND city."Name" @OP @PARAM
```

Depending on modifying terms found in the query as specified in the grammar, the SQL fragment is selected and the parameters are substituted with the appropriate values. A query for accommodation in Innsbruck produces the following fragment.

```
SELECT entity."EID" FROM entity WHERE
entity."CID" = city."CID" AND city."Name" = 'Innsbruck'
```

Finally, the SQL fragments have to be combined to a single SQL statement reflecting the natural language query of the user. The operators combining the SQL fragments are again chosen according to the definitions in the grammar.

3   Data and Examples

For the examples described below, we demonstrate the application of our natural language interface to search for accommodations throughout Austria. In particular, we use a part of the database of the largest Austrian web-based tourism platform *Tiscover* (Pröll et al., 1998), which, as of October 2001, provides access to information about

13,117 accommodations. These are described by a large number of properties including the respective numbers of various room types, different facilities and services provided in the accommodation, or even the type of food.

These accommodations are located in 1,923 towns and cities that are again described by various features, mainly information about possible sports activities, e.g. mountain biking or skiing, but also the number of inhabitants or the sea level. The federal states are the higher-level geographical units. For a part of the data, we integrated the geographical coordinates of the cities and towns to additionally provide information about the distance between places. Therefore, the system can be queried for accommodations *close* to a certain place as will be shown in the second example.

As our first example consider the following English query:

"I am looking for a hotl in St. Abton am Arlberg with sauna and a swiming pool. The hotel should furthermore be suitable for children and pets should be allowed".

As can be seen, the query contains several misspellings such as *"hotl"*, *"Abton"* and *"swiming pool"*. In the case of *"Abton"*, our improved spell checking mechanism does not choose the word *"Baton"*, which is ranked first in the list of suggested corrections, but instead chooses *"Anton"*. This selection is performed because of a previously posed query, where *"St. Anton am Arlberg"* has been spelled correctly.

For our second example we use the following German query:

"Ich brauche ein Einzelzimmer mit Frühstück in einer Pensoin in der Nähe von Insbruck aber nicht in Innsbruck selbst".

This query, again with misspellings, shows the effect of different prepositions modifying a noun. The query states that a pension with breakfast close to Innsbruck but not in the city of Innsbruck is searched for. The first occurrence of *"Innsbruck"* is preceded with *"close to"* and therefore the following SQL fragment is constructed:

```
SELECT entity."EID" FROM entity
WHERE entity."CID" IN
(SELECT b."CID" FROM city AS a, city AS b WHERE
a."DEC_Lat" != 0 AND a."Name" ~* '^(.* )?Innsbruck( .*)?$'
AND (|/(((a."DEC_Lat" - b."DEC_Lat")^2) +
((a."DEC_Long" - b."DEC_Long")^2)) <= 0.13489734))
```

This statement is based on the assumption that *"close to"* means within approximately 15 kilometers. This range can be adapted by the user to her or his particular needs. Regarding the second occurrence of *"Innsbruck"*, the identification of *"in"* before the city name leads to the following SQL fragment:

```
SELECT entity."EID" FROM entity
WHERE entity."CID" = city."CID"
AND city."Name" !~* '^(.* )?Innsbruck( .*)?$'
```

The negation *"not"*, preceding *"in Innsbruck"*, determines the operator that will be applied to merge the two sets of accommodations retrieved by the above SQL fragments. In this particular case, the pensions in the result set will be those close to Innsbruck except those located in Innsbruck directly.

4 Design Considerations for the Web-Based Interface

In Figure 2, a screen shot of our interface is depicted. A simple and easy to use interface was our major design goal. Hence, we provided only short textual descriptions in both German and English, a text area in which the user can enter the query and the submit button. The sample query *"I am looking for a double room in the center of Salzburg with indoor pool."* is the only hint on the capabilities of the interface. The intention was to cover a broad range of accommodation requests and to find out what the user really wants. We wanted to avoid to narrow the user's imagination when formulating a query, admittedly, with the risk of disappointing the user when no or inappropriate results were found.

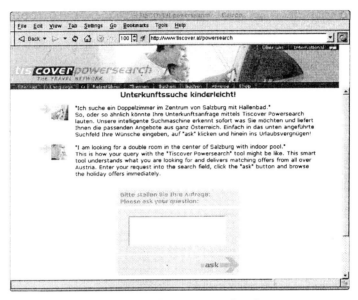

Fig. 2. Natural language query interface

With the conventional interface of *Tiscover* for searching accommodations the area (federal state, region, city) can be chosen either by typing the name directly into the text field or via clicking through the hierarchy of place names. Further criteria are the name of the accommodation, the chain it belongs to and, perhaps, a particular "theme", e.g. family hotel, as well as several amenities the accommodation should provide. Note, this list of amenities is rather small compared to the complete information contained in the *Tiscover* database to keep the interface concise.

We also implemented the look-and-feel of the *Tiscover* design in order to avoid distraction from the user's task. On the result screen (see Figure 3), we provide the original query as well as the concepts identified by the natural language processing to provide the user with feedback regarding the quality of natural language analysis. Below the list of accommodations matching the criteria, we have provided a feedback form where users can enter a comment and rate the quality of the result. After the field test, it turned out that only 3.37% of the queries have either been annotated or rated where the numbers of positive and negative comments were nearly equal. Due to the unsupervised nature of the test without any reward for the test persons, this figure is not surprising thinking of the additional time it takes to assess the quality of the result and then comment on it. At the bottom of the page, the input field prefilled with the posed query is presented to allow for convenient query reformulation or refinement. About 10% of the queries were modified correspondingly.

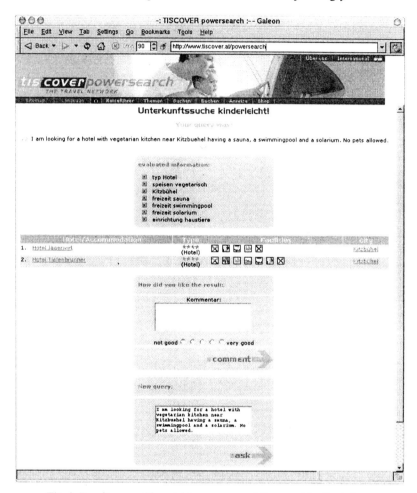

Fig. 3. Result page with matching accommodations and feedback form

5 Field Trial

The field trial was carried out from March 15 to March 25, 2002. During this time our natural language interface was promoted on and linked from the main *Tiscover* page. We obtained 1,425 unique queries through our interface, i.e. equal queries from the same client host have been reduced to one entry in the query log to eliminate a possible bias for our evaluation of the query complexity. Expectedly, most of the queries (39.73%) came from Austrian hosts, followed by hosts from the .net top-level domain, most of which have been identified as German Internet service providers by manual inspection. After the 13.13% of queries from the US commercial domain several European countries can be found. In 20.42% of the queries we were unable to identify the originating country because of non-resolvable domain names.

Of those 1,425 unique queries, 1,213 (85.12%) were German, 120 (8.42%) were English and 92 (6.46%) were not identifiable, e.g. non-sentence queries like *"hotel salzburg"* that are possible in both languages or just nonsense like *"ghsdfkjg"*. Based on the 1,333 identified queries we found 52 queries that were not in the scope of our natural language interface. Among these were, for example, questions about purchasing used cars or, of course, sex among other topics that could not be answered by the system. Obviously, in any kind of publicly available service like this, not all of the people are using it for the intended purpose. However, this number is rather low assuming the rather short description we displayed on the start page to give an idea what kind of information can be queried.

To provide some technical information, for the 1,333 processed queries, the mean processing time was 2.63 seconds with a standard deviation of 1.42 seconds. The median of 2.27 seconds shows that there were only a few outliers with longer processing times. Given these figures, we can say that our system is usable regarding its response time. Even with adding a few seconds for data transmission time over the Internet, the response time still lies below the maximum of ten seconds as suggested by Nielsen (2000). These ten seconds have been measured in usability studies as the approximate maximum attention span of users when waiting for a web page to be loaded before canceling the request.

We will compare the results of two studies analyzing query log files of the large and popular search engines AltaVista and Excite with the results of our analysis, since only few research papers dealing with user behavior in web searches exist. Silverstein et al. (1998) and Jansen et al. (1998) have shown that the average number of words per query is very small, namely 2.35, interestingly the same in both studies. This indicates that most of the people searching for information on the Internet could improve the quality of the results by specifying more query terms. Our field test revealed the amazing result of an average query length of 8.90 words for German queries, and of 6.53 for the English queries. In more than a half (57.05%) of the 1,425 queries, users formulated complete, grammatically correct sentences whereas only 21.69% used our interface like a keyword-based search engine. The remaining set of queries (21.26%) contains partial sentences like *"double room for 2 nights in Vienna"*. This approves our assumption that users accept and are willing to type more than just a few keywords to search for information. Furthermore, the average number of relevant concepts occurring in the German queries is 3.41 with a standard deviation of 1.96, which is still one word per query more than found in the surveys of web-search engine usage as mentioned above. We can thus assume that, by formulating a query in natural language, users are more specific than compared to keyword-based searches.

To analyze the complexity of the queries, we considered the number of concepts and the usage of modifiers like *"and"*, *"or"*, *"not"*, *"near"* and some combinations thereof as quantitative measures. Table 1 shows the distribution of the numbers of concepts per query. For example, consider row four of this table. The entries in this row show the number of queries with three concepts. In particular, we have 310 German and 28 English queries. Note that these figures were derived by manual inspection of the users' original natural language queries. The majority of German queries consist of one to five concepts relevant to the tourism domain with a few outliers of more than 10 concepts. People asking for an accommodation in a specific region by enumerating potentially interesting cities and villages can explain the latter.

Table 1. Total concepts per query (counted by manual inspection of the queries)

concepts	query language		
	ge	en	totals
0	47	5	52
1	77	28	105
2	272	38	310
3	310	28	338
4	245	12	257
5	137	5	142
6	49	2	51
7	38	1	39
8	18	1	19
9	11		11
10	4		4
11	1		1
17	3		3
21	1		1
totals	1213	120	1333

We shall note that most of the concepts not identified, originated from queries falling into the categories of region names, pricing information, room availability and arrival and departure dates. This information, however, was not contained in the data we received from *Tiscover* for inclusion in our database during the field trial.

Another aspect of the complexity of natural language queries are words connecting concepts logically or modifying their meaning. The evaluation of the queries showed that *"and"* is by far the most frequently used modifier and its distribution of occurrences roughly corresponds to the number of concepts.

The second-most frequently used modifier in the queries collected during the field trial was *"near"* expressed in terms like *"around"* or *"close to"*. A common way to use *"near"* is to find accommodations in the surroundings of popular sites, cities or facilities, e.g. *"I am looking for a hotel with sauna and pool in St. Anton near the Galzig-Seilbahn"*. The modifier *"or"* is used far less than *"and"*. *"Or"* is mostly used to provide a set of locations or types of accommodations of interest, e.g. *"I am looking for a farm or an apartment in Tyrol or Salzburg"*. An interesting fact revealed during the field trial is, that the *"not"*-modifier is used in a very small subset of queries. This implies, that the vast majority of users formulate their intentions without the need of excluding concepts. In most of the cases a *"not"* is used to exclude a specific property of a region or an accommodation. For instance, users wanted to avoid places where pets are allowed as well as quiet accommodations without children. Another common use of *"not"* was to exclude one or more cities from a query where an accommodation in a federal state or region was wanted, e.g. *"I am looking for a hotel in Tyrol, but not in Innsbruck and not in Zillertal"*.

In general, queries are formulated on the basis of combining concepts in a simple manner, e.g. *"I am looking for a room with sauna and steam bath in Kirchberg"*.

Only a small subset of queries consist of complex sentence constructs that require a more sophisticated sentence evaluation process, as, for instance, when the scope or type of the modifier cannot be determined correctly. As an example, consider the query *"I am looking for an accommodation in Serfaus, Fiss or Ladis"*. In contrast to the assumption that the default operator of combining concepts is *"and"*, the modifier *"or"* must be used to combine the geographical concepts in this sample query.

For a more detailed report on the complexity of the queries processed during the field trial, we refer to Dittenbach et al. (2002).

6 Conclusions

In this paper we have described a multilingual natural language database interface. At present, the interface allows queries to be formulated in German and English. The language of the query is automatically detected using an n-gram-based text classification approach. A spell checker is used to compensate for orthographic errors. The strategy of word replacement is further improved by taking into account word occurrence statistics from previous queries. After simple syntactic and semantic analysis the concepts addressed by a query are transformed into parameterized SQL fragments. Our analysis of the field trial shows that the level of sentence complexity is rather moderate which suggests that shallow text parsing should be sufficient to analyze the queries emerging in a limited and well-defined domain like tourism. Nevertheless, we found out that regions or local attractions are important information that has to be integrated in such systems. We also noticed that users' queries contained vague or highly subjective criteria like *"romantic"*, *"wellness"*, *"cheap"* or *"within walking distance to"*. These concepts are difficult to model in the knowledge base of information systems and pose a challenge for the future.

References

Cavnar, W. B. and Trenkle, J. M. (1994) N-gram-based text categorization. In *Proc Int'l Symposium on Document Analysis and Information Retrieval (SDAIR'94)*, Las Vegas, NV.

Dittenbach, M., Merkl, D., and Berger, H. (2002) What customers really want to know from tourism information systems but never dared to ask. In *Proc Int'l Conf on Electronic Commerce Research (ICECR-5)*, Montreal, Canada.

Jansen, B. J., Spink, A., Bateman, J., and Saracevic, T. (1998) Real life information retrieval: A study of user queries on the web. *SIGIR Forum*, 32(1).

Levenshtein, V.I. (1966) Binary codes capable of correcting deletions, insertions and reversals. *Soviet Physics Doklady,* 10(8).

Nielsen, J. (2000) *Designing Web Usability: The Practice of Simplicity.* New Riders Publishing.

Philips, L. (1990) Hanging on the metaphone. *Computer Language Magazine,* 7(12).

Pröll, B., Retschitzegger W., Wagner, R. R., and Ebner, A. (1998) Beyond traditional tourism information systems - TIScover. *Information Technology and Tourism,* 1.

Shneiderman, B., Byrd, D., and Croft, W. B. (1998) Sorting out searching. *Comm. ACM,* 41 (4).

Silverstein, C., Henzinger, M., Marais, H., and Moricz, M. (1998) Analysis of a very large AltaVista query log. Technical Report 1998-014, digital Systems Research Center.

Beyond the Trip Planning Problem for Effective Computer-Assisted Customization of Sightseeing Tours.

Jean-Marc Godart

Mathematics & Operations Research Department
Faculté polytechnique de Mons, Belgium
jean-marc@godart.net

Abstract

In two previous articles, it was suggested to use the Trip Planning Problem (TPP) to provide decision support with regard to the planning of sightseeing tours. The purpose of the present article is to suggest a few models that are based on the TPP but which relax some of the assumptions made in the TPP. The implementation of these improvements is discussed and a real-life example is given to illustrate what can actually be done with these kinds of models. Going beyond the TPP, by relaxing or eliminating its main assumptions, make it possible to set up a valuable Decision Support System (DSS) for effective computer-assisted customization of sightseeing tours.

Keywords: Trip Planning Problem (TPP) ; Decision Support System (DSS) ; customization ; sightseeing tours.

1   Introduction

Trips are made of various components : transportation (e.g. by plane or car), lodging (e.g. in hotels or campgrounds), activities (e.g. museums or beaches), etc. Thus, planning a trip consists in selecting and arranging specific travel components in the most appropriate way, as to come up to the tourist's expectations (Godart, 1999). The tourist for whom the trip is being planned either can be identified precisely, as when she/he is sitting in front of the travel agent, or she/he can be any member of a target group, as when a tour operator is setting up a new package. In both cases, actual or implicit expectations of the customer consist of wishes (e.g. about the kind of destination), requirements (e.g. about the duration of the trip), interests (e.g. about the nature of selected activities), etc.

As tourists have learned to become more and more demanding about their trips, their preferences have shifted away from standardised packages, designed by tour operators, to individualised products, specifically selected and arranged to meet customers' expectations (Bennett and Radburn, 1991). This change leads to greater customer involvement in the trip planning process.

Trip planning is a complicated and time-consuming activity. In the past, the main difficulty for tourists was the accessibility of pertinent information (especially about schedules, availability, fares, rates...) which was typically the domain of travel agents. Now, the Internet and other recent information and communication technologies play a powerful role in enabling customers to choose and purchase travel components to plan their own trip (Reinders and Baker, 1998). Yet, little has been done so far to assist users in the trip planning process itself (Hruschka and Mazanec, 1990 ; Godart, 2001).

Given the huge number of products that are made available and the interdependence of their choice for a given trip, it is desirable that some kind of help be provided, by a Decision Support System (DSS), in selecting and arranging those travel components, in order to ease the planning of customized trips. Such a tool will be useful to tourists who set up their own trips as well as to travel agents, tour operators, etc. It will also enable online travel providers (e.g. travel portals, hotel chains...) to offer extra added value to their customers and to take advantage of interesting new opportunities to market their products (Godart, 2000).

This paper will specifically focus on sightseeing tour planning. Sightseeing tours are leisure trips that start and finish at a same place (e.g. an airport), are aimed at visiting several sights (e.g. museums) in a given region, and may have overnight stays (e.g. in hotels) in different locations along the route. The tremendous number of possible combinations makes the planning of sightseeing tours even more difficult than for other types of trips, such as a trip to a major city or a seaside resort.

2   The Trip Planning Problem

Even though trip planning, and in particular sightseeing tour planning, has not been studied much from the point of view of decision support, different approaches coexist. The most usual one relies on expert system concepts, but does not seem to give rise to practical applications. In a previous article (Godart, 1999), an innovative alternative approach was proposed, which consists in using the mathematical tools of combinatorial optimisation. These have already been applied with success in many different fields (routing, scheduling...), and seemed to be appropriate also in the matter of trip planning (Godart, 1999 ; Ladany, 1999 ; Loban, 1997).

Following this new approach, consecutive alterations of the well-known Travelling Salesman Problem (TSP) led to define a model which is more appropriate for the purpose of trip planning and which was thus called "Trip Planning Problem" (TPP) (Godart, 1999).

To keep it rather simple, the principle of the TPP is schematised in figure 1. Trips are built as sequences of different activities and of N, possibly redundant, lodgings that are selected from databases (N is the number of nights as imposed by the tourist). All these sequences start and finish at the same "point of origin and destination" (represented by a black square in figure 1), which is also chosen by the tourist. Knowing the duration of each activity and the time needed to go from a place (point of origin and destination, activity, or lodging), to another, it is ensured that the time elapsed between two consecutive lodgings in the sequence falls between d^- and d^+, which are respectively a lower and an upper bound on the active duration of every day of the trip.

Finally, knowing the cost and the attractiveness of each activity and lodging, and knowing the cost of going from a place to another, all the feasible trips (sequences) can be evaluated as to five conflicting objectives : "minimisation" of the cost of transportation, "minimisation" of the cost of activities, "minimisation" of the cost of lodgings, "maximisation" of the attractiveness of activities and "maximisation" of the attractiveness of lodgings. The TPP consists in selecting one (or a few) of the best compromises with regard to these objectives. As an example, figure 2 gives the spatial representation of what could be a solution of the TPP (Godart, 2001).

The TPP has proven to be an appropriate basis for the development of a DSS for trip planning. Yet, this model relies on rather strong assumptions that limit the range of trips, and particularly sightseeing tours, that it can help to customize. The main hypotheses of the TPP were enumerated in a previous article (Godart, 2001) and the reasonableness of these assumptions was also discussed in relation to the planning of sightseeing tours. The most cumbersome hypotheses were studied further to show how they can be made less strong (relaxed) or even eliminated, in order to enhance the usefulness of the TPP in this matter.

The purpose of the present article is to suggest a few models that are based on the TPP but which relax some of the assumptions made in the TPP. Using these new models, it will be possible to provide decision-makers (tourists) with more effective assistance for sightseeing tour planning. The implementation of these improvements will also be discussed and a real-life example will be given to illustrate what can actually be done with these kinds of models.

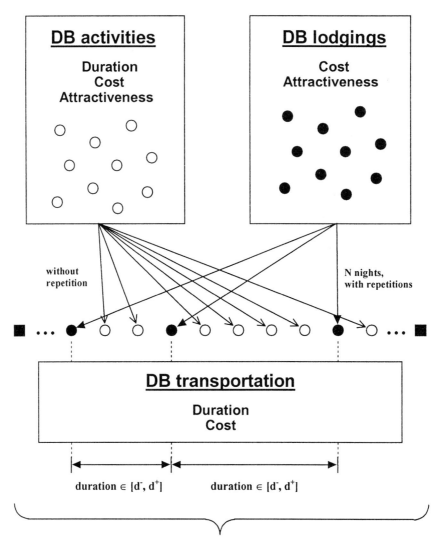

Fig. 1. Principle of the TPP

167

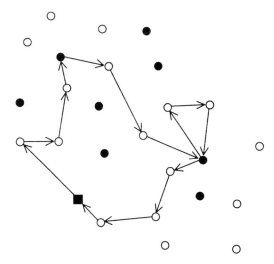

Fig. 2. Example of a solution of the TPP

3   Beyond the Trip Planning Problem

From a practical point of view, there are many ways in which the model based on the TPP can be improved, to be made even more appropriate to trip planning. However, it has to be remembered that the problem underlying the model needs to be solved to find optimal or, at least, close-to-optimal solutions (trips). As a matter of fact, it has to be kept in mind that the TPP is already difficult to solve as a combinatorial optimisation problem and that computation times increase drastically as the instance size (number of activities, lodgings...) gets bigger. Thus, it is important to make sure that good solutions for normal instances of the new problems can still be obtained in a reasonable amount of time.

Moreover, it also has to be taken into account that the data needed for the new model must be available. For data that needs to be input by the decision-maker, it is necessary to make sure that this can be done in real life situations, for example on a B-to-C web site.

At this stage, three main alterations of the TPP-based model have been considered reasonable and studied further. They concern (1) the selection of the origin and the destination of the trip, (2) the inclusion of specific activities, and (3) the inclusion of specific lodgings.

3.1   Selection of origin and destination

In the TPP, it is assumed that the trip starts and finishes at the same place, which is called "point of origin and destination" ; it is also supposed that the tourist knows what is the right point of origin and destination for her/his trip (e.g. Miami International Airport, for a trip in Florida).

In reality, a sightseeing tour is not always a tour as such, since the tourist can decide to start at a place (e.g. Miami International Airport) and finish at another place (e.g. Orlando International Airport). Moreover, in the first stage of the planning process, the tourist may have no idea of where her/his sightseeing tour should start and finish. Hence, it is desirable to set up an improved model where the "point of origin" and the "point of destination" need not be the same place, and where these points are selected endogenously (that is by the model itself) based on what seems to be most appropriate. Yet, it should still be possible for the tourist to decide on the point of origin and/or the point of destination, or require them to be identical.

To improve the TPP model in this respect, two extra databases were added to what was presented in figure 1 : one database of possible points of origin and one database of possible points of destination. The optimisation algorithm was then modified as to select the point of origin and the point of destination of the trip from the respective database. Also, an optional constraint can oblige the algorithm to select the same place in both databases.

3.2   Inclusion of specific activities

Another implicit hypothesis of the TPP model considers that there is no activity that the tourist absolutely requires to be included in the trip. As a matter of fact, any activity of the database of activities can or cannot be selected, depending on what seems to be better, considering the trip as a whole.

Yet, most often the tourist knows as from an early stage in the planning process that an activity has to be part of her/his trip, because she/he considers that activity as a "must-do". For example, a tourist who is planning to go to Florida could consider that she/he absolutely has to go see "Disney World".

To take into account such requirements, the "must-do" activities have to be pinpointed in the database of activities. Furthermore, the algorithm needs to be modified as to include and keep those activities in the trip.

3.3   Inclusion of specific lodgings

The TPP model assumes as well that there is no lodging where the tourist absolutely wants to spend one night (or more) during her/his trip. In the same way as for activities, any lodging of the database of lodgings can or cannot be selected, depending on what seems to be better, considering the trip as a whole.

Again, though, the tourist can demand that one or several nights be spent at a given place. This can be due to a good past experience or to the reputation of the lodging in question. For example, a tourist who is planning to go to Florida could consider that she/he absolutely has to spend a night at the famous "Casa Marina" hotel, in Key West.

This kind of requirements can be dealt with by pinpointing the affected lodgings in the database of lodgings and by indicating the minimum number of nights that have to be spent there. The algorithm then needs to be modified as to include and keep in the trip the required number of copies of those lodgings.

4   Computer-Assisted Customization of Sightseeing Tours

Since the above suggested extensions of the TPP model relax a few strong assumptions on which this model is based, they are a big step towards effective assistance in the trip planning process, and especially for the customization of sightseeing tours.

Of course, as it was mentioned above, the combinatorial optimisation problems on which the models rely still need to be solved. The optimisation algorithms which have been used for the TPP are approximative methods called metaheuristics. These enabled to find close-to-optimal solutions in a reasonable computing time for normal instances of the TPP. Also, as metaheuristics are very flexible methods which makes it possible to tackle a large range of different problems (Osman and Kelly, 1996), it was not too difficult to adapt them to solve the extensions of the TPP model. Consequently, good solutions could also be found for a global model that gathers the new features that were mentioned before.

As an example, using basic databases of 173 activities, 91 lodgings, and 3 origins/destinations (international airports), a sightseeing tour was planned in Florida for a fictitious couple of tourists. This couple is supposed to have the following specific requirements as to what their trip should be like : it does not matter from what airport their tour starts, but they want to end up at Orlando International Airport ; they want to visit Edison's Winter Home, in Fort Myers, and Miami Museum of Science ; they want to spend two nights at the "Key Ambassador Resort Inn" in Key West. Finally, it was assumed that the trip has to be eight days.

The first sightseeing tour that was obtained is represented in figure 3. Travel plans are as follows :

> Origin : Miami International Airport
> Activity : Art Deco District - 0 h. 55 - $ 0.00 / person
> Activity : Little Havana - 0 h. 55 - $ 0.00 / person
> Activity : Villa Vizcaya - 2 h. 00 - $ 10.00 / person
> Activity : Miami Museum of Science - 2 h. 30 - $ 6.00 / person
> Lodging : Key Ambassador Inn - $ 110.00 / room
> Activity : Key West Aquarium - 1 h. 30 - $ 6.50 / person
> Activity : Hemingway's House - 1 h. 00 - $ 6.50 / person
> Activity : Pelican Path - 4 h. 30 - $ 0.00 / person
> Activity : Conch Tour Train - 1 h. 30 - $ 14.00 / person
> Activity : Casa Marina - 0 h. 20 - $ 0.00 / person
> Lodging : Key Ambassador Inn - $ 110.00 / room
> Activity : Coral Castle - 1 h. 20 - $ 7.75 / person
> Activity : Edison's Winter Home - 2 h. 30 - $ 6.00 / person
> Lodging : Admiral's Inn - $ 53.00 / room
> Activity : Sea World - 8 h. 15 - $ 35.95 / person
> Lodging : Plaza International Hotel - $ 69.00 / room
> Activity : Magic Kingdom - 9 h. 00 - $ 37.00 / person
> Lodging : Admiral's Inn - $ 53.00 / room
> Activity : Disney-MGM Studios - 9 h. 00 - $ 37.00 / person
> Lodging : Plaza International Hotel - $ 69.00 / room
> Activity : EPCOT - 9 h. 00 - $ 37.00 / person
> Lodging : Admiral's Inn - $ 53.00 / room
> Activity : Universal Studios - 8 h. 15 - $ 37.00 / person
> Activity : Church Street Station - 0 h. 35 - $ 0.00 / person
> Destination : Orlando International Airport

Based on this first proposal, the couple can review the different parameters of the trip (activities/lodgings to be included, point of origin/destination, duration, etc.) and have another sightseeing tour generated. This way, interactively, a fully customized tour will be designed for the tourists.

This example shows that the newly defined model enables to take account of pretty elaborate demands that are often made by the tourists as to their trip. As such, it makes it possible to consider designing a DSS for effective computer-assisted customization of sightseeing tours.

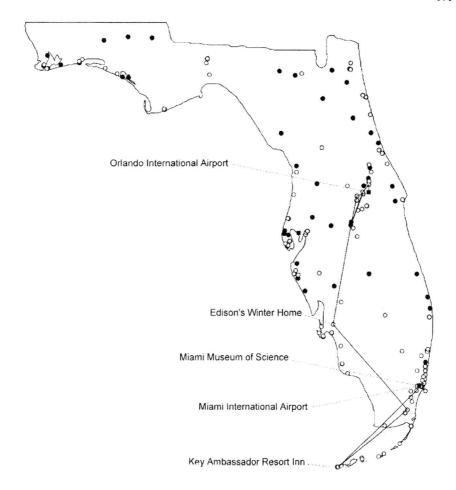

Orlando International Airport

Edison's Winter Home

Miami Museum of Science

Miami International Airport

Key Ambassador Resort Inn

Fig. 3. Example of a sightseeing tour in Florida

5   Conclusions

Using mathematical models, such as the one based on the TPP, for planning sightseeing tours, constitutes plainly a simplification of a very complex process. Thus the output which is obtained from the TPP should not be considered as ready-to-book travel plans but rather as a good starting point for further "manual" improvements. Going beyond the TPP, by relaxing or eliminating its main assumptions, can reduce those necessary manual improvements and make it possible to set up a valuable DSS for sightseeing tour customization.

In the near future, as tools and knowledge become more readily available, it should be possible to relax most hypotheses of the TPP in order to get even closer to real travel situations, and to assist users in the planning not only of sightseeing tours but also of other kinds of trips.

References

Bennett, M., & Radburn, M. (1991). Information Technology in Tourism : The Impact on the Industry and Supply of Holidays. In M.T. Sinclair, & M.J. Stabler (Eds.), *The Tourism Industry, An International Analysis* (pp. 45-65). United Kingdom: C.A.B. International.

Godart, J.-M. (1999). Combinatorial Optimisation Based Decision Support System for Trip Planning. In D. Buhalis, & W. Schertler (Eds.), *International Conference on Information and Communication Technologies in Tourism, proceedings* (pp. 318-327). Austria: Springer-Verlag.

Godart, J.-M. (2000). Sightseeing Tour Planning Based on Combinatorial Optimization : a Tool for Hotel Marketing. *International Journal of Hospitality Information Technology, 1* (2), pp. 21-33.

Godart, J.-M. (2001). Using the Trip Planning Problem for Computer-Assisted Customization of Sightseeing Tours. In P.J. Sheldon, K.W. Wöber, & D.R. Fesenmaier (Eds.), *International Conference on Information Technology and Tourism, proceedings* (pp. 377-386). Austria: Springer-Verlag.

Hruschka, H., & Mazanec, J. (1990). Computer-Assisted Travel Counseling. *Annals of Tourism Research, 17* (2), pp. 208-227.

Ladany, S.P. (1999). Optimal Tourist Bus Tours. *Tourism Economics, 5* (2), pp. 175-190.

Loban, S.R. (1997). A Framework for Computer-Assisted Travel Counseling. *Annals of Tourism Research, 24* (4), pp. 813-834.

Osman, I.H., & Kelly, J.P. (1996). Meta-Heuristics : An Overview. In I.H. Osman, & J.P. Kelly (Eds.), *Meta-Heuristics : Theory & Applications* (pp. 1-21). U.S.A.: Kluwer Academic Publishers.

Reinders, J., & Baker, M. (1998). Electronic Direct Retailing of Travel and Tourism Products. *Progress in Tourism and Hospitality Research, 4* (1), pp. 1-15.

Stille, W. (2001). *Lösungsverfahren für Prize-Collecting Traveling Salesman Subtour Probleme und ihre Anwendung auf die Tourenplanung im Touristeninformationssystem Deep Map.* Student thesis. Germany: Ruprecht-Karls-Universität Heidelberg.

Modeling and Comparing Internet Marketing: A Study of Mainland China Based and Hong Kong Based Hotel Websites

Truman Huang[a]
Rob Law[b]

[a] HRC WORLDWIDE, Guangzhou, P.R. China
truman@hrcworldwide.com

[b] School of Hotel & Tourism Management, The Hong Kong
Polytechnic University Hong Kong
hmroblaw@polyu.edu.hk

Abstract

In spite of the relative ease of website design, many hotels do not know how to fully utilize their websites as a marketing tool, and many hotel websites have been designed to serve as "electronic brochures" instead of "innovative marketing tools". In view of this deficiency, successful factors for conducting web marketing will clearly help hotels improve their efforts on Internet marketing. This research serves two purposes. First, it makes an attempt to develop a framework for hotel Internet marketing. Second, the research performs a comparison study for Hong Kong luxury hotels and Mainland China luxury hotels using the developed framework. Hotels in Hong Kong are selected for direct comparison with their China counterparts because the former are generally regarded as having an international standard. Lastly, this research makes recommendations for hotels to make better use of their websites as an effective marketing tool.

Keywords: Internet; hotel websites; Mainland China; Hong Kong.

1 Introduction

In order to remain competitive in the Internet era, hotels have to develop a web marketing strategy and to find an optimal customer-oriented website solution. Having a successful Internet marketing strategy can provide hotels competitive advantages since websites are a very important component of Internet marketing. This research, therefore, aims at offering a WWW marketing solution for the hotel industry, particularly in China, by comparatively evaluating the performance of hotel websites from a marketing perspective.

The specific objectives of this research are:

- to develop an evaluation framework for measuring hotel website performance from practitioners and customers' perspectives;
- to compare and contrast the website performance of Mainland luxury hotels and Hong Kong (an international tourism and business center) luxury hotels using the developed framework; and

- to provide recommendations for Mainland hotels to improve the effectiveness of their websites.

2 Related Work

Kotler & Armstrong (1996) pointed out that the marketing mix is one of the major concepts in modern marketing. Marketing mix is defined as the set of controllable tactical marketing tools that the firm blends to produce the response it wants in the target market. The marketing mix consists of everything the firm can do to influence the demand for its product, which traditionally refers to 4Ps including product, price, place, and promotion. Morrison (1996) revealed that there are a total of 8Ps in the hospitality industry according to its unique characteristics, and that the other more 4Ps are people, partnership, packaging, and programming.

Kotler & Armstrong (1996) commented that the Ps' theory merely represents the sellers' view of the marketing tools available for influential buyers. From a customers' viewpoint, however, four Cs also need to be stressed which refer to customer needs and wants, cost, convenience, and communication (Lauterborn, 1990).

Many hospitality companies created their own websites for various purposes, but mostly used them as a sales and marketing tool (van Hoof et al., 1999). By using the Internet as a marketing tool, tourism organizations have gained some distinct advantages in cost reduction, revenue growth, marketing research and database development, and customer retention (Morrison et al., 2001). Customer expectations are crucial factors when starting to consider the feasibility of developing a website for business (Bell & Tang, 1998).

This study investigates marketing mixes of 8Ps from suppliers' perspective and 4Cs from customers' perspective in the context of hotel Internet marketing. Website promotion is crucial to lead a hotel to the final success of WWW marketing, and for this reason, website promotion is also included in this study as another dimension of evaluation.

3 Methodology

In this research, 74 luxury hotel websites including 45 Mainland hotels and 29 Hong Kong hotels were evaluated. The Mainland hotels selected for analysis were chosen from the top 100 hotels in terms of revenues in 2000. This study only focused on luxury hotels in the five-star and four-star categories. In total, 45 Mainland luxury hotels including 31 five-star hotels and 14 four-star hotels were selected. Though Hong Kong hotels are classified into three categories as High Tariff A, High Tariff B and Medium Tariff hotels, which differ from the hotel star-rating system used in Mainland, High Tariff A and High Tariff B hotels were considered as luxury hotels in Hong Kong. At last, 29 Hong Kong hotels including 13 High Tariff A hotels and 16 High Tariff B were chosen in study.

As shown in Figure 1, the research process consisted of 3 phases. The website evaluation was developed with opinion from industrial practitioners and users, as well as the promotion of websites. Hence, a framework for hotel website performance was developed. A content-based analysis was then carried out to examine the performance of hotel websites on the basis of the proposed framework.

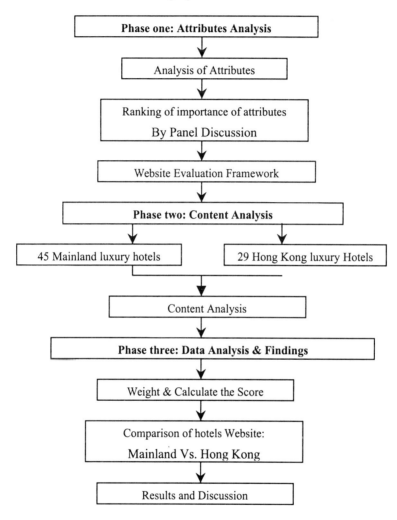

Fig. 1. Research Process

Each component of 8Ps, 4Cs and website promotion was evaluated according to the attributes as listed in Table 1.

Table 1. Website Attributes from the Industry's perspective

Components	Attributes
Product	
Hotel Introduction	Hotel history (background), hotel milestones, maps
Guestroom	Room type, in-room facilities & service, check-in/check out time
Restaurants & Bar	Cuisine type, menu, hours, contact
Convention	Capacity, layout, area, style, service & facilities provided, meeting planner, contact
Entertainment	Introduction, type, hours, contact
Transportation	Vehicle type, service provided, contact
Business Service	Services provided, hours
Shopping Service	Introduction, types, hours
Aesthetic Effect	Photos in each facility, animation aspects
Place	
On-line Distribution	Available product category, distribution for target market
Reservation	Reservation policy, on-line search, on-line available check, on-line reservation form, on-line reservation retrieval, on-line cancellation, real-time processing, create/modify profile and e-mal reservation
Payment	Credit card type, security payment system, payment options
Prices	
Room	Room rate of different types, breakfast price
F & B	Price for banquet, dishes or set menu
Convention	Room rental rate, facilities rental rate
Transportation	Vehicle rental fee
Promotion	
Special Offers	Discount for Internet bookers, special offers for niche market
Advertising	What's new, press release
Multimedia	360 panoramic photo, virtual hotel tour, audio/video on homepage, picture zoom-in/out
People	
On-line Recruitment	Employee opportunity and benefits
Email	Email with own domain suffix
Internal Information	Mission statement, company policy, staff information
Packages	On-line package
Marketing Research	On-line survey

Partnership	
Affiliate Program	Other booking channels, travel service, points with other partners
Linkage	Linkage to travel partners, linkage to other websites
Customer Needs and Wants	
City Tour	City introduction, local attractions, shopping guide, cuisine, entertainment
Local news	Activities, exhibitions, arts
Travel tips	Exchange rate, weather reports, world clock, transportation timetable, custom polices, calendar
Security & Privacy	Credit card security, privacy policy
Cost to customer	
Room Rate	Comparison of Internet rate and walk-in rate
Frequenter Plan (Members Club)	on-line enrollment, restricted frequent guest area, on-line account review, create/modify customer profile, members benefit, special offer for members
Convenience	
Loading Speed	Homepage download time
Language	Number of language choices, currencies, date formats
Design Skill	Navigation, site map, fact sheets
Communication	
Interactivity	FAQs, Feedback forms, chat rooms, on-line talking, e-mail and automated response, e-mail updates on latest development, on-line request, comments from satisfied customers, update date, visitor number
Customized Web Pages	Customer registration, personal web page, address web visitors by name, welcome message, and guest book
Contact Information	Telephone number, fax number, e-mail address
Website Promotion	
Search Engines	Registration in main engines
Links	Links with other sites
Useful skills	Bookmark, email newsletter, 'email this to a friend' notices, announce site update, hit counter
Management	Refresh website weekly

At the initial stage of framework development, a panel discussion was adopted to rank the importance of each component in 8Ps, 4Cs, and website dimensions. The discussion group consists of 4 respondents including senior hotel managers and Internet users. Next, the average weighted scores were calculated for each component based on a proportion of 100 percent using to the following approach:

$D_1 = 8Ps$ $D_2 = 4Cs$ $D_3 =$ Website Promotion
$S_{D_iR_j} =$ Important Score given by respondent $i=1,2,3$ R = Respondent $j = 1, 2, 3, 4$

W_{Di} = Average weighted score of marketing mix; i = 1, 2, 3

$$W_{Di} = \frac{\sum_{j=1}^{4} S_{Di}R_j}{4}$$

$$\text{Total Score} = \sum_{i=1}^{3} \left(\frac{\sum_{j=1}^{4} S_{Di}R_j}{4} \right) \qquad (\text{Total Score} \leqslant 100)$$

Next, the average weighted score of each element of the marketing mix is calculated based on the given score (W_{Di}) of the first step. The calculation results use the following approach:

P_i = Elements of 8Ps, i = 1,2,3, ...7
$S_{pi}R_j$ = Importance Score for Elements of 8Ps given by respondent j=1,2,3,4
C_i = Elements of 4Cs, i = 1,2,3,4
$S_{ci}R_j$ = Importance Score for Elements of 4Cs given by respondent j=1,2,3,4
$S_{Pr}R_j$ = Importance Score for Web Promotion given by respondent j=1,2,3,4
W_E = Average weighted score of each element

$$W_{D1} = \sum_{i=1}^{7} \left(\frac{\sum_{j=1}^{4} S_{Pi}R_j}{4} \right)$$

$$W_{D2} = \sum_{i=1}^{4} \left(\frac{\sum_{j=1}^{4} S_{ci}R_j}{4} \right)$$

$$W_{D3} = \frac{\sum_{j=1}^{4} S_{Pr}R_j}{4}$$

Having calculated the average weighted score for each component, a five-point judgmental scale was used to evaluate the attribute performance of each component in the category as depicted in Table 2.

Table 2. A five-point judgmental performance scale

1	2	3	4	5
Very bad/Not Avail.	Poor	Neutral	Good	Very Good
Performance				
No listed attributes available	Possess only a few listed attributes	Possess half of the listed attributes or Possess most of the attributes but bad organization and Aesthetic Effect	Possess most of the attributes or Possess all the attributes but neutral organization and Aesthetic Effect	Possess all the attributes with Good organization and Aesthetic Effect

The performance mean score for each component was then calculated by combining the average weighted score and the attribute's score using a five-point judgmental scale. To directly reflect the website performance, the mean score was transferred to a performance percentage score. In other words, the performance percentage score was obtained by dividing the average weighted score with the performance mean score according to the following approach:

P_C = Performance score of component

W_{EC} = Average Weighted score of component

A_s = Score of attributes by a 5-point judgmental scale

$P_C = W_{EC} * A_s$

A 100-point judgmental scale was then used to rate the score's performance. The performance rating was divided into 5 categories ranged from very unsatisfactory to very satisfactory (See Table 3).

Table 3. A 100-point judgmental performance scale

Point	0-20	20-40	40-60	60-80	80-100
Performance Category	Very unsatisfactory	Unsatisfactory	Average	Satisfactory	Very Satisfactory

An independent-samples t-test was used to test the significant difference between Mainland hotel websites and Hong Kong hotel websites.

4   Findings

Having evaluated the performance of websites using attributes from the developed framework, significant differences were found between the Mainland hotel websites and the Hong Kong hotel websites.

In terms of evaluation from the 8Ps, Hong Kong hotels received higher performance scores than Mainland hotels. For Hong Kong hotels, Product, Place, Price, Promotion, and Partnership all performed satisfactory; whereas Packaging & Programming, and People had an average performance. For Mainland hotels, the Product and Price dimensions performed satisfactory but Place, Promotion, People, and Partnership only achieved average performance scores. Unfortunately, Packaging & Programming fell in the unsatisfactory level. Table 4 shows the results for the 8Ps.

Table 4. Overall Perceived Performance Score of 8Ps

Components	Expectation Weighted Score	Overall Performance Percent	Perceived Performance Score				Sig.
			Mainland		Hong Kong		
			Mean	Percent	Mean	Percent	
Product	12	70.27	8.12400	67.70	8.91052	74.25	0.031*
Place	7	54.13	3.14222	44.89	4.79310	68.47	0.000*
Price	9	64.80	5.52556	61.40	6.30862	70.10	0.016*
Promotion	6	57.49	2.90167	48.36	4.30000	71.67	0.000*
People	5	46.19	2.10889	42.18	2.62069	52.41	0.018*
Packaging and Programming	4	40.14	1.28889	32.22	2.09655	52.41	0.000*
Partnership	5	49.05	2.04444	40.89	3.08621	61.72	0.002*

*significant at = 0.05

In the 4Cs dimension, Hong Kong hotels also received higher performance scores than Mainland hotels. For Hong Kong hotels, Convenience was rated satisfactory and others were rated average. For Mainland hotels, Convenience and Communication were in the same categories as their Hong Kong counterparts. However, Customer needs and wants as well as Cost to customers were rated unsatisfactory. Table 5 presents the 4Cs findings.

Table 5. Overall Performance Score of 4Cs: Mainland vs. Hong Kong

Components	Expectation Weighted Score	Overall Performance Percent	Perceived Performance Score				Sig.
			Mainland		Hong Kong		
			Mean	Percent	Mean	Percent	
Customer Needs & Wants	10	33.84	2.50889	25.09	4.74138	47.41	0.000*
Cost to Customer	7	39.11	2.28000	32.57	3.44828	49.26	0.003*
Convenience	8	71.44	5.81333	72.67	5.56207	69.53	0.280
Communication	6	57.03	3.32444	55.41	3.57241	59.54	0.152

*significant at = 0.05

Lastly, as shown in Table 6, Hong Kong hotels received higher performance scores than the Mainland hotels in the dimension of website promotion.

Table 6. Performance Score of Website Promotion: Mainland vs. Hong Kong

Components	Expectation Weighted Score	Overall Performance Percent	Perceived Performance Score				Sig.
			Mainland		Hong Kong		
			Mean	Percent	Mean	Percent	
Search Engines	6.5	77.84	4.82444	74.22	5.42414	83.45	0.104
Management	5.75	61.08	2.70889	47.11	4.75862	82.76	0.000*
Links	5.75	35.14	1.66111	28.89	2.57759	44.83	0.014*
Useful Skills	3	22.61	0.64000	21.33	0.73793	24.60	0.163

* significant at = 0.05

5 Conclusion

Empirical findings in this research showed that the website performance of Hong Kong hotels scored higher than Mainland hotels, and statistically significant differences exist between hotels in these two regions. The findings typically reflect that the websites of Mainland hotels mainly remain as "electronic brochures". Hong Kong hotels, however, obtain higher scores in most attributes, which reflect that the websites of Hong Kong hotels can serve as "marketing tools".

The performance of 4Cs on the sites was not satisfactory in both Hong Kong hotels and Mainland hotels. In reality, the Internet can boast its character of interactivity. Without the participation of customers, the website is not really interactive. The website is designed for customers, but not for the industry. Therefore, customer needs and wants, cost to customer, convenience and communication should be focused during the website development stage.

Although each hotel website has its own style in location, facilities, star-rating, management, and target market, the marketing mix of 8Ps and 4Cs should be applicable to most, if not all, hotels when conducting marketing activities.

In the future, a structure model of website marketing based on marketing mix can be developed. The model consists of four functional media including Information Medium, Purchase Medium, Reception Medium, and Traffic Medium. Information Medium helps visitors to get enough understanding about the hotel such as product, price, external environment (customer needs & wants), people and partnership. By Purchase Medium, visitors could select products including hot product (Promotion, Packaging & Programming) and then make an online purchase. If visitors have any questions, they may come to Reception Medium and use various interactive tools to get help. The Reception Medium also helps visitors join some frequent plans to enjoy personalized service. Traffic Medium helps a hotel generate traffic on the website. Specified attributes of each component on those four media should be taken into consideration during website development.

Acknowledgement

An earlier and abridged version of this paper will appear in JICC2002 Proceedings.

References

Bell H. & Tang N. (1998). The Effectiveness of Commercial Internet Websites: a User's Perspective. *Internet Research: Electronic Networking Applications and Policy* 8(3): 219-228.

Van Hoof H., Ruys H. & Combrink T. (1999). Global Hoteliers and the Internet: Use and Perceptions. *International Journal of Hospitality Information Technology* 1(1).

Kotler P. & Armstrong G. (1996). *Principles of Marketing*, Prentice-Hall International, Inc.

Law R. & Leung R. (2000) A Study of Airlines' Online Reservation Services on the Internet. *Journal of Travel Research* 39, Nov: 202-211.

Lauterborn R. (1990). New Marketing Litany: Four P's Passé; C-Words Take Over, *Advertising Age*, Oct. 1: 26.

Morrison A. (1996). *Hospitality and Travel Marketing*, Delmar Publishers, 1996.

Morrison A., Jing S., O'Leary J. & Cai L. (2001). Predicting Usage of the Internet for Travel.

Gaining Competitive Advantage for the Libyan Tourism Industry through E-marketing

Said Al-Hasan[a]
Mokhtar Jwaili[b]
Brychan Thomas[b]

[a]Management Division, Business School, University of Glamorgan, UK
[b]Welsh Enterprise Institute, Business School, University of Glamorgan,
{majwaili; salhasan; bcthomas} @glam.ac.uk

Abstract

The paper investigates enterprises in the Libyan tourism industry and their perception of the role of e-marketing in developing the industry. It considers enterprises that do not apply e-marketing, the problems they face and the benefits they would gain from an e-marketing programme. In relation to this the importance of e-marketing and the extent to which theory is followed is examined. The paper explores what should be done to develop e-marketing for tourist enterprises and how to implement strategic plans for the industry to increase its market share to enable it to compete in the international market place. The tourism industry is important to Libya's economy and greater consideration should be given to its development and, in particular, to e-marketing. This is especially the case for small and medium-sized tourism enterprises (SMTEs) since most industrial sectors depend on these composing 50% privately owned and the rest full or part government owned.

Keywords: competitive advantage, Libyan tourism, e-marketing

1   Introduction

The race for survival amongst enterprises is one of the main issues owner/managers are concerned with. Consequently greater focus is being targeted at the management of e-marketing since it is one of the most influential aspects in the overall performance and competitive advantage of organisations which lead to their growth or decline. In Libya many enterprises consider e-marketing as an extra cost, arguing that if they sell their products why should they consider an activity that will increase costs. E-marketing has been defined as marketing over the Internet. In fact, it covers a wide range of activities including 'advertising, customer communications, branding and relationship-building efforts, loyalty and retention programs – all conducted over the Internet. Much more than creating a Web site, e-marketing focuses more on communicating on line' (CISCO, 2000). The use of the World Wide Web grants SMEs the opportunity to compete with larger organisations on a global level (Quelch and Klein, 1996). Quelch and Klein identified Internet advertising as a means of overcoming some of the problems facing SMEs, as it will allow them to reduce the cost of advertising and overcome entry obstacles. They concluded that lower costs in communications on the Internet might give firms with limited financial resources the opportunity to become global marketers very quickly (Buhalis and Main, 1998). Werthner and Klein (1999) state that tourism is considered one of the most important application domains on the World Wide Web (WWW). Wahab and Cooper (2001) point out that a Global marketing strategy has opened the door for SMEs to compete in the international market with similar opportunities to the big players. Strassel (1997) estimates that 33% and 50% of Internet transactions are tourism based. It is estimated that within the next 10 years 30% of tourism business will be Internet based (Werthner and Klein, 1999). Buhalis (1999) lists the costs and benefits for developing Internet presence for SMTEs, as follows in Table 1:

Table 1. Cost and benefit analysis for developing Internet presence for SMTEs

COSTS
<ul><li>Cost of purchasing hardware, software and communication package</li><li>Training cost of users</li><li>Design and construction of internet presence</li><li>Cost of hosting the site on a reliable server</li><li>On-going maintenance and regular updating</li><li>Marketing the internet service and registration of domain</li><li>Commissions for purchases on line by intermediaries</li><li>Advertising fees for representation in search engines and other sites</li><li>Interconnectivity with travel intermediaries such as TravelWeb, ITN, Expedia</li></ul>
BENEFITS
<ul><li>Direct bookings, often intermediaries and commission free</li><li>Global distribution of multimedia information and promotional material</li><li>Low cost of providing and distributing timely updates of information</li><li>Global presence on the internet, 24 hours a day, 365 days a year</li><li>Durability of promotion (in comparison to limited life of printed advertising in press)</li><li>Reduction in promotional cost and reduction of brochure waste</li><li>Great degree of attention by visitors to web site</li><li>Reduction of time required for transactions and ability to offer last minute promotions</li><li>Low marginal cost of providing information to additional users</li><li>Support of marketing intelligence and product design functions</li><li>Development of targeted mailing lists through people who actively request information</li><li>Great interactivity with prospective customers</li><li>Niche marketing to prospective consumers who request to receive information</li><li>Interactivity with local partners and provision of added value products at destinations</li><li>Ability to generate a community feel for current users and prospective customers.</li></ul>

Source: Buhalis (1999) in Buhalis and Schertler (Eds.), 1999: 224

In the USA travellers using online services for their travel arrangements rose from 11% in 1996 to 28% in 1997, and in 1997 the number of people who preferred to make travel reservations on the internet increased by 19% (Evans and Peacock, 1999). Jupiter communications' (1999) projection for world wide travel sales in 2003 is in the region of $16.60 billions. More recently a WTOBC report has predicted that 'internet transactions may account for more than 25 per cent of all tourism sales over the next three to four years' (e-Tourism Europe, 2002).

The aims of the research are:

- To determine the factors involved in e-marketing for the development of Libyan Small and Medium-sized Tourist Enterprises (SMTEs).
- To assess the perceived relevance of 'best practice' in terms of assisting Libyan SMTEs in e-marketing.
- To develop a model that measures the importance of the factors involved in e-marketing.

The research question, which ties back directly to the above aims, asks - if the factors (which are identified through the survey work and 'best practice' cases) are properly integrated in the operation of e-marketing will this be conducive to the efficient marketing of Libyan Small and Medium-sized Tourist Enterprises?

2   Theory/Issues

One of the reasons for researching e-marketing for small businesses in the tourism industry is that they have been neglected in the past, and if more attention were given to the sector this would embrace the industry as well as the economy through the Internet at a national and international level. Friel (1998) states that a small tourism and hospitality firm may be said to be a business entity operating within the tourism and hospitality industry. This has neither the resources to have its own discrete e-marketing department with specialised job roles nor the funds to hire the services of an external e-marketing agency. While resources dedicated to e-marketing in such firms may be limited, this does not necessarily imply that they cannot or should not undertake e-marketing activities; nor does it imply that they are any less effective in these undertakings than larger firms.

The Marketing Association has defined marketing as 'the process of planning and executing the conception, pricing, promotion, distribution of ideas, goods and services to create exchanges that satisfy individual and organisational objectives'. Marketing management is 'the process of scanning the environment, analysing the market opportunities, designing marketing strategies, and effectively implementing and controlling marketing practices' (Cravens et al, 1987). Also, 'marketing is specifically concerned with how transactions are created, stimulated, facilitated and valued'. (Kotler, 1972). Indeed, the main purpose of marketing is to create and distribute values among the market parties through the process of transactions and market relationships (Sheth et al, 1988). The marketing interface is generally concerned with the process of finding out how SMEs perceive and undertake marketing in their business. SME managers gather information by using different approaches, which do not seem to connect together, and clarify the vague picture of marketing information that would serve as a base for their future acts. The interface contention is that marketing for the small firm should be relevant and appropriate both in respect to the problems that it seeks to address and the relative position of the firm in its life cycle. In particular, it is not against formal planning (Day, 2000). Indeed, marketing and entrepreneurship have essentially three key areas of interface: they are both change focused, opportunistic in nature and innovative in their approach to management (Collinson et al, 2001). Carson et al (1995) perceive the central focus of the interface as being change focused, essentially process based and market driven.

It has been reported that the marketing interface is strongly interdependent although it is not harmonious (Dewsnap and Jobber, 2000). The literature describes the relationship as having little cohesion, limited co-ordination, lacking co-operation and involving conflict (Shocker, Srivastava and Ruekert, 1994, Urbanski, 1987, Wellman, 1995 and Wood and Tandon, 1994). Bradley (1998) has defined the marketing interface as "a fundamental variety of forces within and beyond the organisation". Through e-marketing it is possible to marshal these forces beyond the organisation. According to O'Sullivan (1998) the marketing interface claims "the status of enduring shapers and creators of human experience". Accordingly, the e-marketing interface exists between the SME and external social entities. Due to the absence of models to illustrate this the paper attempts to develop a model that measures the importance of the factors involved in e-marketing in this interface. A problem that can hinder the e-marketing interface is that barriers can arise that affect the operation of the interface (Cepedes, 1994). It is therefore important to manage the interface in relation to overcoming barriers to Libyan tourist enterprises.

3 Methods/Procedures

The research has used the following methods:

- Secondary data to assess the existing environment which exists in the Tourist industry in Libya
- Quantitative methods to determine the different factors involved in the development of e-marketing
- Semi-structured qualitative methods to examine, in detail, the nature and importance of these factors and the relationships which currently exist to facilitate their implementation.

It is one of the primary aims of the research to understand 'best practice' in e-marketing. This approach is important to small and medium-sized tourist firms, since emulation amongst firms enables them to compare themselves with the leading firms in their field and is an effective way of propagating 'good practice'.

The research has been carried out in two stages. Stage 1 has set out the existing tourist environment, which currently exists within Libya and described in the previous section. It has drawn primarily on existing research, secondary data sources and interviews with key policy makers. Secondary data sources include existing literature in the area, which consist of both published material and 'grey' literature (including reports from national bodies, universities and consultants). In addition, there has been a series of interviews with key tourism policy makers in Libya.

Stage 2 of the research has consisted of three main sub-tasks. The first has established a population of suitable SMTEs, using the guidelines established in stage 1. From this a sample of firms has been drawn up which have been stratified by size and type. The second sub-task has been the development of a questionnaire. The questionnaire has gauged the significance of e-marketing. The problem of the non-response rate to questionnaires, especially SMEs, has been attended to by developing a strategy to deal with non response bias (NRB) during design and prior to distribution (Armstrong and Overton, 1977). The final part has been data collection using the questionnaire developed. Data cleaning and analysis have followed this.

A starting point for an SMTE to move towards more efficient e-marketing is to comprehend the importance of the e-marketing interface. If they perceive factors affecting the interface as important then these need to be included in the e-marketing evaluation process. It will be the Internet customer who determines the relative quality of an SMTE's products and services and they will determine which factors are significant. For an enterprise to improve its relative quality it is important for customers' views to be understood with regard to three critical aspects:

- Which factors determine choice of a SMTE?
- How important is each factor in the consumption of the tourism product or service?
- How does the enterprise perform in relation to each factor compared to its competitors?

Information on this can be determined through a profiling exercise for SMTEs. Moreover, profiling can involve three stages in the evaluation of the e-marketing interface. These three stages are:

Stage 1. Which factors are important in the choice of the tourism provider on the Internet?
Stage 2. How important is each factor in the purchasing of tourism products and services on the Internet?
Stage 3. How well does each SMTE perform with regard to each factor relative to the enterprise's main competitors?

Table 2 (Al-Hasan et al, 2002) illustrates the types of factors that can assist SMTEs in the development of the e-marketing interface. These are based upon the findings of the research so far, which can be developed and drawn up into a profile to compare customers' views on performance (Greenan et al, 1997) for an enterprise with its main competitors.

Table 2. Factors assisting SMTEs in the development of the e-marketing interface

Factors	Weighting	Percentage
Accessibility of Web site	X 2	10
Appropriateness of the Web site	X 1	5
Suitability for distributed use	X 1	5
Maintenance of the Web site	X 2	10
Availability in terms of cost	X 2	10
Information provided	X 2	10
Multi-lingual	X 3	15
Delivery platforms	X 2	10
Quality of service	X 2	10
Security	X 3	15
Index	20	100

Although it may be argued that an SMTE's individual influence on a tourism economy may be small; when grouped together they will have considerable influence for emerging economies such as Libya. A way forward concerns electronic networking and this can involve creating linkages between SMTEs and institutions. Useful links may be between SMTEs and local universities and between SMTEs and public bodies. From past evidence Humphreys and Garvin (1995) have shown that only a small number of enterprises (14%) consider using universities to help them develop and implement effective practices. By taking the above stages and linking them together as a process it is possible to develop a model to measure the factors involved in the efficient operation of the e-marketing interface for SMTEs, as illustrated in Figure 1.

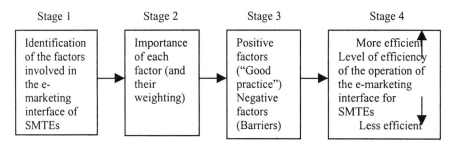

Fig. 1. A model to measure the factors involved in the efficient operation of the e-marketing interface for SMTEs

Through the application and development of the above preliminary model for the evaluation of factors involved in the operation of the e-marketing interface for SMTEs it will be possible to improve the competitiveness of enterprises in the Libyan tourism industry. The problems facing the Libyan SMTE tourism e-marketing interface, can be attended to through 'good practice'. These problems are divided into two categories, the first category are general problems which face many tourism industries in developing countries, and the second category are more specific problems faced by most parties involved in the tourism industry in the country. The general problems are:

- Lack of knowledge and awareness of the potential of e-marketing. There is a lack of awareness of the economic importance of e-marketing. This is true for its positive impact as a potential source of foreign exchange and employment.
- Lack of e-marketing-related infrastructures. The country lacks the sufficient infrastructures necessary for the development of e-marketing.
- Lack of technical know-how and weak promotional activity.
- Lack of e-marketing investments. Investment in e-marketing is considered a risk, and foreign investors especially private companies are still demanding more assurances from the government for their investments in the country.
-

Private sector (specific) problems are often brought about by the public sector, which in turn faces its own financing problems. Table 3 (Jwaili et al, 2002) illustrates the types of problems affecting the e-marketing interface for Libyan SMTEs. If these problems are addressed, and the role of e-marketing for SMTEs is recognised, this will have positive benefits for Libyan tourism development.

Table 3. Problems affecting the e-marketing interface for Libyan SMTEs

Problem	Description
1	Lack of co-ordination between NBT and tourism companies concerning e-marketing.
2	E-marketing has not been developed enough to cope with the increasing demand in all the various types of tourism activities offered by the industry i.e. desert tourism.
3	Due to the lack of e-marketing there is a big disadvantage in relation to international markets.
4	Lack of e-marketing facilities for places of interest to the tourist.
5	Lack of preparation and training of all personnel in e-marketing.
6	Lack of co-ordination to develop a competitive e-marketing policy.
7	Lack of regulations to encourage e-marketing.
8	The delay by NBT in establishing e-marketing.
9	There is no comprehensive regulation for e-marketing for the tourism industry that allows it to benefit from development in the international tourism market and international tourism research.

4   Results

The questionnaire circulated to SMTEs in Libya, as part of stage 2 of the research during August 2002, composed of four principal questions regarding the gaining of competitive advantage through e-marketing. These included questions concerning the extent to which information technology (IT) facilities are used and what they are used for, whether the business has a Web site, and what information is provided on the site. The percentage of the sample of SMTEs in Libya who used IT facilities was 100%. In relation to this another question asked what business functions IT was used for by the SMTEs who had IT facilities. In response to this 100% said they use it for business correspondence, 91% use it for accounts and 93% apply it to customer data and personnel records. Also, 89% of firms use it for reservations, 50% apply IT in marketing analysis and business planning and 77% use it for brochure design. In response to a question whether SMTEs have a Web-site, 86% said they have one, 5% replied they did not, and 9% are working on one.

In another question businesses were asked what e-marketing information and activities their Web-site provided. Table 4 shows that for those SMTEs with Web sites 100% said it provides information on the company, 95% use it to give details of packages provided, only 63% said the Web-site gives prices of all services provided and 34% said the site allows direct e-booking.

Table 4. E-Marketing information and activities provided by SMTEs' Web sites

DOES THE WEB-SITE ALLOW YOU TO:	YES	NO	DON'T KNOW
Provide information on the company	100%	-	-
Give details of packages provided	95%	5%	-
Give prices for all services provided	63%	37%	-
Allow direct e-booking	34%	66%	-
Provide feedback on web site visitors and requests	5%	89%	5%
Give access to contact company staff and partners	5%	95%	-

A small proportion said that their site provides feedback on Web-site visitors and requests. A similar proportion of 5% stated that the Web-site gives access to contact company staff and partners.

5   Conclusions

The importance of e-marketing to the operation of SMTEs in Libya and how it is anticipated that this will affect business in the future has been determined. The results of the research show that e-marketing is viewed by SMTEs as important both now and in the future although this view is yet to be fully embraced by the authorities in terms of support. The significance of the obstacles to advancing SMTEs' business through this lack of e-marketing support has also been assessed. This has been found from the extent to which information technology and electronic facilities have been used within SMTEs. This is supported by the findings of the research that shows 100% of the sample of SMTEs use IT. In addition to the study of micro e-marketing aspects at the level of the SMTE the macro aspects have also been considered with regard to policy. This has included the extent to which tourist offices use e-marketing and the role of the Libyan National Board of Tourism with regard to this. It is contended that in order for the Libyan tourism industry to become more e-marketing based tourist offices and the Libyan National Board for Tourism need to promote its use. In these terms the results of the research are of both academic and practical significance, contributing to the body of understanding on the processes involved in e-marketing applicable to Libyan SMTEs, as well as the development of specific policy measures to aid this process. The work has extended previous studies identified in the literature and provides knowledge about the different factors applicable to SMTEs to develop e-marketing. This may result in the development of specific initiatives in which SMTEs in Libya will have easier access to the benefits of e-marketing at the most relevant level. A policy towards greater marketing over the Internet for the Libyan tourism industry will:

- Enable SMTEs to extend their business beyond their present reach to connect to customers and tourists globally;
- Lower costs;
- Automate formerly time consuming marketing tasks;
- Provide tools to measure the efficiency of marketing procedures;
- Enable an integrated Web based service to be provided for SMTEs to market their services to prospective customers and tourists.

An e-marketing policy will enable SMTEs to have greater promotion and communication online and will significantly complement their marketing mix to lower cost (through e-mail), provide faster turnaround (by on line promotions). There will also be a quicker response to marketing efforts (through online marketing) and the opportunity for personalisation (by tailoring online promotions for each customer). The operational area of e-marketing will therefore cover the range of activities and prospects in the electronic marketplace, and this will range from e-business to e-service and information (Osterreich Werbung, 2002). In relation to the research question, it is apparent that if the factors (identified through the survey work and 'best practice' cases) are properly integrated in the operation of e-marketing this will be conducive to the efficient marketing of Libyan Small and Medium-sized Enterprises. It will also enable them to gain competitive advantage.

References

Al-Hasan, S., Jwaili, M. and Thomas, B. (2002) Evaluating the competitiveness of SMEs in the Libyan Tourism Industry, *Tourism Research 2002: An Interdisciplinary Conference in Wales, Cardiff,* 4th-7th September 2002.

Armstrong, J.S. and Overton, T.S. (1977) Estimating Non-response Bias in Mail Surveys, Journal of Marketing Research, XIV, August, pp. 396-402.

Bradley, F. (1998) From Clashmore Containers to value relationships in the business system, *Irish Marketing Review,* 11, 1, pp. 78-84.

Buhalis, D., The cost and benefits of information technology and the Internet for small and medium-sized tourism enterprises. In D. Buhalis & W. Schertler (Eds), *Information and Communication Technologies in Tourism1999* (pp224). Innsbruck: Austria.

Buhalis, D. and Main, H. (1998) Information Technology in peripheral small and medium hospitality enterprises: strategic analysis and critical factors, *International Journal of Contemporary Hospitality Management,* 10,5, pp.198-202.

Carson, D. Cromie, S. McGowan, P. and Hill, J. (1995) *Marketing and Entrepreneurship in SMEs: An Innovative Approach,* London, Prentice Hall.

Cepedes, F.V. (1994) Industrial marketing: Managing new requirements, *Sloan Management Review,* 35, 3, pp. 45-63.

CISCO (2000) *The Easy Guide to E-Marketing, A Guide for Promoting Your Business in the New Internet Economy,* Cisco Systems.

Collinson, E. and Shaw, E. (2001) Entrepreneurial marketing – a historical perspective on development and practice, *Management Decision,* 39, pp. 761-766.

Cravens, D.W., Hill, G.E. and Woodruff, R.B. (1987) *Marketing Management*.

Day, J. (2000) The value and importance of the small firm to the world economy, *European Journal of Marketing*, 34, p. 2033.

Dewsnap, B. and Jobber, D. (2000). The sales-marketing interface in consumer packaged-goods companies, *The Journal of Personal Selling & Sales Management*, 20, 2, pp. 109-119.

E-Tourism Europe (2002) *Strategies and Alliances for Multi-Channel Tourism Sales and Distribution*, 7-8 October, Marriott, Amsterdam.

Evans, G. and Peacock, M., A comparative study of ICT and tourism and hospitality SMEs in Europe. In D. Buhalis & W. Schertler (Eds), *Information and Communication Technologies in Tourism 1999* (pp248). Innsbruck: Austria.

Friel, M.(1998). Marketing. In R. Thomas. (Ed), *The Management of Small Tourism and Hospitality Firms*, London, Cassell, pp. 117-137.

Greenan, K., Humphreys, P. & McIvor, R. (1997) The green initiative: improving the quality and competitiveness for European SMEs, *European Business Review*,Vol.97, No.5,pp.208-214.

Humphreys, P. and Garvin, J. (1995) Environmental management in Northern Ireland businesses: the green initiative, *Irish Business and Administration Research*, Vol. 16, pp.32-43.

Jupiter Communications. (1999) In *Marketing Tourism Destinations Online: Strategy for the Information Age*. World Tourism Organisation. Madrid. pp.49.

Jwaili, M, Thomas, B. and Al-Hasan, S. (2002) The Problems faced by small firms and the role of policy makers in the Libyan Tourism Industry: The national and international context, *25th ISBA National Small Firms Conference: Competing Perspectives of Small Business and Entrepreneurship*, Brighton, 13th-15th November.

O'Sullivan, P. (1998) It's not what you make, it's the way that you say it: Reflections on the design-marketing interface, *Irish Marketing Review*, 11, 1, pp. 69-77.

Osterreich Werbung (2002) *Customer Orientation and First Hand Service*, Austria Tourism.

Quelch, J. A. and Klein, L. R. (1996) The Internet and international marketing, *Sloan Management review*, Spring, 60-75.

Sheth, J.N., Gardener, D.M. and Garrett, D.E. (1998) *Marketing Theory: Evolution and evolution*, New York, John Wiley and Sons.

Shocker, A.D., Srivatava, R.K. and Reukert, R.W. (1994) Challenges and Opportunities facing Brand Management: An Introduction to the Special Issue, *Journal of Marketing Research*, 31, May, pp. 49-158.

Strassel, K. A. (1997) *E-Commerce can be elusive*. Convergence, 3/3, 1997

Urbanski, A. (1987) Repackaging the Brand Manager, *Sales and Marketing Management*, April, pp. 42-45.

Wahab, S. and Cooper, C. (2001) *Tourism in the Age of Globalisation*. London, Routledge.

Wellman, D. (1995) Brand Management Report: People management, *Food and Beverage Marketing*, January, pp. 16-18.

Werthner, H. and Klein, S. (1999). *Information Technology and Tourism - A Challenging Relationship*. Springer-Verlag Wien, Austria.

Wood, V.R. and Tandon, S. (1994) Key Components in Product Management Success (and Failure): A Model of Product Managers' Job Performance and Job Satisfaction in the Turbulent 1990s and Beyond, *Journal of Product and Brand Management*, 3, 1, pp. 19-38.

Evaluating the Electronic Market Hypothesis
in the Airline Distribution Chain

Marianna Sigala

The Scottish Hotel School
University of Strathclyde, UK
M.Sigala@strath.ac.uk

Abstract

The airline distribution chain is an excellent example of the impact of ICT and particularly of the Internet on industry structures and interorganisational relations. However, although the opposing effects of ICT on market structure have been summarized and tested into the Electronic Market Hypothesis (EMH), none study has examined yet the impact of Internet advances on EMH's hypotheses. By gathering data from the airline distribution chain, this study tests EMH's validity in the light of e-commerce developments. An enhanced model of EMH is proposed and implications for both researchers and professionals are provided.

Keywords: airlines; distribution; Global Distribution Systems; Electronic Market Hypothesis

1 Introduction

The transaction cost theory predicts two conflicting impacts of ICT. By reducing transaction and coordination costs, ICT foster electronic marketplaces, but electronic networks may also have an opposing effect on market structure, as they can be used to strengthen existing commercial relationships and lock-in partners by increasing the switching costs. The opposing effects of ICT on market structure are summarised into the Electronic Market Hypothesis (EMH) (Malone et al., 1987), which has been widely tested (Barret & Konsynsky, 1982; Venkatraman & Zaheer, 1990; Christiaanse & Venkatraman, 1996). However, since its development, ICT advances had been dramatic, e.g. the Internet affects all players in the distribution chain.

The airline distribution chain is an excellent example of the impact of ICT and particularly of the Internet on industry structures and interorganisational relations. Airlines, GDSs and travel agents (cyberintermediaries or existing players) have all created their own Internet storefronts, while competing airlines also combine forces on a common website (opodo.com, orbits.com). However, e-commerce has left the future of airline distribution systems and markets uncertain and so, any framework that can help to forecast future possibilities is very fruitful for both academic and industry professionals. The EMH provides a good starting point for understanding how industries may evolve in the future, but due to environmental and ICT changes, further research needs to test and/or amend its hypotheses. This paper aims to develop a model for predicting future developments in the structure of the airline distribution by undertaking the following steps: a) review the EMH and the response that it has generated in the academic literature; b) consider the developments occurring in the airline electronic distribution and use these to determine the current validity of the EMH; and c) extend the EMH by proposing a model that predicts the impact of Internet on the structure of the airline distribution market by including current developments. Because of the sensitivity of the data required, data are gathered through in-depth interviews with key players from airlines, GDS and travel agencies.

2 Analysis & debate of the Electronic Market Hypothesis (EMH)

Traditional economics describe the two basic mechanisms for coordinating the flow of products/services between buyers - suppliers; markets and hierarchies. In markets, buyers can search goods from many suppliers and market forces of supply and demand define the design, price, quality and delivery schedule of the items. As buyers must search all suppliers, they incur transaction and co-ordination costs. So, if the coordination and transaction costs are large, hierarchical coordination between buyers and suppliers are preferable. In their EMH, Malone et al. (1987: 486) stated that: "*by reducing costs of coordination and transactions, IT will lead to an overall shift toward proportionately more use of markets – rather than hierarchies*". However, Malone et al. (1987) went further arguing that inter-organisational electronic networks can improve coordination between firms in two contrasting ways. ICT have an electronic brokerage and market effect when electronic networks are used to reduce search costs. By connecting different buyers and sellers through a shared network and providing some searching tools, electronic networks help buyers to quickly, conveniently and inexpensively evaluate the offerings of various suppliers and seamlessly and efficiently finalise and conduct any transactions. However, when ICT are used to more tightly couple buyers and suppliers, an electronic integration effect takes place which implies a crucial change in their relationship.

Grover & Ramansal (1999) also highlighted the opposing ICT impacts on market structure by identifying the counter-myth of the following five ICT impacts: ICT reduces transaction costs, perceived complexity of products, asset specificity and free information flow. Specifically, they provided evidence that: sellers use ICT for product customisation for exploiting and locking-in buyers; ICT fosters outsourcing but it reinforces the seller's monopoly by sustaining higher prices; suppliers exploit open ICT architectures for creating captive buyer networks to sustain higher prices; linking multiple market centres using ICT networks can fragment markets benefiting suppliers; expansion of customer bases via networks lets suppliers to exploit buyers.

However, in the information era, more importance is given to knowledge-based ICT advantages/resources. So, although traditional transaction economics have mostly focused on tangible ICT assets, some studies have begun to recognise the importance of intangible assets reflecting knowledge and expertise. Thus, previous studies on the ICT impact on market structure focused on the role of physical asset specificity due to the presence of dedicated ICT (Venkatraman & Zaheer, 1990; Barret & Konsynsky, 1982), on the role of systems in creating process specificity (Zaheer & Venkatraman,

1994), but recent research is also investigating the impact of information acquisition, dissemination and exploitation. Christiaanse & Venkatraman (1998) proposed the concept of expertise exploitation to further understand ICT's role in vertical electronic channels. Expertise exploitation refers to the capabilities a firm develops to monitor and influence downstream channel members' behaviour by the utilisation of knowledge assets in exchange relationships. It also refers to learning effects related with the use information as an asset by exploiting critical information gathered by distribution members. Their study revealed that firms developing expertise exploitation through electronic channels create hierarchical relationships with their partners. Firms exploiting ICT generated information also gain greater strategic benefits in the development of yield management (Sigala et al., 2001), e-marketing strategies (Sigala, 2001) and the materialisation of ICT benefits (Sigala, 2002).

Whether ICT has a brokerage or integration effect is influenced by: the attributes (specificity and complexity) of the products; the information networks used (open/ proprietary) and their locus of control; the business environment in which networks are deployed and the pre-existing relationships among suppliers-buyers. Williamson (1991) distinguishes six types of asset specificity: site, physical asset, human asset, dedicated assets, brand name capital and temporal specificity. Lastminute.com is an example of ICT exploitation for creating time specificity. Product complexity refers to the amount of information needed to specify product attributes in enough detail to allow buyers to select. Low product asset specificity and low complexity are compatible with a market relationship whilst the greater the product asset specificity and the complexity, the more likely it is to favour a hierarchical structure. However, product complexity and specificity may interact with ICT tools, e.g. buyers can use shopping agents to compare complex products in various dimensions simultaneously.

Networks are more open when they allow easy communication with a new customer, supplier or other entity over the network. Networks become more open if they use public protocols, if many individuals/firms already use the network, and if the financial and behavioural costs of acquiring and deploying the required ICT are low. Open networks also foster electronic markets because the cost to a customer to search for potential suppliers is low, while the cost for a supplier to add an additional customer is also very low. However, open networks are only a prerequisite and not a guarantee for electronic markets, e.g. encryption can be used to control network access. Locus of control also impact on inter-firm relationships. Networks run by third parties are more likely to lead to market-like relationships, as their return is based on the volume of network transactions. On a seller controlled network, sellers will be motivated to exclude other sellers in order to retain customer loyalty (e.g. American Hospitality Supply). Similarly, on a buyer controlled network, buyers may be motivated to exclude other buyers to capture the efficiency benefits that the network provides (e.g. Avendra, the e-procurement hub for Marriott, Hyatt, Bass, ClubCorp and Fairmont hotels). Under some circumstances, buyers may exclude some sellers to reduce their search costs and to convince a restricted number of sellers to engage in relationship specific investments (Bakos & Brynjolffson, 1993).

Electronic network effect will have its effects primarily when the power relationships among sellers, among buyers and between buyers and sellers are unequal. If one seller dominates a market, then the seller will perceive opening their networks as threatening its market share. If no seller/buyer has dominant market share, then the benefits of network externalities (by increasing network size) will overwhelm the disadvantages of sharing efficiency gains and so, sellers/buyers would like to open their network to others. Network control and business context can also influence the evolution of network services. In a seller-initiated network, when one seller dominates a market, not only are efforts to open the network resisted, but a service evolution strategy is pursued to raise switching costs and lock-in buyers (Steinfield et al., 1997). Overall, the EMH forecasted that industries will follow an evolutionary path in their transformation towards electronic markets, in which the final stage will always be an unbiased market (buyers access all suppliers), while there will frequently been an intermediate stage, a biased market, in which the buyer can access a limited number of suppliers. Malone et al. (1987) also predicted that ICT may lead a personalised market in which software decision aids are used to compare product attributes.

3   Aims and methodology of research

This paper aimed to apply and test the EMH to explore the Internet impact on the market structure of airline distribution. As the emphasis was on discovery and understanding, a qualitative research was used, with the EMH and its extensions as sensitising concepts. Primary data were collected through in-depth interviews with major airline distribution players from the Global Distribution Systems (GDSs), travel agent and airline sectors. Due to the sensitivity of the data, access to key informants and disclosure of information was possible because of the personal contacts of the researcher. Twelve e-business managers were interviewed (4 managers of European travel agents, 5 managers of European airlines, 3 managers representing 3 GDSs).

Semi-structured interviews were judged most appropriate for exploratory research, as it allowed understanding to increase incrementally throughout the interviews. Each interview was informed by preceding interviews, topics and questions were revised, while respondents' opinions were also sought on previous respondents' ideas. General demographic questions asked initially to facilitate familiarity with firms, but no attempt was made to build a rich description of them, as the emphasis was on overall structure and dynamics of the distribution chain rather than on comparisons between individual firms. Interviews were not built around the conceptual implications of the EMH to avoid difficulties of explaining and misrepresenting complex theories/terms and imposing theoretical constructs on respondents. Although this approach may result in some inherent misinterpretations if interviews are treated as conservations, it is supported by Burguess (1984: 110) and *"merely indicates the kinds of topics, themes and questions that might be covered rather than the actual questions that were used"*. Thus, interviews varied in the time and attention devoted to each topic, reflecting Patton's (1980: 203) belief that *"to understand the holistic world view of a group of people it is not necessary to collect the same information from each person"*.

A topic list gave interviews a high degree of commonality and helped in collating responses into broad data categories (Burguess, 1984), which represented the major areas of commonality that interviewees found particularly important. The list grouped questions into three major areas: ICT used for electronic distribution before and after Internet; exploitation of ICT tools for affecting relationships with customers and partners before and after the Internet; and the perceived impact of the two previous discussed issues. In analysing the data, the author began to look for patterns and the logic behind these. The resulting 'hermeneutic circle' principle in moving back and forth between the data and the theory enabled the understanding of the large data sets. This approach greatly helped in identifying the links and promoting the development of theoretical arguments that tightly reflected respondents' concerns. Findings were further corroborated and supplemented by the analysis of secondary data, e.g. studies, press news. This triangulation of data and sources of data largely increased validity.

4 Analysis and discussion of findings

GDSs represent proprietary ICT networks that emerged from the convergence and development of CRS first developed by major airlines. CRS-owner airlines had the locus of control, but they opened their networks to smaller airlines to benefit from network externalities (i.e. create one-stop shop), get more transactions and revenues. However, the CRS opening did not lead to electronic markets, as their airline owners engaged in several practices aiming to increase their market power and competitors' costs (e.g. differential pricing policies for competing airlines, weighting fee structures inequitably between distributors and subscribers, charging excessive fees for services and structuring prices so that high initial costs prevented smaller carriers from accessing them, Pemberton et al., 2001). Two other techniques illustrate more directly EMH's claim regarding the impact of product information/features and network attributes on market structure. *"Screen bias"* was a ploy to display a CRS-owner airline's flight information ahead of that of its competitors. Feldman (1987) reported that 70 – 90% of bookings came from first screen displayed. *"Halo effect"* refers to the bookings that owner airlines achieved due to their status as network providers.

Although the provision of an ICT platform has been a crucial factor for locking-up smaller airlines (i.e. creating hierarchies through asset specificity), the Internet gave smaller and new carriers the opportunity to develop their own inventory, distribution and marketing systems. All respondents highlighted the low cost carriers that have managed to cut down distribution fees to offer competitive fares. The Internet enabled low cost airlines to also reduce operational and technology costs as well as their operational dependence on GDS. Traditional carriers are still technologically locked-in in the GDS, because most of their operations and corporate functions are based on GDSs (sometimes the GDS is also their corporate system, e.g. Lufthansa). This was stressed by all GDS and airline respondents and confirms EMH's claim, i.e. operations' specificity leading to hierarchical inter-organisational relations. To decrease their distribution costs and face the increased online fare competition, airline respondents highlighted the attempt of several airlines to develop their own websites. However, it was felt that airlines' direct, online sales are constrained because: online consumers prefer systems whereby they can compare prices of several carriers; and airline websites do not always provide the cheapest fares due to the deals that airlines give to certain travel agents. To address these, two carriers reported that their websites were enhanced to sell tickets of competing airlines (Lufthansa, British Airways), in order to make airline marketplaces whereby airlines are the controller (Sigala, 2002). Jarach (2002) also argued Internet's impact on the strategy/operations of carriers.

Various regulatory initiatives have somewhat managed to minimise GDSs' power by eliminating the impact of unfair practices (Barber et al., 1998). Recently, CRS-owner airlines aim to strengthen their power by gathering and exploiting GDS generated information at the expense of no-system owners' airlines (i.e. expertise exploitation). All airline respondents mentioned the discriminatory practices of CRS-owner airlines about GDS information dissemination and accessibility. While non-participating GDS airlines receive information in microfilm /fiche on a monthly basis, CRS-owner airlines reported to receive information on magnetic tape or cartridge on an "as requested" basis. The format of information provision facilitates or impedes its use for gaining competitive intelligence and advantage. Magnetic or electronic media allows swift access to data due to the inherent compatibility of the medium with computers. Microfilm/fiche must be converted to paper, and then entered into computers. When asked about these differences, one GDS respondent replied that CRS-owner airlines actually get data on the format requested, rather than having to accept the standard format offered to participating airlines. Another claimed that they could not access on the system information of other airlines' business on competing routes. Stonehouse et al. (2001) revealed that CRS-owner airlines had an advantage in building competitive intelligence by: accessing data of other participating airlines; restricting frequency of data provision to participating carriers; charging airlines high fees for information; providing only selected GDS data to other airlines; managing GDS data more effectively than participating airlines. Recognising the importance of information exploitation, respondents reported that airlines now try to gather and utilise customers' data from different channels. So, both airlines and GDS aim to create knowledge-based marketplaces and build power via personalisation and targeting.

The Internet created another competitor for GDSs, as not only airlines but also third parties sell airline tickets. GDSs also created their Internet storefront to gain a market share and direct access to travellers, but new .com companies develop innovative business models (e.g. auctions), whose success accelerate competition. According to respondents successful online models would need to provide "one stop shop" solution and exploit customer data for implementing customer relationship management. In their efforts to regain control of marketplaces, airlines have recently joined forces. Orbits.com is a marketplace created by USA airlines. European airlines also followed the co-opetition business model by launching opodo.com (a marketplace with regional interfaces in UK, France and Germany, opodo.co.uk/fr/de). Respondents claimed that the latter's success vitally depends on their ability to successfully gather and use customer information. In response to these initiatives, GDS respondents claimed that GDSs are trying to get low cost airlines in their systems. In this way, GDSs aim to become controllers of a marketplace that creates vicious circles of great network externalities, i.e. enhance their product base, which will boost their client database, transactions etc. Internet's ubiquity also offers further opportunities and threats to airline distributors. Two GDS and one airline respondent claimed that they are already looking on m-commerce activities but they are constrained by the lack of knowledge regarding the new services, platforms, mobile devices and market. However, they all recognised that successful m-commerce players would need to exploit the location and person specific mobile devices for providing personalised and localised services.

Concerning travel agents, their relationship with GDS and airlines has been very complex since the CRS introduction. A mutual dependency has characterised their relationships, as airlines have been the leading network and information provider for travel agents for nearly two decades. GDSs also exploited asset/business operation specificity to influence and control travel agents. As, travel agent respondents claimed their major reason for adopting GDS was to computerise their business operations. They also recognised that computerisation is not a major issue for small travel agencies, as their operations can be efficiently managed in traditional ways. To that respect, one GDS has modified its strategy to offer its services through the Internet. In this way, the only investment required by travel agents is a Internet connected. GDSs have also been exploiting GDS data for gaining market power over travel agencies. An interesting example stressed by respondents was SMART, a system developed by American Airlines (AA), owner of Sabre (GDS), that by consolidating GDS information on sales and revenue of travel agents, it gives AA's sales force specific information about their clients enabling them to customise marketing strategies.

Overall, despite the Internet an unbiased airline distribution marketplace as predicted by the EMH did not happen, as every distribution player has initiated and tries to control a marketplace by exploiting ICT generated data to lock-in customers/partners. So, electronic hierarchies (biased markets) rather than electronic markets are created with an emphasis in (Figure 1): a) forward/consumer biased marketplaces, e.g. reversed/auction markets; b) regional and/or backward (supplier or distributor) biased marketplaces; c) global marketplaces. Most of these marketplaces are personalising their product/services to lock-in customers (e.g. *my expedia.com*). Soon, Internet ubiquity and accessibility from mobile devices will lead to totally one-to-one personalised marketplaces offering services/products customised to each person, his/her location and circumstances at any time. Findings also revealed that Internet's openness is not the only condition for creating electronic markets. The development and use of ICT tools helping in the collection, transmission and utilisation of ICT generated information also play a vital role (Figure 2). To gain market power in electronic networks, firms need to support and foster the acquisition, transmission and exploitation of ICT information. But as ICT's role moves from a transaction processing to a decision making/knowledge management tool, ICT networks usually lead to electronic hierarchies than electronic markets.

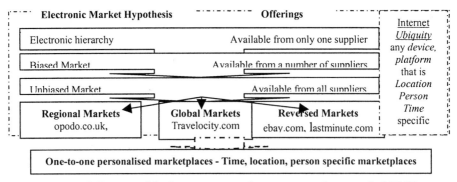

Figure 1. Impact of the Internet on market structure: the amended EMH

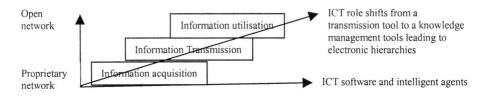

Figure 2. Network characteristics & exploitation & inter-organisational relationships

5 Conclusions and suggestions

Findings revealed that the EMH has to be amended to reflect the impact of Internet advances on airline distribution. As similar phenomena to that described are taking place in other industries (e.g. banking, insurance, music industry), further research is required to investigate whether the proposed models can be generalised to other settings. So, although findings contributed to the general building of new knowledge, the proposed models can be further validated, enhanced and refined from insights in other sectors. It was also shown that by relying solely on ICT, firms cannot gain sustainable advantages, but technological innovations require complementary assets. GDS had significant complementary assets (their industry and ICT expertise, customer/partner base, knowledge and relationships), which helped their reintermediation and reinvigoration of their business model. As it is impossible for one firm alone (specifically for new .com firms) to create or control relevant complementary assets to its ICT innovations, strategic alliances may emerge. Coopetition distribution models among distribution players are being developed, but research can further investigate alternative cooperative strategies for creating complementary assets. Regarding the practical implications, managers should realise that a focus on proprietary systems does no longer provide competitive advantages as knowledge-based expertise exploitation is another vital factor. Leveraging knowledge and being able to quickly react and adapt to changes in channel structure is key. Firms also need to renew their skills, resources and functional competencies to sustain the advantages they build. As ICT advance, current innovation must be the wellspring for further innovation. This must become systemic, but it requires more tightly coupled business processes. Strategies to achieve these can be the focus of future research.

References

Burguess, R.G. (1984) *In the field.* London: Routledge

Bakos, J.Y. & Treacy, M.E. (1986). Information technology and corporate strategy: a research perspective. *MIS Quarterly,* June, 107 – 119

Barret, S. & Konsynski, B. (1982). Inter-organisational IS. *MIS Quarterly,* 8 (2), Special Issue

Grover, V. & Ramanlal, P. (1999). Six myths of information and markets: IT networks, electronic commerce and the battle for consumer surplus. *MIS Quarterly,* 23 (4), 465 – 495

Christiaanse, E. & Venkatraman, N. (1996). Expertise exploitation of electronic channels. Best paper. Proceedings of the Academy of Management Conference, August 1996, Cincinnati

Christiaanse, E. & Venkatraman, N. (1998) *Beyond Sabre: an empirical test of expertise exploitation in electronic channels.* Working paper, University of Amsterdam

Feldman, J. (1987) CRS in USA: Determining competition. *Travel Tourism Analyst,* 3, 3 – 14

Jarach, D. (2002) The digitalization of market relationships in the airline business: the impact and prospects of e-business. *Journal of Air Transport Management,* 8, 115 – 120

Malone, T., Yates, J. & Benjamin, R. (1987). Electronic markets and hierarchies: effects of IT on market structure and corporate strategies. *Communications of the ACM,* 30 (6), 484-497

Pemberton, J.D., Stonehouse, G.H. & Barber, C.E. (2001) Competing with CRS-generated information in the airline industry. *Journal of Strategic Information System,* 10, 59 – 76

Sigala M. (2001). Modelling e-marketing strategies: Internet presence and exploitation of Greek hotels, *Journal of Travel and Tourism Marketing,* 2 (3), 83 – 104

Sigala, M. (2002). *Assessing the productivity impact of Information and Communication technologies in the hotel sector.* PhD Thesis, University of Surrey, UK

Sigala M., Lockwood A. & Jones P. (2001). Gaining advantage from Yield Management: strategic implementation in the rapidly developing world of IT. *International Journal of Contemporary Hospitality Management,* 17 (3), 364-371

Steinfield, C., Kraut, R. & Plummer, A. (1997). The impact of interorganisational networks on buyer-seller relationships. *Journal of Computer Mediated Communications,* 1 (3)

Stonehouse, G., Pemberton, J. D. & Barber, C. (2001) The role of knowledge facilitators and inhibitors: lessons from airline reservations systems. *Long Range Planning,* 34, 115 – 138

Venkatraman, N. & Zaheer, A. (1990). Electronic integration and strategic advantage: a quasi-experimental design in the insurance industry. *Information Systems Research,* 1 (4), 377-393

Williamson, O.E. (1991) Comparative economics organisation: the analysis of discrete structural alternatives. *Administrative Science Quarterly,* 36, 269-296

Zaheer, A. & Venkatraman, N. (1994) Determinants of electronic integration in the insurance industry: an empirical test. *Management Science,* 40, 5, 549 – 567

Destination Management Systems Utilisation in England

Catherine Collins
Dimitrios Buhalis

Centre for eTourism Research (CeTR)
School of Management, University of Surrey, Guildford, GU2 7XH,
UK
{c.collins; d.buhalis} @surrey.ac.uk

Abstract

A framework is being developed in England to support tourism businesses and destinations to become more competitive and profitable. The aim of the EnglandNet project is to create a national system through full integration of information between national, regional and local Destination Management Systems (DMS) to provide benefits to both consumer and the tourism industry. This research provides a snapshot of the degree of implementation of DMS in England and the types of business models used. This study of 73 Destination Marketing Organisations (DMOs) illustrated that DMS have been extensively implemented to support the strategic objectives of DMOs. Although most destinations examined employed ICTs extensively and 52% utilise DMS, they have yet to realise the full potential and benefits to be obtained through co-operation and partnerships with the local industry. Funding is acknowledged as a critical factor for successful DMS, and most DMS are currently moving towards the direction of operating as commercial entities.

Keywords: Destination Marketing Organisations, Destination Management Systems, ICT, England

1.   DMS as strategic tools for DMOs

Destination Marketing Organisations (DMOs) can act as facilitators to achieve the strategic objectives of the destination. Buhalis (2000) addressed these as enhancing the long-term viability of the local population, provision of visitor satisfaction and maximising profits for Small Medium-Sized Tourism Enterprises (SMTEs). This is particularly important in the UK as approximately 80% of tourism businesses are SMTEs (ETC, 2002) and many lack the skills required for promoting themselves and their destination (O'Connor, 1999). With new information technologies having escalated over the last decade it is of paramount importance DMOs in England re-engineer their business process, develop new business models and take advantage of these new tools (Gretzel, Yuan & Fesenmaier, 2000). Destination Management Systems (DMS) can be described as the IT infrastructure of the DMO (Sheldon, 1997). DMS should be able to act as an enabling mechanism to integrate the different services and products from the tourism industry. They should not only be capable of handling both pre-trip, post arrival information requests, but they should also integrate an availability and booking service too (Buhalis, 1997; Frew & O'Connor, 1998). DMS can increase visitor traffic, attract the right market segment with the provision of an accurate and up to date comprehensive electronic database (Sheldon, 1997; Pollock, 1995). DMS may also create more efficient internal and external networks, which can have long-term positive effects on the local economy in achieving competitive advantage (Fischer, 1998; WTO, 2001). DMOs can also support the wide distribution of destination information online. WTO (2001) has indicated that a number of new electronic distribution channels are emerging through online travel agencies, search directories and destination portals. Therefore an important technical consideration in the design of DMS is the development of open systems so that links can be developed with alternative distribution channels (ADCs) (Sheldon, 1993). ADCs are required to support DMS to interface and distribute information through websites, TICs, call centres, kiosks and traditional marketing channels (O'Connor, 2002). Therefore DMOs need to realise that DMS can act as an enabler in sustaining competitive advantage (Lewis, 2002; Gretzel et al, 2000).

2.   Towards DMS business models

The financial viability and profitability is becoming more of an important issue for DMOs as they often determine their success. The two types of DMS business models, which are available to DMOs, namely non-revenue generating/information only or revenue generating/fully transactional DMS. There are several factors, which influence this, namely, funding and finance, industry links, technology expertise and availability of technology. DMS can be implemented at national, regional or local level. Presently within the UK, DMS are implemented at regional and local level. This is because funding is derived from this level rather than at national level (Rapke & Appleford, 1998). The majority of DMS in the UK are often funded publicly, by DMO budgets or EU supported programs (O'Connor & Rafferty, 1997) with few being privately owned (Baker, Hayzelden & Sussmann, 1996). This has implications as according to Archdale (1994) those DMS that are totally dependent on public funding may be at a particularly high risk for failure because of ownership issues and technology.

The future for DMOs implementing either type of DMS business model is seen as a challenging one, as funding constantly being reduced, which leads to an increased onus on DMS to raise their own revenue (Connell & Reynolds, 1999). However, a solution to this is to encourage the private sector to provide capital for the development and implementation of DMS as higher returns on investment may be achieved (WTO, 2001). Moreover, efficient partnerships and co-operation between the public and private sector is critical for DMS to become effective, efficient and dynamic reservation systems. By offering reservation facilities on their own platform the transformation from eMarketing to eCommerce is realised (Werthner & Klein, 1999). However, this can also result in the strategic direction of the DMOs changing from one of sustaining economic regional development to being a commercial entity. This type of DMS may not be as accurate and as fully comprehensive as non-commercial DMS (Buhalis & Spada, 2000). Potential problems may also arise if the private sector is not provided with evidence of prospective tangible benefits (Frew &

O'Connor, 1998). DMOs, which are operating commercially also need to ensure that conflict is avoided with the local industry who may see this as unfair competition (Tunnard & Haines, 1999; Mistillis & Daniele, 2001). Although several business models gradually emerge in the global marketplace for DMS, there is still uncertainty on the most appropriate one. Therefore DMOs around the world adopt a combination of commercial and non-commercial features that fit best their funding and operational requirements.

3.   Methodology

The aim of the research was to identify the extent of information technology usage and DMS implementation in DMOs in England. This research was undertaken in close collaboration with the ETC and reflected some of the key issues that they need to address towards the design and development of EnglandNet. The primary research for this study consisted of three parts. Firstly a focus group was held at the University of Surrey on 24th April 2002, to examine the critical issues arising for DMOs implementing DMS. These clearly placed less restriction on the expression of the data that was more difficult to analyse. Secondly a qualitative questionnaire that enabled the researcher to collect opinions of DMO representatives, DMS suppliers and academic staff to identify the key success factors of DMS implementation. Thirdly, following the exploratory research a number of variables were established for the mail survey. Prior consideration of the relevant literature and theory was vital in determining the types of questions asked. The quantitative questionnaire consisted of seven sections, with a framework of fifty mainly closed and ordinal level of measurement through Likert scales. The former enabled respondents to provide straightforward quantitative answers whilst the latter required the respondents to indicate the order of importance for a list of attributes related to IT and DMS implementation. The Regional Tourism Boards in England and the English Tourism Council (ETC) were also involved in the development of the quantitative questionnaire. The final population sample consisted of 250 DMOs throughout England. It was not possible to target all of the UK DMOs due to time constraints and the limited time period in which to send and receive back the questionnaires.

In partnership with the ETC the questionnaire was e-mailed to ten Regional Tourist Boards (RTBs) throughout England, who in turn distributed them via e-mail to a total of 250 local tourist boards and DMOs. Respondents were asked to respond either electronically or via a Freepost address which was provided by the ETC. The response rate was 29.2% (73 questionnaires returned), therefore providing a valid sample size for the subsequent statistical analysis carried out using SPSS. For this exploratory research, descriptive statistics were used for estimating frequencies, initially to understand the rating of responses and to use for further discussion in the data analysis.

4.   Findings and Discussion

From the analysis it was discovered that 52% of DMOs that replied have implemented DMS. The majority of those DMOs that have not yet implemented DMS realise the need for one and are currently in the process of developing a DMS as illustrated in Table 1. There are few destinations that do not intend to implement DMS at all. Only 3% of the respondents have stated they will only implement DMS when all the systems become interoperable on a national basis through the EnglandNet Project.

Table 1 Time Scale for implementation of DMSs in England

Implementation	No. of Respondents	Out of total Respondents (73)	Out of total respondents not implemented DMSs
		%	%
DMS Implemented	38	52	
Implement 2-3yrs	3	4	9
Implement when funding agreed	5	7	14
Do not know	7	10	20
Total of no DMS implemented	35	48	100
Total	73	100	

4.1 Geographical Distribution of DMS in England

Most of the well-established destinations as those in the South of England use DMS widely. The results illustrated in Figure 1, shows South West Tourism (SWT), South East England Tourist Board (SEETB) and Southern Tourist Board (STB) have the highest implementation of DMS. However, the majority of destinations within RTBs who do not yet have DMS intend to implement one within the next three years or when funding has been agreed. However this data is only representative of the 73 DMOs who responded to the questionnaire and does not reflect the real situation currently for each geographical region.

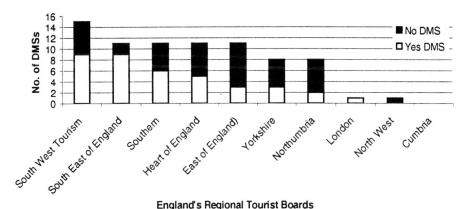

England's Regional Tourist Boards

Figure 1 Number of DMS implemented by DMOs within Regional Tourist Boards

4.2   Transactional/Information Vs Non Transactional/Reservation DMS

A key factor DMOs need to decide upon is whether they should be facilitators of information or be involved in the reservation process. A total of 15 respondents (39%) of those who have DMS, have a non-transactional one and 23 respondents (61%) have a fully transactional one. The Focus group discussion identified several factors that may influence DMOs decision to procure either a non-transactional or transactional DMS. The respondents were asked to rate and rank in order of importance the factors that would influence their decision and are illustrated in Tables 2 and 3. For non-transactional DMS, little demand from local industry was considered the most important factor by 60% of respondents. Some 40% of respondents agreed that high cost implications were important and another 40% have not implemented a transactional DMS because they maintain the technology has not yet been proven for a secure reservation process. Only 33% agreed that no demand from the consumer was an extremely important factor to consider.

Table 2 Decision Factors for implementation of non-transactional DMS

	N	Minimum	Maximum	Mean	Std. Deviation
little demand from local industry	15	2	5	3.73	.884
high cost implication	15	2	5	3.67	.976
technology not yet proven	15	1	4	3.07	1.033
Factor- Info only -no demand frm consumer	15	1	4	3.07	.884
not a priority at present	1	1	1	1.00	

Most of the respondents who have implemented transactional DMS (61%) agreed that having the right technology available was the most important factor for implementing transactional DMS. Being able to obtain funding was the second most important factor for 36% of respondents. High consumer demand was considered as an important decision factor, and overall 48% agreed that if DMOs realised consumer demand was high for implementation of transactional DMS, then they would see this as an important factor to consider. The focus group demonstrated that DMOs will eventually have to adopt transactional DMS and facilitate eCommerce as more consumers purchase tourism products online.

Table 3 Decision factors used for implementing a fully transactional DMS

	N	Minimum	Maximum	Mean	Std. Deviation
technology available	25	3	5	4.32	.627
funding available	25	2	5	3.96	.889
transaction-high consumer demand	25	3	5	3.88	.726
demand from local industry	25	2	5	3.64	.907

It is surprising that quite a small percentage (32%) considered demand from local industry to be an influencing and important factor for implementing transactional DMS. Even 8% of respondents disagreed that it was an influencing decision factor. DMOs felt confident that they would be able to provide sufficient inventory online to cover the market needs.

4.3 Development Options and Suppliers

DMOs have 3 options available, either to purchase, build, or outsource DMS. A very high percentage of DMOs (87%) have purchased DMS from a DMS supplier as indicated in Table 4. The disadvantages are that DMS purchased may not have all the specifications that are required by the DMO. There may also be problems of integrating existing DMO databases. Although 32% of respondents who did purchase DMS from suppliers, had their system built to their specifications.

Table 4 Options for DMSs development in specific regions

Region	Bought	Built	Outsource	Non-transactional DMS	Transactional DMS
South of England	13		1	6	8
South East England	6		1	3	4
England's North Country	5	1		3	2
Heart of England	4		1	1	4
South West England	4	1		2	3
London	1				1
Total	33	2	3	15	23

Figure 2 illustrates those suppliers who supply the highest number of DMS to DMOs. They are Touch Vision, Belmont Solutions, New Media, Integra, CTV and Velvet Software. 20% of the respondents did not specify who their supplier was. Touch Vision also supply the highest number of transactional DMS followed by CTV, Integra and ICL.

Figure 2 Type of DMSs supplied by DMSs Suppliers

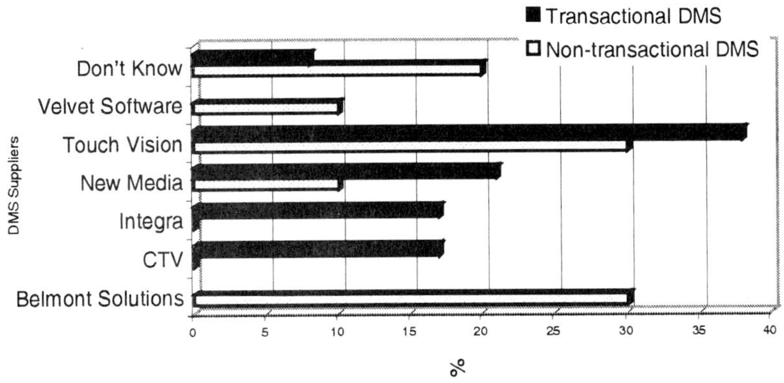

Very few respondents actually built their own DMS (5%). However 67% of respondents completed this section and indicated that having flexibility, quality control and less risk were considered the most important factors to build DMS. Corporate policy was considered important by 50% of the respondents. Only 33% of the valid responses considered cost the least important of all the variables to build DMS. In contrast 75% of the valid responses maintained reducing costs was an important factor when considering outsourcing. Lack of internal technology skills would influence 56% to outsource DMS. Whilst 37% of respondents agreed that having additional skills through partnerships and outsourcing was important. Surprisingly, 33% of the respondents stated that obtaining competitive advantage was the least important factor.

4.4   Distribution and Booking Mechanisms

The management of information and the reservation/booking functions are the most significant features of DMS (51%). Although 58% of DMOs are using e-mail to make accommodation bookings, this does not necessarily mean they operate a DMS. However, results on Table 5 illustrate that only 10% of DMOs are using real time online booking and confirmation for accommodation. This also indicates that not all-transactional DMS (30%) are fully utilising this function, as e-mail, fax forms, counter and telephone bookings are predominately used to make accommodation bookings.

Table 5 DMOs Accommodation Booking Methods

Counter bookings	67%
Telephone bookings	58%
Email reservation forms	46%
Fax reservation	12%
Real time online bookings	10%

The analysis also showed that the majority of transactional DMS and who use real time online accommodation bookings and confirmations are located in the South East of England and London tourist board (LTB). This may be because of the high number of visitors that come to this area and accommodation providers provide allocations to systems as well as to the fact that these regions concentrate on large properties. The most commonly used databases on DMS are accommodation (80%), attractions (80%), local services and events (60%) and restaurant databases (53%). Non-transactional DMS place a stronger emphasis on updating systems on a weekly, monthly and yearly basis and transactional DMS place priority on updating more than once a day. Results indicated that 30% of the respondents receive accommodation updates via the extranet. Although a vast majority (56%), are still using traditional methods i.e. telephone, fax and forms returned by suppliers for inventory to be updated on DMS. None of the accommodation suppliers have yet to use SMS text messaging. Most respondents (83%) agreed that linking to alternative distribution channels was important whilst 27% of respondents used Regional and National Tourist Boards. A significant number of respondents (44%) use DMS to feed kiosks and electronic information points. Transactional DMS (67%) more so than non-transactional DMS have implemented a consumer database and use this database mainly for general direct mail promotional campaigns. Presently a very small percentage (22%) of DMS customer databases is used for e-mail campaigns.

4.5   Funding for DMS

It is evident that EU supported funding (36%) funds the majority of DMS. A significantly high percentage allocated this funding (34%) implemented a transactional DMS (see Figure 3). Private sector funding is critical, particularly for the development of transactional DMS. However the results indicate that no funding has been allocated to DMOs for any type of DMS development. This may be attributed to the fact that investors from this sector want immediate commercial benefits. Funding received from Local Authority ICT budgets may have been quite substantial as 13% had the finance to implement transactional DMS. Funding allocated to DMOs from the public and private sector enabled both types of business models to be developed. It is critical that DMS are self-funded for their establishment and growth. Therefore it is highly unlikely DMOs will obtain funding for the operational costs incurred by DMS. DMOs therefore have to generate revenue to supplement these costs. A total of 22% of the respondents fund the operational costs of DMS through commission/booking fees. 15% provide an overall DMO marketing package, 8% do not charge local industry and 4% charge membership fees. However charging fees to the local industry suppliers for representation on DMS conflict may arise as suppliers may be reluctant to pay because they do not see the potential benefits that can be gained.

Figure 3 Type of Funding allocated to Non-Transactional & Transactional DMSs

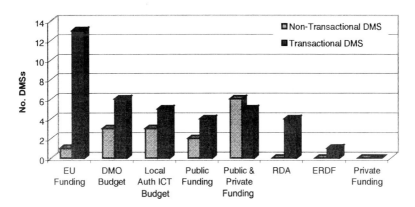

5   Conclusion

The research demonstrates that DMOs within England do realise the importance of implementing DMS and that they can provide the destination with competitive advantage and long-term benefits. A significant number of DMOs are not outsourcing DMS regardless of the advantages that can be gained. The majority of DMOs still prefer to purchase the system and few proceed to build their own DMS, lack of funding and technological expertise are seen as two of the major barriers. Respondents have indicated that DMS require the support of the local industry in order to operate a comprehensive system. However, it seems there is still a lack of co-operation between DMOs and accommodation providers as the results showed that accommodation providers are still reluctant to give allocation to DMS. The analysis also indicated that real-time online reservations on DMS are still not fully explored. Hence DMS are not being used to their full capability. However, various issues may account for this, which may also be related to the lack of infrastructure, required to perform real time inventory searches and to have an accurate picture of availability on a real time basis. Secondly, training DMO employees on how to effectively use DMS may be another issue. It could also be that customer's still research online and use off-line reservation techniques. The analysis has also demonstrated that the local industry is still using the traditional methods for updating information. Therefore training is required to assist the local industry in utilising information technology. DMOs need to better communicate the benefits and justification of implementing DMS to funding bodies and the local industry. It is also evident that closer collaboration is also required with the private sector and DMOs to obtain funding to finance DMS implementation. More transactional DMS business models have been implemented than non-transactional, which indicates that DMOs are moving their strategic direction to that of a commercial entity. Interestingly those DMOs who have not implemented transactional DMS state that it is because technology has not yet been proven. It is evident they are still in the early stages of the evolutionary steps of DMS development. Alternatively they do realise that the technology is available but they do not yet have the adequate funding to implement transactional DMS or lack the expertise required, but blame the technology. However most DMOs appreciate that consumers require these information and reservation services and are therefore keen to provide them. The respondents are still not fully utilising Alternative Distribution Channels for onward distribution and for the creation of partnerships. The majority of those respondents with a consumer database on the DMS generally use it for general direct mail promotional campaigns. Utilisation of one-to-one or general e-mail campaigns would be more useful for customer relationship marketing (CRM) and would enable DMOs to identify more specific target market segments. DMS are perhaps the most important technological tool for the future of DMOs and destinations. Therefore it is critical for DMOs to realise this and work in partnership with all stakeholders towards the successful development and implementation of DMS.

Acknowledgement: This research was undertaken with kind support from the English Tourism Council and we would like to acknowledge contribution by Andrew Duff and Andrew Daines to the development and distribution of research instruments.

References

Archdale, G., 1993, Computer reservation systems and public tourist offices, Tourism Management, 14 (1), 3-14.

Baker, M., Hayzelden, C., and Sussmann, S., 1996, Can Destination Management Systems give you competitive advantage? Progress in Tourism &Hospitality Research, 1(3), 89–94.

Buhalis, D., & Spada, A. 2000, Destination Management Systems: criteria for success an exploratory research, Information Technology in Tourism, 3, 41-58.

Buhalis, D., 1997, Information Technology as a Strategic Tool for Economic, Social, Cultural & Environmental Benefits Enhancement of Tourism at Destination Regions, Progress in Tourism and Hospitality Research, 7 (3), 71-93.

Buhalis, D., 2000, Marketing the Competitive Destination of the Future, Tourism Management, 21 (1), 97-116.

Connell, J. and Reynolds, P., 1999, The implications of technological developments on Tourist Information Centres, Tourism Management, 20(4), 501-509

English Tourism Council (ETC), 2002, E-Tourism in England: A strategy for modernising English tourism through e-business, English Tourism Council.

Fischer, D., 1998, Virtual Enterprise – Impact and challenges for Destination Management, Presentation at ENTER Conference 1998, Istanbul.

Frew, A.J., and O'Connor, P., 1998, "A Comparative Examination of the Implementation of Destination Marketing System Strategies: Scotland and Ireland" in Buhalis, D., and Schertler, W., (eds), Information and Communication Technologies in Tourism Springer Wien, New York.

Gretzel, U., Yuan, Y.L. and Fesenmaier, D.R., 2000, Preparing for the New economy: Marketing Strategies and Change in Destination Marketing Organisations, Journal of Travel Research, 39, 146-156.

Lewis, R.D., 2002, Modelling Tourism impacts using IT based DMS, Fesenmaier, D, Klein, S & Buhalis, D., (eds.) Information and Communication Technologies in Tourism, Springer Wien-New York, 97- 104.

Mistilis, N. and Daniele, R., 2001, Does the public sector have a role in developing destination marketing systems? in Sheldon, P.J., Wober, K.W., Fesenmaier, D., (eds.) Information and Communication Technologies in Tourism, Springer Wien-New York, 353-364.

O'Connor, P., 2002, The Changing Face of Destination Management Systems, Travel and Tourism Analyst, No.2, April, 1-25.

O'Connor, P., 1999, Electronic Information Distribution in Tourism and Hospitality, CAB International: New York.

Pollock, A., 1995, The impact of information technology on destination marketing, Travel and Tourism Analyst, 22 (3), 66-83.

Rapke, K., and Appleford, M., 1999, The Evolution of a Modular Destination Management System in Lake Placid, Insights, C-9 – 14.

O'Connor, P., and Rafferty, J., 1997, Gulliver – Distributing Irish Tourism Electronically, Electronic Markets, 7 (2), 40-46.

Sheldon, P.J., 1993, Destination Information Systems, Annals of Tourism Research, 20, (4), 633-649.

Sheldon, P., 1997, Tourism Information Technology, CAB International: New York.

Tunnard, C.R., and Haines, P., 1999, Destination Marketing Systems – A new role for tourist board marketing in the information age, Journal of Vacation Marketing, 1 (4).

Werthner, H., and Klein, S., 1999, Information Technology and Tourism, A Challenging Relationship, Austria, Springer, Wien New York.

World Tourism Organisation, 2001, E-Business for Tourism, Practical Guide for Destinations and Businesses, World Tourism Organisation Business Council, (WTOBC), Madrid.

Evaluation of DMO Web Sites
Through Interregional Tourism Portals:
A European Cities Tourism Case Example.

Karl W. Wöber

Institute for Tourism and Leisure Studies
University of Economics and Business Administration, Austria
karl.woeber@wu-wien.ac.at

Abstract

This paper demonstrates the applicability of tourism portals as an analytical market research tool and a source for highly relevant comparative management information for destination marketing organizations (DMOs). The case example outlined here focuses on the European Cities Tourism Portal, the B2C site developed by European Cities Tourism, a pan-European organization of more than 80 European city tourism organizations currently representing 31 European countries. The use of web content mining and web usage mining tools for the development of inter-regional tourism portals is proposed. Along with a comprehensive description of the conceptual approach, the paper provides detailed information about the technical realization and experiences with the prototype version of the system. Finally it highlights a number of research issues that need further attention in the future.

Keywords: Tourism; web portals; web effectiveness, web content mining; web usage mining.

1 Introduction

Due to the vital role of tourism in many countries and regions in Europe a number of programs concerning tourism promotion have been installed. Government and private tourism organizations have been established in order to strengthen a tourism destination. These organizations, commonly referred to as destination marketing organisations (DMOs), are the main suppliers of up-to-date information for people who travel to a certain destination. The Internet provides unprecedented opportunities for destination marketing organisations, as there are few other economic activities where the generation, gathering, processing, application, and communication of information is so important for day-to-day operations (WTO 2001). In fact, the web has revolutionized the conception of communication and interaction for destination marketing organisations. It offers new ways of business-to-business and business-to-consumer transactions, new mechanisms for person-to-person communication, new means of discovering and obtaining information, services and products electronically.

The volume of web data increases daily and so does its usage. Potential tourists use the web more and more for obtaining information about their prospect destination and it therefore becomes an increasingly important source of information. Both web contents and web usage data are potential bearers of precious knowledge. However, the web grows freely, is not subject to discipline and contains information whose quality can be excellent, dubious, unacceptable or simply unknown. The complexity and inscrutability of the net makes it more and more difficult for destination marketing organisations to make their offer visible and accessible for potential tourists or actual visitors who want to prepare themselves before or during their stay. As a consequence research on the effectiveness of tourism web sites has gained significant popularity recently.

2 Studies on the effectiveness of tourism web sites

Web site evaluation studies rely either on qualitative methods by collecting and analysing user or expert opinions (e.g. Jeong and Lambert 2001, Benbunan-Fich 2001) or on applying quantitative measures (e.g. Olsina and Rossi 2001, Ivory et. al. 2000, Wöber et al. 2002). A combination of quantitative and qualitative assessment for the evaluation of European city tourism and national tourism organisations' web sites was introduced by Bauernfeind et al. (2002). This quantitative assessment was mainly based on data retrieved by an automated web site analysing tool, to obtain information about navigation, interactivity, layout and textual features. The authors complemented the web content mining data with qualitative methods to obtain a more comprehensive picture of the sites. Finally the variables were combined with success indicators such as number of visits or inquiries retrieved by a survey among the managers of the corresponding tourism organisations.

The strengths of this computer supported type of evaluation lies in the inclusion of more objective measures, and by means of larger samples, in the chance to draw more general conclusions compared to solely qualitative studies. Weaknesses remain concerning the comparability and significance of the success measures and the fact that the actual user needs are not considered in the evaluation process (Jansen et al. 2000). For example, a high volume of page views recorded by a tourism web site is rather a sign of a management's success in effectively marketing an organisation's web site than an indicator for a demand-oriented and informative content which is available through a well-designed, easy-to-use system. As discussed and demonstrated in the following, interregional-web portals, which follow a collaborative information provision concept, promise to provide a solution to this major drawback in tourism web site evaluation research.

3 Web portals in tourism

A tourism web portal is a web site or service that offers a broad array of resources and services, such as search engines, e-mail lists, forums for a number of content providers who usually are located elsewhere and maintained by other parties. The main objective is to serve potential or actual visitors (undecided or decided customers) who want to access one or more official tourism web sites in order to satisfy their information needs. In addition to fulfilling customers actual needs, tourism portals provide a platform for tourism organisations to bundle their destinations' web sites in order to allow joint marketing initiatives. Finally, this concept opens valuable opportunities to gain insights into the consumers decision making process by tracking the users' information needs.

3.1   Service for the customer

Tourism web portals directly interact with customers, helping them to find information about destinations or services they may like to visit or purchase. In this sense, tourism web portals act as domain specific data query systems, which distinguishes them from the more general search engines, like Alta Vista or Yahoo. By summarising data from only official information providers (e.g. destination management organisations), they guarantee that the most accurate and up-to-date information does not get lost on the web. For the traveller, or a potential tourist, tourism web portals provide a minimum level of security to receive the most reliable information available.

Tourism web portals are similar to, but also different from recommendation and reservation or booking systems. In contrast to recommendation or booking systems, tourism web portals do not suggest specific products to the customers, however, they support the user by identifying his/her needs and redirecting him/her to the appropriate travel web site. They are particularly suitable for supporting travellers (= decided customers) in their final travel preparations and for information seekers (= undecided customers) who want to inform themselves about particular offers (sport activities, cultural offers, events etc.) in various destinations.

In both cases traditional recommendation systems usually provide only very limited service. Since tourism web portals are less demanding concerning their data repository, they are able to grow very fast and scope with more destinations (and services) than traditional recommendation and booking systems.

3.2   Network for tourism organizations

A network of tourism organisations where all members are able to exchange information on the success of their systems will not only help to learn and improve present tourism web sites, but also contribute to the standardisation and harmonisation of design and data models. Inter-organisational relations among tourism organisations have become an interest of tourism practitioners and academics (Selin and Beason 1990). The effectiveness of these relations depends on the ability of each participating organisation to generate any additional value for its customers. One of the main drawbacks for a customer who browses various tourism web sites is the lack of standardisation which exists. This heterogeneity refers not only to the content that is actually provided by the different organisations, but also to the style of presentation, and the way how users are navigated through a system.

Furthermore, the information overload on the Internet hampers users, who are interested in the most accurate and relevant travel information, to identify official tourist offices web sites among the many sources usually presented by traditional search engines. To address this information and cognitive overload problem, research has been conducted in developing techniques and tools to analyse, categorise, and visualise specialised collections of web pages. In turn, a variety of tools have been developed to assist searching, gathering, monitoring and analysing information on the Internet. A prominent example is the European City Tourism web portal.

4 Case study: The European City Tourism Web Portal

The following case example is an initiative of European Cities Tourism, the pan European organisation of currently more than 80 European city tourism organisations (see www.europeancitiestourism.com) representing 31 European countries, and was developed by the author. It demonstrates the applicability of tourism portals as an analytical market research tool and a source for highly relevant comparative management information.

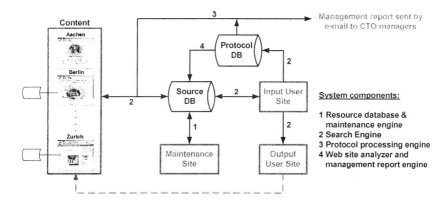

Fig. 1. The Conceptual Approach of an Intelligent Tourism Web Portal

Web content mining and web usage mining are two important and active areas of research which are combined for the implementation of the system. Web content mining describes the process of information or resource discovery from various sources across the World Wide Web (Scharl 2000). Web usage mining, is the process of mining web access logs or other user information user browsing and access patterns on one or more web localities (Masand and Spiliopoulou 1999, Spiliopoulou 2000).

Figure 1 illustrates the general architecture of the European City Tourism Web Portal. The system consists of four components: (1) Resource database and maintenance engine; (2) Search engine (incl. web content miner); (3) Protocol processing engine, and (4) Web site analyser and management report engine.

4.1 Resource database and maintenance engine

The resource database and maintenance engine is the principal component which defines the domain and the system's capabilities. The core element of the resource database is the data repository which consists of one or more addresses of tourism related web sites and additional information which is required for common services offered by the web portal. Normally, the official tourist board's web site will be the only data source represented for each individual city in the data repository. However, since some tourism organisations have outsourced all or parts of their services (e.g. booking services) to various content providers, the web portal also offers the opportunity to define several web sites which jointly represent the web content provided by the official tourism authority. Additional data stored in the resource database refers to structured information which is relatively static in nature and is available for all cities. Examples are a short description or slogan of the city, longitude and latitude information to identify city on a dynamically generated map, name of the country the city belongs to, size of population, currency, passport and visa requirement, official language(s), a standard picture etc. Although this information is centrally stored by the web portal's server it is decentralised maintained by a city board's web site operator.

4.2 Search engine

The search engine is the main system component which is used by the visitor who enters the tourism web portal. In difference to the well-known general search engines like AltaVista, InfoSeek or Yahoo, the search engine implemented in the European City tourism portal is a special purpose domain-specific search engine which builds its index by scanning only consumer relevant information from the net from a pre-defined list of (tourism) web sites. The search engine consists of following components:

- A **domain-specific web crawler** which collects web pages from all in the resource database specified web sites (i.e. servers);

- An **indexer** which compares keywords extracted from each individual web page with information commonly requested by users and generated by the protocol processing engine (see below). Congruencies are analysed and indices are stored into the database.

- The **user interface** allows users to query the database and customise their searches. Users can search documents using keywords together with appropriate Boolean operators and by voluntarily specifying one or several cities.

Fig. 2. The European Cities Tourism B2C Web Portal

If the keywords specified by the user are available by the current index, the system will retrieve and present all available web pages ranked by its relevance (see Figure 2). In case the user has specified a keyword which is unknown by the system, the word will be stored and marked for the next web mining procedure. By this means the web portal learns from actual system usage and adapts itself automatically to the customers' preferences.

4.3 Protocol processing engine

The web portal monitors all information entered by the user and stores it into a protocol database. Data stored in the protocol database does not only comprises the actual text entered by the user, but also the name of the city and its web page which was finally selected. Analysis of the protocols are necessary whenever the web mining procedure, which looks-up for previously unknown (and unidentified) words, is launched (see previous paragraph). Moreover, the protocols are regularly analysed and a list of highly interesting key-indicators is send in form of a management report automatically by e-mail to all participating city tourism organisations (i.e. members of European Cities Tourism). Information which is extracted from the system protocols can be classified into three categories:

(1) Information which describes system usage as an indicator for the success in marketing the tourism web portal as an effective tool for undecided and decided city break travellers. Typical key-indicators in this category are the total number of visitations and queries made to the system and the distribution of languages selected by the users.

(2) Information which describes the users' interests is available from the text he/she enters in the fields 'keyword' and 'city'. Information collected here is

comparable to the unaided response frequently applied in travel surveys where customers are asked where they want to spend their vacation and/or what they are interested in. A great number of valuable information can be generated here including the most frequently asked keywords and cities, and the distribution of query styles applied by users (i.e. number of times users have specified only a keyword, a city name, or both), which can be used to estimate the user's stage in the travel decision making process.

(3) Relative measures for each participating city based on comparative information and analysis. The analysis of 'city bundles', i.e. groups of cities which are commonly requested by a single web user, is a precondition for effective collaborations in city tourism management and therefore highly interesting information from a practitioner's point of view. Another example for comparative information now available to the managers is the user's interest for a particular city, measured the number of times the city's web site was clicked divided by the number of times the web site was actually offered on the screen.

Overall, the protocol database is an outmost valuable data source not only from a managerial but also from a consumer behaviour research point of view since it prevents some of the severe contortions (e.g. interviewer bias) frequently found in traditional surveys.

4.4   Web site analyzer and management report engine

In addition to the web usage mining procedures introduced in the previous paragraph a web site analyser automatically accesses and analyses the city tourism organisations web sites, always respecting the wishes of site owners with regard to limits where robots may search. To minimise the number of time consuming parsing operations, the content mining procedure is launched whenever the maintenance routine for updating the keyword index is started. From a methodological perspective, the web miner generates two main types of information: syntactic and semantic.

Syntactic information describes noncontextual information about content, such as total number of documents, distinct file types, or the number of embedded images. Due to the multi-dimensional nature of hypertext (embedding content, visual, structural and navigational information), the mark-up tags can be used to extract information about all of these dimensions simultaneously. For example, the number of distinct images identified has a visual impact on the site, the use of tables partially determines the structure of documents, internal and external links define the navigational system, and the (unformatted) raw text contains a significant part of the content.

Although interesting from web developer's point of view, syntactic metadata offers no insight into a document's meaning. By contrast, semantic metadata describes domain-specific information about the content. For this purpose the textual raw data extracted from the hypertext documents is used to investigate the textual content. The first step of any lexical analysis is the conversion of a stream of characters (= the original text) into a stream of coding units (= words). Identical words are then grouped together by counting their occurrences in order to create a word list that is usually sorted in order of decreasing frequency. By comparing these words with the users' interests found by the protocol-processing engine, web pages are evaluated and classified according to their relevance. In addition, the data is used to study attributes such as textual richness, and the number of available languages.

Statistics generated by the protocol-processing engine is then combined with information collected by the web mining tool and send to the participating managers in form of a comprehensive management report. Two areas of investigations are currently covered by the European Cities Tourism system: *uniqueness of offer* and *usefulness of information*. Uniqueness of offer refers to the content provided by a certain destination's web site compared to content provided by other destinations in the database. Usefulness of information refers to the content requested by the users versus content actually provided by the DMO's web site. For the first time tourism managers are able to monitor not only the overall success of their web appearance, but also to control if the Internet communication strategy they pursue is successful. By continuously revising their content according to the figures in the management report they are able to adapt their web site to actual user needs and industry standards.

5   Conclusions

The advances in information and Internet technology, the growing complexity of works, the increasing competition, and the search for higher-quality lives have pushed the demands for new tools or systems to remove the burdens of human beings from those tedious and repeated tasks, so that they can focus on those higher value-added tasks. This paper provides a comprehensive description of the conceptual approach of a market research oriented tourism web portal, and contains detailed information about the technical realisation and experiences with the prototype version of a city tourism application. The European cities tourism portal consists of a number of features that support the integrity and accurateness of the presented information. These features include automatic removal and alert functions for web sites with broken links, and an automatic identification of cities on maps based on geographical co-ordinates. Moreover, the system consists of sophisticated features that support the concept of decentralised maintenance for all participating destination management organisations.

The innovative part of the European city tourism web portal, however, are the content mining and web usage mining tools which are incorporated in order to access and analyse the effectiveness of the members' web sites and the portal itself. By associating meaning - as expressed by the users - with content, the web portal facilitates not only the search of the consumer, but also provides valuable information for the operator of a tourism web site. A number of measures that demonstrate the type of insights someone can gain through the comparative assessment offered by the content mining and usage-mining tools integrated into a web portal have been introduced.

There are a number of problems that remain unsolved and will evoke new research initiatives. One problem refers to the content that is not visible on the web (Sherman and Price 2001). Certain file types, especially those for images and audio cannot be reached by traditional search engines. Textual information which is stored in pdf (portable document format), postscript files, or compressed files can be crawled and indexed, but they present technical challenges that increase costs and effort.

Tourism web sites that exist as layer upon layer of static HTML (hypertext markup language) pages soon become labyrinthine and terrible unwieldy. In response, many webmasters implemented database-driven web sites that generate pages only when queried. This structure promises quicker response times, in part by enabling a web site's visible portion to remain small and nimble. The web site itself may remain visible to search engines, but the individual pieces of information encompassed lie beyond the view of the crawler. When a crawler rummages around a database-driven web site, it will typically get as far as a query template, but the treasure behind it lies hidden. Technological improvements among the search engines will not necessarily render all things visible on the web.

Another example for a challenging research field refers to the extraction and categorisation of semantics associated with information requested by the user and presented by the web sites. Today's search engines do not consider the query's context. They cannot know whether the search term "golf" is a Middle East ocean, a product (type of car manufactured by Volkswagen), or an activity a tourist wants to enjoy during their holiday. Although the city tourism web portal's search engine limits the search to certain relevant web servers, this alone does not guarantee that the user will find what he/she is looking for. In this context many advanced and exciting capabilities of semantic metadata extraction of structured and semistructured text are currently researched in information retrieval, artificial intelligence, database management, and knowledge representation (Fensel et al. 2002).

References

Bauernfeind, U., Wöber, K.W., Scharl, A., Bauer, C., Natter, M., and A. Taudes (2002) The Evalua-tion of European Cities' Tourism Offices' Web Sites. In K.W. Wöber (Ed.) City Tourism 2002. Proceedings of the International City Tourism Conference in Vienna, Austria, Vienna-New York: Springer, 323-334.

Benbunan-Fich, R. (2001). Using Protocol Analysis to Evaluate the Usability of a Commercial Web Site. Information & Management 39(2): 151-163.

Fensel, D., Hendler, J., Lieberman, H., and W. Wahlster (2002). Spinning the Semantic Web. MIT Press, Cambridge, MA.

Ivory, M.Y., Sinha, R.R. and M.A. Hearst (2000). Preliminary Findings on Quantitative Measures for Distinguishing Highly Rated Information-Centric Web Pages. Proceedings of 6th Conference on Human Factors and the Web. Austin, USA.

Jansen, B.J., Spink, A. and T. Saracevic (2000). Real Life, Real Users and Real Needs: A Study and Analysis of User Queries on the Web. Information Processing & Management 36(2): 207-227.

Jeong, M. and C.U. Lambert (2001). Adaptation of an Information Quality Framework to Measure Customers' Behavioral Intentions to Use Lodging Web Sites. Hospitality Management 20: 129-146.

Masand, B. and M. Spiliopoulou (1999) Web Usage Analysis and User Profiling. Berlin, Springer.

Olsina, L. and G. Rossi (2001). A Quantitative Method for Quality Evaluation of Web Applica-tions. Proceedings of the Argentinian Symposium on Software Engineering (ASSE '2001). Bs. As., Argentina.

Scharl, A. (2000). Evolutionary Web Development. London, Springer.

Selin, S. and K. Beason (1990). Interorganizational Relations in Tourism. Annals of Tourism Research, 18(4), 639-652.

Sherman, C., and G. Price (2001). The Invisible Web: Uncovering Information Sources Search Engines Can't See. Information Today, Inc. Medford, NJ.

Spiliopoulou, M. (2000). Web Usage Mining for Web Site Evaluation. Communication of the ACM (Association for Computing Machinery) 43(8): 127-134.

Wöber, K.W., Scharl, A., Natter, M., and A. Taudes (2002) Success Factors of European Tourism Web Sites In K.W. Wöber, A. Frew, M. Hitz (Eds.) Information and Communication Technologies in Tourism. ENTER 2002. Proceedings of the International Conference in Innsbruck, Aus-tria, Vienna-New York: Springer, 397-406.

World Tourism Organization (2001). E-Business for Tourism. Pratical Guidelines for Destina-tions and Businesses. Madrid.

Third-Generation Mobile Services and the Needs of mTravellers

Riyam Ghandour
Dimitrios Buhalis

Centre for eTourism Research (CeTR)
School of Management, University of Surrey, Guildford, UK
riyam.ghandour@world.net and d.buhalis@surrey.ac.uk

Abstract

Tourists are increasingly using various sources in order to collect tourism information before and during their trip. The upcoming third-generation (3G) mobile services will offer targeted information according to the location that will support travellers' flexible behaviour. This study identifies the travel information needed by travellers using mobile phones and wireless devices (mTravellers) during their visit to a tourism destination. The needs of mTravellers are first extracted through an exploratory survey, followed by a quantitative study to assess the usefulness and the best offering (price and presentation) of the location-based travel and tourism services. Finally, an attempt is made to detect the influence of these services on destination information providers, mainly Destination Management Systems (DMS). It is concluded that the value of mobile travel solutions lies in increasing the offering across the travel activity chain by providing wireless real-time useful information and services, combining tourism products, transportation networks and navigation system.

Keywords: location-based services, third-generation tourism, mobile services, DMS

1   Introduction

Mobile Internet is the combination of two major technologies: the mobile telephony and the Internet. The convergence of the Internet and the broadband wireless communication will combine high-speed broadband, real time interactivity and always-on interconnectivity. The new third-generation mobile services (3G) will transform the way we buy and search for tourism services from brick and mortar tourism distribution channels (Travel agents, tour operators, tourist offices, etc.) to 'on the move' mobile travel and tourism providers (Clapton, 2001). Hitherto, mobile travel applications are being developed by mobile operators who concentrate on technological aspects, rather than the information and services needed by consumers (Barnes, 2002). This paper investigates the tourism location-based services that are of interest to: travellers on the move, (mTravellers), and the impact of the new mobile services on destination information providers, mainly Destination Management System (DMS).

2   Theory

Mobile technology had gone through tremendous evolution between the first generation and the upcoming third generation wireless services. The upcoming

Universal Mobile Telecommunication System (UMTS) networks, the name developed for the 3G mobile telephone standard, are to provide better support to push data to devices by establishing 'always on' session that remain active as long as the device is turned on and within coverage area (Singhal, 2001; Warren, 2001). This will enable the development of several types of 3G services including purchasing goods and services through the mobile devices (m-commerce). It will also facilitate receiving information according to location (location-based services); or simply accessing the Internet on the move (Mobile Internet). All of these applications are expected to transform the way we buy and search for information, products and services.

Mobile generation evolution incorporated the migration from analogue systems (1G) to digital systems (2G and beyond). Each mobile generation was coupled with technological infrastructure developments to support its services and applications, as illustrated in Table 1. Even though mobile applications have evolved through time, they have not yet met consumers' expectation due to a set of problems such as limited data transmission rate, high power consumption of devices and inadequate mobile interfaces (Barnes, 2000). The upcoming third-generation mobile service (3G), promises the delivery of content rich next generation services which are expected to revolutionise the telecommunication industry by providing 'always connected' networks.

There are several types of 3G applications and services, the most revolutionary is expected to be the Location-Based Services (LBS). LBS, which will provide location-aware information and services to consumer through push mechanisms, whereby information, will be provided based on mobiles in a particular geographic location. Another major 3G application will be Mobile Commerce or M-commerce which is defined as "any transaction with a monetary value either direct or indirect- that is conducted over a wireless telecommunication network " (Barnes, 2002 p.2). Some analysts predicted that by 2003, the number of people accessing the Internet through mobile phones will exceed the number of PCs (Raisinghani, 2001). Hence more innovative 3G services will emerge to cater for the need of consumers 'on the move'.

The main driver behind the deployment of 3G services is to satisfy the current changes in lifestyle, including high mobility and increase in demand for information and Internet usage. Travellers are expected to be among the highest users of the new wireless services, given their need to access information while on the move. Travel and tourism services are already among the most used on the net and account to 28% of the e-commerce market (Hultkrantz, 2002). Some analysts expect "more than 23 million Europeans will use their mobile phones to buy travel products and services by 2005" (Leonardi, 2002). Hitherto, wireless travel and tourism services are being offered across the tourism industry. However services are limited to SMS and WAP platforms and are still far from providing enough added value. However they are considered a stepping stone before the emergence of the entire range of 3G services.

Table 1 Mobile Generations Evolution

Generation	Main Service(s)	Dominant Applications	Network & Technologies
First Generation: 1G *(Year: Early 1990)*	Voice traffic only		• Network: GSM (Global System from Mobile Communication) • Devices: Good only for voice traffic.
Second Generation: 2G *(Year: 2000)*	Voice and limited data services	• Short Message Services (SMS)	• Network: GSM • Devices: Small, limited processing power and short battery life.
Second and the half Generation: 2.5G *(Year: 2000-2002)*	• Personalised data services • Auction bidding • Stock trading • Email • Direction aid and Intranet access.	• Wireless Application Protocol (WAP) in Europe • I-mode in Japan	• Network: GPRS (General Packet Radio Service) • Devices: More memory and longer battery life.
Third Generation: 3G *(Year: end 2002-2004)*	• Location-based and time specific services • Video • Large file transfer • B2B services • Laptop replacement	• Multimedia Message Services (MMS) • M-commerce • Location-Based Systems (LBS) • Mobile entertainment	• Network: UMTS • Bluetooth: Identify location and enable LBS. • IPV6: Enable Mobile Internet. • Devices: HTML data, colour screens, long battery life, high processing speed.

Source: Adapted from Bechtolsheim et al., 2001

The upcoming mobile technology will create opportunities for developing new services for tourists who use mobile devices (Zipf and Malaka, 2001). The value of mobile travel solutions lies in increasing the offerings across the travel activity chain. Therefore the integration and partnership among travel providers is vital since no single industry alone can establish the online digital economy (Barnes, 2000). At this point, mobile networks and providers who are faced with serious technical challenges are developing tourism mobile applications. However, without useful and reliable content and robust business models even the most sophisticated applications will not succeed. Hence, the tourism and the telecommunication industries should join forces to deliver useful tourism information and innovative services in a flexible way.

There is great potential for 3G tourism services on the destination level, given that location-based services and information can increase customer satisfaction and influence the traveller experience (Hultkrantz, 2002; Schmidt-Belz, Makelainen, Nick, Poslad, 2002). The new mobile technology will open the door for new entrants that were previously non-existing in the tourism market. Examples include location-based city guide systems (e.g. www.Wcities.com) and wireless content aggregators (e.g. www.Kizoom.com), which will challenge Destination Management Systems (DMS). In the future Mobile services will not replace web services. However they will offer updated information for sophisticated tourists when they want it, where they need it, and in the media they prefer (Gjesdal et al., 2002). As tourism content will be critical there is a clear issue for DMS to repackage travel information for mobile devices and to create partnerships for distributing it through a wide variety of channels. This research examines the characteristics of the m-traveller, the distribution channel and the information they search before and during their trip. It also studies the market for location-based travel services and identifies which information and services travellers need most on their mobile. It also explores which form they would like to receive (text, voice, etc), how (randomly or per request) and how much are they willing to pay for it. Success in the new world depends on creating customer-centric tourism applications that fulfil the needs of mTravellers and enhance their experience.

3 Methods

The principle aim of this study is to provide an understanding of:
- The market for location-based tourism services: What LBS the mTravellers need, when and in what form they would like to receive it, and how much are they willing to pay for it.
- The impact of the new mobile services on destination information providers and Destination management Systems (DMS).

This research was based on secondary data as discussed in the literature review and primary data collected through qualitative and quantitative methods. The research was conducted between April and September 2002. The research execution began with an exploratory face-to-face research to extract the needs of leisure travellers. The qualitative data collection involved 22 interviews in the London area. It employed a questionnaire with a mix type of close and open-ended questions, targeted only to travellers using mobile phones. Since a lot of tourists had no experience of the new wireless technology and couldn't imagine how it would work, a brief presentation about Location-based mobile services was included in the interviews. Each interview lasted approximately around 15 minutes. During these interviews the researcher generated ideas and new variables that were included in the quantitative survey.

A quantitative study followed three-weeks after the completion of the quantitative study. Data from the qualitative study was analysed and guided the development of a structured questionnaire for the quantititative study. The quantitative research was conducted through self-administered questionnaires distributed to travellers that use mobile phones and were travelling on the Eurostar train (London/Paris route) or visiting the London Eye. The quantitative study was conducted to test which tourism services and information are valued most by travellers to be received and booked on their mobile devices. The quantitative research also provided an understanding of the multiple distribution channels mTravellers used and the information needed before and during their journey. The questionnaire was handed out randomly to tourists using mobile phones at the two locations. A total of 133 questionnaires were completed. The study used non-probability, convenience sampling for both the qualitative and quantitative surveys, given that the population is not known. The data was analysed by using descriptive statistics, cross tabulations and correlation tests.

4   Results

The majority of the respondents of this study are Europeans, who have planned their trip on average 1-4 weeks in advance. Most have already visited the destination more than ten times and are travelling for leisure purposes, for a short stay (1-3 nights), independently (not with a package tour). When asked the question " Will you use such a device (3G) in the future?" most of the subject responded "Maybe" (47.7%) followed by "Definitely" (39.4%), which indicates that 87.3% of the respondents were rather inclined to use 3G services. The majority of the respondent in this study (55.6%) found location-based travel information of "average usefulness". Around twenty nine percent (29.3%) of the respondents found 3G tourism services 'useful', and only 9% found 'not useful' and 6% 'not useful at all', whilst none found it 'very useful', as illustrated in Figure 1.The results are more encouraging than disappointing. One can expect that in the future, once the users have tried location-based services, their attitude will be even more positive.

The survey identifies which specific travel and tourism information are of interest to the users. As pointed out in Table 2, the most useful travel and tourism information are Transportation, Flights, Maps, Guiding system, Restaurant, accommodation, road condition and weather location-based services.

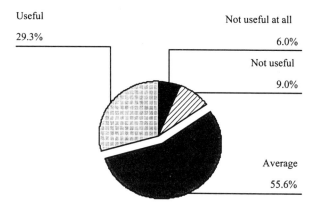

Useful
29.3%

Not useful at all
6.0%

Not useful
9.0%

Average
55.6%

Fig. 1. Usefulness of Mobile tourism services (N=133)

The research also examined whether the travellers are interested in 3G merely to search for information or to conduct actual transactions – mainly booking. The majority of the respondents (41.7%) found it useful to book tourism services through mobile phones and 12.1% found it very useful (much more useful than just searching for information). Whilst, only a small percentage indicated that it is not useful (9.1%) or not useful at all (6.8%). The most useful m-commerce tourism transactions were identified to be transportation, flights and accommodation booking as illustrated in Table 3.

Table 2 Usefulness of Location-Based Tourism Information

		Location-based Information	Mean Scale: 1= not useful at all; 5= very useful
Useful Mean: 3.5 to 4.4	1.	Transportation	3.99
	2.	Flights	3.98
	3.	Interactive maps	3.82
	4.	Guiding system	3.78
	5.	Restaurants	3.53
	6.	Accommodation	3.52
Average Mean: 2.5 to 3.4	7.	Road condition	3.48
	8.	Weather	3.45
	9.	Shopping	3.38
	10.	Services	3.23
	11.	Attractions	3.21
	12.	Events	3.13
	13.	Tips/advises	3.10
	14.	Promotions & offers	2.93
	15.	Nightlife	2.93
	16.	Off the beaten track	2.88
	17.	Car Rental	2.69
Not useful Mean: 1.5 to 2.4	18.	Finding like minded travellers	2.34
	19.	Sports' activities	2.15
	20.	Health/beauty activities	1.95
	21.	Children activities	1.77

Table 3 Usefulness of Booking Tourism Mobile Services

	Mean Scale: 1= not useful at all; 5= very useful
Transportation	3.75
Flights	3.71
Accommodation	3.50
Events	3.39
Car rental	3.07
Daily Excursion	2.90

Not all travellers will adopt 3G services at the same time. Therefore, the study goes further in dividing the respondents into three categories: the innovators, the sceptical ones and the laggards, following Kotler (1999) segmentation according to products' adoption process. The study identifies each category based on the question *"Will you use such a device in the future?"*. *"Innovators"* are the adventurous and the opinion leaders in a community. They are the early adopters who "Definitely" want to use 3G devices. They account to 39.4% (n=52) of the sample, which is positive outcome amid all the doubts about the success of 3G services after the failure of WAP. *"Sceptical consumers"* answered "maybe". They are sceptical to use of new technologies such as 3G devices. They represent 47.7% (n=63) of the sample in this research. *"Laggards"* are suspicious of changes and not interested at all to use 3G devices in the future. Although they answered "no", things may change once the usage of 3G services becomes commonplace. This segment represents 12.9 % (n=17) of the sample. The study demonstrates that the innovators need a lot of information and want this information now, 'anytime and anywhere' explaining their interest in 3G services. Scepticals are more relaxed about their information needs. The majority of the respondents needed better information about transportation and restaurants in the destination (Figure 2).

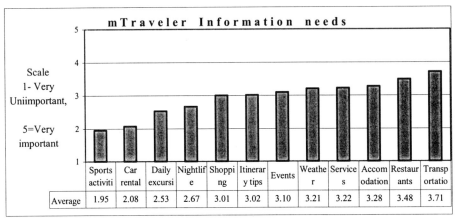

Fig. 2. Information Needs

Furthermore, after conducting correlation tests and descriptive statistics, the study identified specific traits per adoption of 3G devices, as illustrated in Table 4. The importance of the new technology lies in creating useful customer-centric applications. Good content was identified in this study as the most important factor that will encourage users to embrace mobile services (Table 5). Interestingly health issues have ranked last in the mTravellers assessment, who assured that their devices are safe to use. Hence, it is very important for application providers to offer above all useful content to end-users.

The research attempted to identify the useful travel content, which should be taken into consideration when developing mobile tourism services. Good content, presented in an interactive and interesting way will attract the attention of the public. Users expect a similar 3G service to be displayed in the familiar web services presentation. The majority of the respondents (37%) are interested in receiving text messages, and preferably interactive text presented in a similar way with the Internet. Good services and cost are also important factors for embracing the new technology. The majority of the respondents (51.5%) are willing to pay for 3G services a price similar to mobile phone calls. Only 15.2% are not willing to pay for 3G services, whilst 9.8% are willing to pay more than they pay for their mobile phone calls, if it offers useful content as previously indicated. The rest of the respondents (23.5%) are interested to pay for 3G services less than the amount they pay for their calls on mobile phones.

Table 4 Profile Per Adoption Process

	Characteristics per adoption of 3G devices
Age	Younger people are more likely to adopt the 3G devices quicker than elder one. The sample identifies the 20s age as the innovator, whilst the 30s, 40s, and 50s the sceptical ones
Education	The more educated the person is, the more likely 3G services.
Profession	Managers and professionals are more likely to use 3G services than people working in administrative support positions. Students were identified as the most interested group in 3G devices.
Planning Time	The lesser travellers plan their trip the more likely they are to use 3G services. People on the move are interested in seamless connectivity given that they have lesser time to plan things ahead of time. This suggests that last minute offers should have a potential market on the 3G platforms.
ICT Usage	Travellers who were more familiar with technology and used more the Internet, Mobile, PDA, Laptop, SMS, e-commerce were likely to be prime adopters of 3G services.

Table 5 Factors influence the adoption of 3G services (N=133)

Frequency (%)	Content	Cost	Good Service	Security issues	Health Issues
Not Important at all	11.0	9.3	12.4	11.0	46.5
Not Important	9.4	9.3	10.9	9.4	11.8
Average	18.1	27.9	11.6	18.1	16.5
Important	23.6	18.6	35.7	23.6	3.1
Very Important	37.8	34.9	29.5	11.0	22.0
Total	*100.0*	*100.0*	*100.0*	*100.0*	*100.0*
Mean	**3.68**	**3.60**	**3.59**	**3.06**	**2.43**

The study also tested the concerns of the users on 'pushing' information to their mobile devices. Based on the interviews, respondents mentioned uneasiness toward the fact that information would be pushed to their devices, fearing that they will be invaded by junky mobile messages. They expressed their interest in receiving information on request only, or to some extent according to location, if filtering tools are available on future mobile devices. In the quantitative survey, the majority of the respondents (59.5%) would prefer to receive information only at request. Few (7.6%) were interested to receive all information randomly on location. However, a substantial number of respondents (28.2%) were interested in having filtering tools on their mobile devices which will enable them to receive only specified information randomly and the rest on request. Such tools, based on intelligent agent software, are under development (for example the European Commission Crumpet project) and are to be embedded in future mobile devices (Shmidtz-Belz, 2002).

Last but not least, the study examined if the upcoming tourism mobile services will influence the traveller experience at the destination. Based on the correlation test conducted (p=0.000 and r=0.479) a positive relationship exists between using 3G devices and its influence on the travel experience at the destination. This implies that the new 3G services will impact travellers' experience. More specifically they will provide better information where needed (about trains, restaurants and services). New

3G services are capable of answering questions like " when is the next train?" "What are the nearest Indian restaurants?" "Is there an Internet café around? Or a bank machine?" Those questions will provide travellers with the information they need during their trip, which could not be covered in guidebooks, identified as the most important source of information during the journey by the majority of the respondents. Therefore, it is expected that the upcoming Tourism Mobile Services will influence the travellers' experience at the destination by offering better information mainly about transportation, accommodation, services, weather, events, and guiding systems and by enabling real-time bookings for transportation, flights, and events.

5   Conclusion and implications for destinations

Mobile services are evolving to better cater for the need of consumers. Increasing numbers of consumers are currently searching for information remotely via the Internet or mobile phones. SMS has been a tremendous success worldwide, whilst advanced digital services, like I-mode, are a success story in Japan. Upcoming 3G service operators are banking on these consumer trends. 3G mobile technologies is the combination of two successful technologies: mobile telephony and the Internet. Among the 3G applications, Location-Based Services (LBS) will provide the user with information according to his location and preferences. This is expected to revolutionise the way people search and buy tourism products and services. There is great potential for the upcoming mobile services in the tourist sector, especially when tourists are on their way to or at the destination. The majority of the people will be skeptical in the beginning in adopting the innovative 3G services. However, if the new mobile platform provides them with useful and affordable content that enhances their experience then it will eventually be adopted.

This research ascertains that location-based services will create a new way for travellers to buy and search for tourism products and services, and will influence traveller's experience, particularly when they are at the destination. Ideally, travellers would like to have received reliable information on a device providing them with useful information, namely transportation, flights, restaurants, hotels, services and weather forecast. Currently, the main destination information aggregators are Destination Management Systems (DMS) and Destination Management Organisations (DMOs). The position of DMS in the industry can be strengthened if they move forward and enhance their offering by integrating useful location-based information in order to fulfil information needs of travellers. Application providers can build the tourism databases themselves (which currently Location-based companies are actually doing) and will integrate important services, such as trains' information, by dealing directly with the suppliers (which Kizoom in the UK is currently doing with British Rail).

If DMS enhance their offering and bring in all the useful mobile tourism information together, then they will create an one-stop shop for reliable tourism information for mobile content providers. This will also help DMOs in marketing better their destination and enhance their products and services by directly interacting with tourists and get real-time feedback. The mobile industry should pay particular attention to users' needs in setting the price for 3G services in the same range as mobile phone calls and in integrating filtering techniques in the upcoming 3G devices in order to enable users to select the information they want to receive. The research demonstrates that for 3G services to be successful the needs of the mTravellers will need to be addressed and therefore a close collaboration between all stakeholders is a prerequisite.

References

Barnes, C. (2002). The mobile commerce value chain: Analysis and future developments. International Journal of Information Management [Online serial]. Article in Press.

Bechtolsheim, M., Muller P., Kranz S., and Loth B. (2001). Travel and Tourism: M-Commerce for the Mobile Business. Arthur D. Little. http://www.adlittle.com/management/services/e-business/articles04/mT&TChallengesFinal.pdf

Clapton, A. (2001) Future mobile networks: 3G and beyond. London: BT Exact Technologies.

Ericsson, O. (2002) Location Based Destination Information for mobile tourist. In Wober, K.W., Frew, A.J., and Hitz, M. (eds.), Information and Communication Technologies in Tourism, *Proceedings of ENTER 2002 International Conference*, 254-264. Innsbruck, Austria.

Gjesdal, O., Sulebak, J., and Borge, M. (2002). Market research in the boat tourism segment. In Wober, K.W., Frew, A.J. and Hitz, M. (eds.), Information and Communication Technologies in Tourism, *Proceedings of ENTER 2002 International Conference*, 320-328. Innsbruck, Austria.

Hultkrantz, L. (2002). Will there be a unified wireless marketplace for tourism? Commentary. *Current Issues in Tourism*, 5 (2), 149-161.

Leonardi, C. (2002). Online Travel Services Set to Boost Mobile Commerce. IDC Press Release. http://www.idc.com/getdoc.jhtml?containerId=pr2002_01_23_150830

Raisinghani, M. (2001). WAP: Transitional technology for m-commerce. Information System Management, 18(3), 8-9.

Schmidt-Belz, B., Makelainen M., Nick A. and Poslad S. (2002). Intelligent brokering of tourism services for mobile users. In Wober, K.W., Frew, A.J. and Hitz, M. (eds.), Information and Communication Technologies in Tourism, *Proceedings of ENTER 2002 International Conference*, 275-284. Innsbruck, Austria.

Singhal, S. (2001). The wireless Application Protocol: Writing application for the mobile Internet, North Carolina: Edward Arnold.

Warren, P. (2001). 3G in Europe: Expensive but essential. London: Yankee Group.

Zipf, A. and Malaka, R. (2001). Developing Location Based Services for Tourism. Enter 2001, Sheldon, P., Wober, K., Fesenmaier, D., Springer Verlag, Wien-New York, pp. 83-92in Tourism, *Proceedings of ENTER 2002 International Conference*, 329-328. Innsbruck, Austria.

DIETORECS: Travel Advisory for Multiple Decision Styles

Daniel R. Fesenmaier [a], Francesco Ricci [b,] Erwin Schaumlechner [c,] Karl Wöber [d], and Cristiano Zanella [e]

[a] National Laboratory for Tourism and eCommerce
University of Illinois at Urbana Champaign, USA
drfez@uic.edu

[b] Electronic Commerce and Tourism Research Laboratory
ITC-irst, Italy
ricci@itc.it

[c] Tiscover AG - Travel Information Systems
Softwarepark Hagenberg, Austria
erwin.schaumlechner@tiscover.com

[d] Institute for Tourism and Leisure Studies
Vienna University of Economics and Business Administration, Austria
woeber@isis.wu-wien.ac.at

[e] Azienda Promozione Turistica del Trentino, Italy

Abstract

This paper presents Dietorecs, a novel case-based travel planning recommender system. Dietorecs has been designed by incorporating a human decision model that stresses individual differences in decision styles. Dietorecs supports decision styles by means of an adaptive behavior that is learned exploiting a case base of recommendation sessions that are stored by the systems. Users can enter the system through three main functional doors that fit groups of decision styles, but they can eventually switch the type of support required. The dialogue (questions) is personalized using both the user model (cases) and statistics over the data available in the virtual catalogues provided by two DMOs.

Keywords: recommender systems; travel planning; decision modeling, personalization.

1 Introduction

There is a growing number of web sites that support a traveller in the selection of travel destinations or travel products (e.g., flight or hotel). Typically, the user is required to input product constraints or preferences that are matched by the system in an electronic catalogue. Major eCommerce web sites dedicated to tourism such as Expedia, Travelocity, and Tiscover have started to cope with travel planning by incorporating recommender systems, i.e., applications that provide advice to users about products (Schafer et al., 2001). Recommender systems for travel planning try to mimic the interactivity observed in traditional counselling sessions with travel agents (Delgado and Davidson, 2002). The current generation of travel recommender systems focus on destination selection and do not support the user through a personalized interaction in bundling a tailor-made trip made of one or more locations to visit, an accommodation and additional attractions (museum, theatre, etc.).

The Dietorecs system extends current recommender systems by incorporating a human choice model extracted from both the literature and the empirical analysis of the traveller's behaviour. Dietorecs supports the selection of travel products (e.g., a hotel or a visit to a museum or a climbing school) and building a 'travel bag'; that is, a coherent (from the user point of view) bundling of products. Dietorecs also supports multiple decision styles by letting the user 'enter' the system through three main 'doors': iterative single item selection, complete travel selection and inspiration driven selection. The first door enables the most experienced user to efficiently navigate in the potentially overwhelming information provided by the two integrated databases (Tiscover and APT Trentino). The user is allowed to select whatever products he/she likes and in the preferred order, using the selections done up to a certain point (and in the past) to personalize the next stage. The second door enables the user to select a personalized trip that bundles together items available in the catalogue.

The personalized plan is constructed by "reusing" the structure and main content of trips either built by other users or available from some provider. The third door allows an inspiration-seeking user to choose a complete trip by exploiting a simpler user interface (icon based) as well an interaction that is kept at the minimum length as possible. It must be stressed that all these decision styles are supported in a uniform and seamless way by means of a graphical user interface. Hence, switching from one style to another is always possible and easy to do.

The next two sections describe our conceptual approach to travel planning and illustrate how the notion of decision styles emerged from the research. Then, we briefly describe the fundamental element of the designed application and its technological implementation. We end the paper by summarizing the state of the work of this research project.

2 Conceptual Approaches for Solving the Application Problem

An in-depth understanding of the destination choice process, those factors considered during travel planning, and the interaction between human and computer mediated environments are an important foundation for any destination recommendation system (Hwang, Gretzel and Fesenmaier 2002). Hence, developers of a travel advisory

234

system cannot focus solely on computer science theory but must also consider the findings and achievements in travel decision theory. In tourism research, travel decision theory is one of the most comprehensively investigated areas with a relatively long tradition (see Mayo and Jarvis 1981 for one of the first textbooks devoted to the psychology of leisure travel).

Many conceptual approaches to understanding travel decision making have been proposed. In general, these approaches can be conceptualised into four different frameworks: (1) Choice set models (e.g. Crompton and Ankomah 1993; Um and Crompton 1990); (2) General travel models (e.g. Woodside and Lysonski 1989); (3) Decision net models (Fesenmaier and Jeng 2000); and, (4) Multi-destination travel models (e.g. Lue, Crompton and Fesenmaier 1993).

Accumulated literatures on travel destination choice indicate that the variables used to explain and predict one's destination choice can be classified into two broad categories: (1) personal and (2) travel characteristics. Personal characteristics encompass socio-economic characteristics as well as one's psychological/cognitive traits. Travel characteristics include situational factors that make the travel distinguishable from other travel.

Although there exists a rich literature explaining each of these conceptual approaches, only a few contributions focus attention on those factors that need to be integrated into human-centric destination recommendation systems. Another important limitation concerning the applicability of travel decision theory relates to its demand-driven focus, hence requesting a universal supplier database. The majority of models are based upon in-depth observational studies investigating the needs and/or benefits of consumers and largely ignore the characteristics and constraints associated with a specific media. Traditionally, research initiatives have focused on what is 'ideal' but not on what is 'optimal' within a certain decision-making environment. It is, therefore, not surprising that IT developers and an increasing number of tourism researchers who are involved in building intelligent travel advisory systems question the practical value of many of the traditional (purely conceptual) models.

3   Travel Decision Styles

Decomposition of a behavioural framework linking destination choice, information search and human-computer interaction enables one to develop a set of propositions that guide the design of a travel destination recommendation system (Hwang, Gretzel and Fesenmaier 2002): (1) Destinations recommended should vary according to the nature of each trip and user characteristics; (2) Recommendations should reflect personal differences in the nature of the information sought and processing styles; and, (3) The destination recommendation system should facilitate experiential aspects of the decision and information searching process. In order to efficiently match travellers' needs with the offers in the databases a travel destination recommendation system should adapt itself to the user in terms of: (1) Hierarchy of questions (sequence of sub-decisions); (2) Mode of representation (information search vs. experiential); and, (3) Degree of recommendation vs. search functions (user or system driven sessions). With these objectives in mind, it becomes clear that the dynamics of decision-making and information retrieval processes need to be examined. Therefore, an observational study in three design variations was conducted in order to investigate the individual trip planning and information behaviour process from a dynamic point of view (Grabler and Zins 2002).

The exploratory approach resulted in six different decision styles which were suggested to be useful when setting up a new generation IT-supported travel recommendation system: (1) Highly pre-defined users; (2) Accommodation-oriented users; (3) Recommendation-oriented users; (4) Geography-oriented users; (5) Price-oriented users; and, (6) The individual traveller. The decision styles link the various issues raised in the consumer behaviour literature with an operational perspective in order to integrate the key factors (socioeconomic, psychological/cognitive variables and trip characteristics) into a travel recommendation system.

Ideally, the system must enable a (new) user entering the travel destination recommendation system to be classified quickly in order to provide him/her the optimal navigational path and mode of presentation (type and sequence of questions, graphical widgets selection, length of interaction etc.). The user should be able to influence the dialogue management component by explicitly volunteering information useful to determine his/her decision style. This can be achieved by self selection of the decision style, for instance, by presenting iconic descriptions of the styles. Although the study by Grabler and Zins (2002) recommends that information on a user's decision style should be acquired at a very early stage in the session, the system should provide the possibility of switching between different interface styles. Again, this can be achieved by self-selection of the user or derived from a pattern of user interaction with the computer. Information presented at a later stage of the user session should be structured differently according to the requirements of the respective decision styles. However, the following three stages are used to decompose the dialogue:

(1) Filtering: The user must be able to enter the primary variables/constraints that describe his/her decision style; however, not all information categories are required at the beginning of the dialogue.

(2) Specification: Additional information related to the responses provided in first stage are presented to the user. An important aspect of this stage is that the user believes that only personally 'important' features are asked.

(3) Selection/sorting: The user must be able to make his/her final decision based on a short number of alternatives presented. This list may be sorted by one (or more)

of the key factors associated with the identified decision style and/or specified during the decision process by the user.

As the number of possible offers should be counted and provided to the user at any time, it is possible that a specific user session does not require all three stages as the number of matched offers may be limited. Additional information which appears to be important in describing the characteristics of the decision process and therefore needs to be monitored and stored during a session include: Usage type, sequence of sub-decisions, flexibility/rigidity of trip characteristics, degree of pre-specification, number of alternatives, decision style, and experiential proneness.

4   The proposed Approach

The theoretical considerations provided above sets the requirements for a travel planning recommender system that must support, by means of an adaptive behaviour, rather different decision styles and must personalize the suggestions on the base of both personal and travel characteristics. In this section we shall describe the basic design choices of the Dietorecs recommender system and illustrate a typical man/machine interaction. The Dietorecs system is based on the following elements:

- **Bundling a mix-and-match travel.** Dietorecs basically supports the user in building a personalized travel plan that can either comprise a pre-packaged offer or can be obtained by iteratively selecting travel components (items) such as locations to visit, accommodations, attractions, services. Item selection dialogue is driven by the personal and travel characteristics that are structurally decomposed into what are referred to as "general" and "detailed travel wishes".

- **Allow the user to enter through three functional doors.** The user can build (configure) his travel plan by means of three top system functions that acts as different doors to enter the system. These doors enable users to provide in whatever order and amount he/she likes general and detailed travel wishes. The first door allows the user to select whatever products he likes and in the preferred order. The second door enables the user to select a personalized trip that bundles together items available in the catalogues. The third door enables an inspiration seeking user to choose a complete trip exploiting (selecting and modifying) examples of travels shown by means of a user interface which is strongly based on images and that minimizes the interaction length.

- **Decision styles and functional doors.** Decision styles are initially mapped (probabilistically) to these functional doors. This means that Dietorecs is bootstrapped with a default assignment of decision styles to doors (for instance the "highly predefined user" to the first door), but user activity logs, stored as cases, will provide data for training the Dietorecs classifier to (1) identify the decision style (2) suggesting the user to switch to another door.

- **Register the user interaction session as a case.** The adaptive behaviour of Dietorecs is based on a structured representation of the interaction session that is stored as a case in case-based reasoning system (Ricci et al, 2002a). A case includes general and specific travel wishes acquired during the interaction, items included in the 'travel bag', feed-back provided by the user on the items selected, ordered list of the system functions called during the interaction (activity log).

- **Personalise the questions posed to the user using cases and catalogue analysis.** After having acquired some travel wishes from the user, Dietorecs poses in-context questions that either try to further specify the travel wishes, whose effect is to tighten the search, either to relax conditions that cannot be satisfied. The identification of those travel wishes that could be asked or should be relaxed rely on: (1) the analysis of the users behaviours stored in the cases (statistics over user explicit wishes), (2) constraint relaxation techniques, and (3) information theory indicators, such entropy, computed on the catalogues of products (Ricci et al., 2002b).

- **Personalise the recommendations using collaboration filtering through case similarity.** The items (and the complete trips)) suggested by Dietorecs are ranked according to a collaboration-via-content based approach (Pazzani, 1999). In fact, item filtered according to the user travel wishes are then sorted such that those contained (or more similar to) in similar recommendation sessions (cases) are scored best. With this respect Dietorecs is a hybrid recommender system that overcomes classical problems of pure collaborative-based approaches such as huge amounts of registered user logs data needed to deliver recommendations.

In the rest of this section we provide a sample session with Dietorecs. For lack of space we cannot cover all the potential interactions, initiated by entering into the three doors mentioned above. When starting to use the system the tourist must first decide how to search through the travel cases managed by the system. If the tourist is more recommendation-driven, then the selection process will be supported from the system in the form of pictorial representations of former cases (third door). The user can qualify the suggestions by labelling them as interesting versus not interesting, which should result in a case near enough to the user's travel imaginations to be selected. Otherwise, information-driven tourists with more concrete ideas such as travel destination or accommodation category normally prefer a more straightforward access to the travel cases as shown in Fig. 1 (first door).

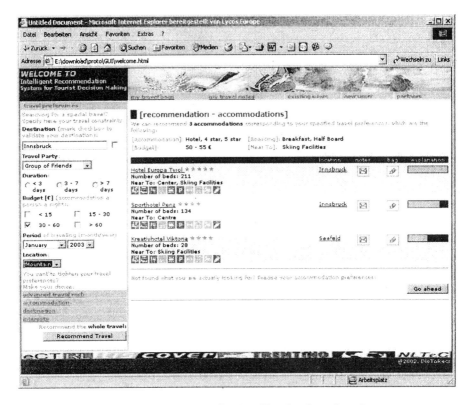

Fig. 1. Dietorecs Application Sketch – Item Search

The left frame contains all travel preferences, clustered into general travel preferences and so-called item preferences, represented by accommodation, destination and interest items. These are the elementary components of a case, namely of the 'travel bag'. Based on these preferences it is up to the system to support the user with suitable recommendations. After retrieving relevant cases from the case base and comparing their content with real items contained in the virtual catalogues of an integrated tourist information system, Dietorecs offers the user a list of recommendations. The recommendations are sorted according to the degree of similarity between found items, satisfying travel preferences, and items contained in past cases. If a user decides that one of the listed items fits, he/she can add it to his/her current travel bag by clicking the button in the column "to bag". A travel bag collects the selections made by the user from different item types and can be stored by assigning a label. Users of the system if they are registered can manage as many travel bags as supported by the system.

Advanced functionalities include the possibility to complete an already created travel bag by finding the most similar case or the chance to find similar items from within the current travel bag and allow a comfortable and flexible creation of suitable travel recommendations.

5 Technical Solutions and Software Architecture

The system exploits a case base of travel bags that is built by the community of Dietorecs users as well as catalogues provided by Tiscover AG and APT Trentino (DMO). The proposed case structure is hierarchical (Ricci et al., 2002b) and implemented as an XML view over a relational data base. Dietorecs integrates case-based reasoning with interactive query management. When asked to retrieve a travel product, Dietorecs tries to cope strictly with user needs and, if this is not possible, it suggests query changes that will produce acceptable results. Similarity-based retrieval is exploited: (1) when a complete travel is searched (second and third door) and (2) when the single products (as the result set of the user's query) must be ranked (first door).

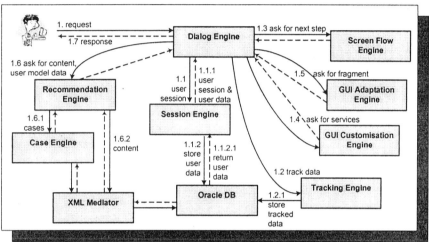

Fig. 2. Cooperation among engines in the Dietorecs system.

The Dietorecs system is developed as component-based architecture, which consists of several components that interact with each other. The units encapsulate functionalities essential for managing the Dietorecs system and are described in the following, whereby Fig. 2 shows the cooperation among those engines. The Dialog Engine serves as a central management component to support the interaction between user and system, to read the input, to validate it, to manage the workflow and to return the response back to the user. The Session Engine is responsible for authenticating and authorizing users according to their identification and permission rights as well as guaranteeing that a user's session will not be lost during interaction. It overcomes stateless HTTP connections by keeping sessions alive. The Tracking Engine manages tracking of interesting data such as user input or user behaviour. Tracked data is logged in the local database and provides methods to retrieve once tracked and logged data to support the recommendation process.

The Screen Flow Engine informs state-dependent engines about workflow sequences. It is responsible for delivering information about what is the next step that should be delivered to the user. The GUI Customisation Engine provides interaction services independent from content that have to be customised for an individual user. The GUI Adaptation Engine handles all relevant actions for supporting the Dialog Engine with essential layout information for the next required step that has to be visualized and delivered to the user. It assembles a Web page through the adaptation of single fragments described in HTML, XML or JSP.

The Recommendation Engine and Case Engine manage personalized recommendations by considering user-specific travel wish preferences. Together, they are responsible for retrieving ranked travel item or travel bundle recommendations, finding similar travel items or bundles, and adapting travels of previous recommendations to available ones supported by the two tourism information systems APT and Tiscover. The XML Mediator provides access to this data and serves as bridge for Recommendation and Case Engine to get use of this data.

6   Conclusions

The proposed Dietorecs system represents a new generation of travel recommender systems that can cope with individual differences in travel wishes and decision styles. We plan to empirically validate the system by A/B comparisons with more traditional approaches. The validation will be conducted in Europe (Italy and Austria) and in the US and will provide real time recommendation sessions to be used as training cases for the adaptive behaviour of Dietorecs.

Acknowledgment: This work has been partially funded by the European Union's Fifth RTD Framework Programme (under contract DIETORECS IST-2000-29474).

References

Crompton, J. L., and P. K. Ankomah. (1993). Choice set propositions in destination decisions. Annals of Tourism Research, 20: 461-476.

Delgado, J. and Davidson, R. (2002). Knowledge bases and user profiling in travel and hospitality recommender systems. In *Proceedings of the ENTER 2002 Conference*, pages 1-16, Innsbruck, Austria. Springer Verlag.

Fesenmaier, D. R., and J. Jeng. (2000). Assessing structure in the pleasure trip planning process. Tourism Analysis, 5: 13-27.

Grabler, K. and Zins, A.H. (2002) Vacation Trip Decision Styles as Basis for an Automated Recommendation System: Lessons from Observational Studies. In: Wöber, K.W., Frew, A.J. and Hitz, M. (eds.) Information and Communication Technologies in Tourism. Springer: Wien, 458-469.

Hwang, Y.-H., Gretzel, U. and Fesenmaier, D.R. (2002) Behavioural Foundations for Human-Centric Travel Decision-Aid Systems. In: Wöber, K.W., Frew, A.J. and Hitz, M. (eds.) Information and Communication Technologies in Tourism. Springer: Wien, 356-365.

Lue, C. C., J. L. Crompton, and D. R. Fesenmaier. (1993). Conceptualization of multi-destination pleasure trip decisions. Annals of Tourism Research, 20: 289-301.

Mayo, E.J. and Jarvis, L.P. (1981) The Psychology of Leisure Travel. Boston: CBI Publishing.

Pazzani, M. J. (1999). A framework for collaborative, content-based and demographic filtering. *Artificial Intelligence Review*, 13:393-408.

Ricci, F., Arslan, B., Mirzadeh, N. and Venturini, A.(2002a). ITR: a case-based travel advisory system. In Craw, S. and Preece, A. (eds.) 6th European Conference on Case Based Reasoning, ECCBR 2002, Springer: Berlin. 613-627.

Ricci, F., Mirzadeh, N., and Venturini, A. (2002b). Intelligent query managment in a mediator architecture. In *2002 First International IEEE Symposium "Intelligent Systems"*, pages 221-226, Varna, Bulgaria.

Schafer, J. B., Konstan, J. A., and Riedl, J. (2001). E-commerce recommendation applications. *Data Mining and Knowledge Discovery*, 5(1/2):115-153.

Um, S., and J. L. Crompton. (1990). Attitude determinants in tourism destination choice. Annals of Tourism Research, 17: 432-448.

Woodside, A. G., and S. Lysonski (1989). A general model of traveler destination choice. Journal of Travel Research, 27: 8-14.

Travel Information Search on the Internet:
A Preliminary Analysis

Bing Pan[a]
Daniel R. Fesenmaier[a]

[a] Department of Leisure Studies
University of Illinois at Urbana-Champaign, USA
{bingpan, drfez}@email.edu

Abstract

Travel information search is a complex, dynamic and contingent process. Travel information search and planning on the Internet becomes more and more prevalent nowadays along with the development of information technology. However, the usability of the Internet as a travel information source still remains as a problem. This research intends to explore the interaction process between a travel information searcher and the Internet based upon the assumption that travel information searchers search through the tourism information space following their semantic mental models and the mismatch between their semantic mental models and the conceptual model of online conceptual tourism information space contributes to the futile and unsatisfactory travel information search process. An experiment of travel planning on the Internet was conducted and the results confirmed that the overall tourism information space is huge, travelers appear to have a different mind sets as compared with the information on the Internet, and this discrepancy contributed substantially to the usability problem of the Internet.

Keywords: Semantic Network; Travel Information Search; Internet; Mental Model.

1   Introduction

While more and more tourists are using the Internet for travel planning purpose, the usability of the Internet still remains as a problem (Pan & Fesenmaier, 2000). Travelers have trouble finding the information they are looking for as the amount of online travel related information increases. In addition, Internet navigation requires different skills as it is fundamentally different from traditional linear reading. Searching through the Internet is an interactive process between the information searchers and the hypertext (the Internet) implemented through a computer and a web browser. The understanding of travel information searcher's behavior on the Internet is essential to the design of useful Internet-based technology. This research explores the interaction process between a travel information searcher and the Internet for travel planning purpose through the perspective of tourism semantics.

2   Conceptual Development

Tourism research has shown that travel planning involves many sub-decisions and can be viewed as a hierarchical, dynamic, multi-stage and contingent process where the

the central decisions are made at the beginning of travel planning and usually hard to change (Fesenmaier & Jeng, 2000; Jeng, 1999). Travel planning implies a dynamic and contingent travel information search whereby the goals and objectives of the potential traveler adapt to changes in his/her information environment. Recent research in human-computer interaction (HCI) has demonstrated that the discrepancy between the mental models of computer users and the information system is the major element which determines the usability of computer system (Nielson, 1993). Research in cognitive information retrieval (Ingwersen, 1996) has also revealed that the differences between the mental models of the information searchers and the conceptual space of information system are crucial in determining the results of information search. Mental models can be viewed either as the different metaphors or affordance of the computer system (Norman, 1988), or as the inter-relationship between different concepts in information user's mind and information space (Carley & Palmquist, 1993). The Internet is an interactive hypertext system where information nodes are "hyperlinked" according to their semantic relevance (Boechler, 2001). For information searchers, traversing through the Web space involves information processing and learning, and judgments of semantic relevance according to information searchers' knowledge and search goals.

Semantic information plays an important role in the information search process. In this study semantic mental models, which capture the travel information searcher's background knowledge and the understanding of the Internet, were used to explore the usability problem in the travel information searching process. It is argues that the incongruity between the semantics in travel information space as well as in the travelers' mind contributes to the futility and frustration of travel information search process. For example, travelers may consider "What to do" is the most important element in travel planning process; however, the tourism destination marketers may display hotel and motel information on the first page of their web sites. Semantic mental models can be represented by semantic networks, which is a keyword matrix (represents matrix of concepts) with different weight of relevance between them. Semantic network analysis then can be used to assess the nature and structure of relationships between the respective semantic models (Doerfel, 1998).

It is also posited that the travel information search process can be represented as a network of goals where each goal can be represented by a cluster of concepts as defined by keywords related with it, which may include nouns, adjectives, and verbs. When a traveler searches for information on the Internet, his/her choices of links are determined by the value of relevance of the link anchors (linked texts, pictures or contextual information), in other words, the value of "information scent" determines which link the information searcher is going to traverse (Card, Pirolli, Van Der Wege and etc., 2001). According to Kim and Hirtle (1995), information seeking on the Web involves reading/understanding and navigating, and the two processes happen simultaneously. The destination page is usually the "content" page (Nakayama, Kato, & Yamane, 2000) and the web pages the searchers quickly clicked through are "index" pages, where the content is marginal in the travel information searcher's semantic mental model. For example, a travel planner with a central concept of "Disneyland" will click through "South", "Florida", and "Theme Parks" quickly in order to reach the "Disneyland" page, and then read that page in depth.

It is also argued that the travel information search process can be broken into smaller goal-directed sequences or "episodes". Each episode will have one destination page (reading page), which is related to the solution of a sub-goal. In the navigation process the semantic mental model of travel information searcher and the representative semantic network are continually changing. For example, after destination choice has been made, their semantic network may change accordingly. After a certain period time of information search, either all the goals will be satisfied, or the travel planner encounters some obstacles (fatigue, no relevant information, or time constraint), and the navigation process will terminate. Those web pages on which the travel planner stays for a longer period of time are essential to his/her decision making in that these reading pages (usually content page) constitute a "sub-space" of the overall travel information space and thus, are the result of the interaction between the semantic mental model of travel information searcher and the semantic model of travel information space.

Finally, it is proposed that the congruence between the semantic mental models of travel information searcher and the semantic model of travel information space will determine the level of satisfaction in the travel information search process. If the travel information space contains the information that the travel information searchers are looking for and the interrelationships of those concepts are similar, the information search process will be more efficient and satisfactory. Information searchers' travel experience and the experience of using the Internet, as an information source will determine the degree of congruence between travel information searcher's initial semantic mental model and the semantic model of the travel information space.

3 Methodology

In order to analyze the travel information search process, a travel planning experiment using "thinking-aloud" method (Ericsson and Simon, 1993) was carried out with 15 subjects from a major university located in the Midwest United States during August of 2002. The subjects were asked to plan a trip to a specified US destination (San Diego, California) on the Internet using computers located in the National Laboratory for Tourism and eCommerce, University of Illinois at Urbana-Champaign. Demographic information, background knowledge about the destination, and behavioral data were collected using questionnaire, voice recorder, online camcorder, Internet monitor software, and microphone while the subjects searched information on the Internet. The data was transcribed and analyzed using semantic network analysis and protocol analysis (Hsieh-Yee, 2001). The following are the detailed data collection methods.

3.1   Experimental Scenario

Upon arriving at the Lab, the subject was introduced to the goal and procedures of the experiment. The subjects were then informed that one subject would be randomly selected to win two round-trip airline tickets to San Diego, California in the fall of 2002. The subject was asked to plan a weekend trip vacation to San Diego on October 19th and 20th of 2002 (a Saturday and Sunday) under the assumption that they were actually awarded the two round-trip airline tickets to that city. The subject needed to plan where to visit during a two day trip and where to stay overnight under the assumption that the budget of the trip would be determined by the subject's own financial situation.

3.2   Data Collection

The experiment started with a survey regarding the subject's travel experience and computer use experience. In addition, a pre-experiment interview was conducted to measure the semantic mental model of the information searcher prior to his/her information search, which represents their search tasks, background knowledge, and their understanding of travel information space (Hsieh-Yee, 2001). The interview followed a semi-structured format in order to elicit different semantic mental models of travel information searchers. The subject was then asked to construct a travel plan to San Diego using the Internet (along with pen, paper, and printer to facilitate the information search and travel planning process) with a maximum time of an hour, and during this process, the subject was asked to think-aloud (Ericsson and Simon, 1993). An online camcorder (TechSmith® Camtasia™) was used to record everything happening on the screen; another program (iOpus® STARR™) was used to record the subject's keystrokes and links he/she clicked on; a microphone was also used to record verbalization of the subjects. After the experiment, the subject was asked to write down their travel plan and to complete a survey regarding their levels of satisfaction toward their travel information search and travel planning process.

3.3   Data Analysis

Three types of semantic analysis were conducted; semantic mental model of each travel information searcher prior to their information search; conceptual model of travel information space; and the semantic model of each interaction space, which is constituted of those web pages a travel information searcher stayed for a longer period of time to process information. The semantic mental models of the subjects were obtained through semantic network analysis on the transcript of each interview. Conceptual models of travel information space were generated through semantic network analysis on full texts of the first pages of the commonly used web sites. The semantic model of interaction space was generated through semantic network analysis on the full texts of each interaction space. QAP analysis (Krackhardt, 1987, 1988) was then used to determine the relationships between these three semantic models and also their relationships with satisfaction of travel information search, travel information searcher's travel experience and Internet use experience. It was hypothesized that the interaction space is the results of the interaction between each individual semantic mental model and the conceptual model of travel information space; the degree of their congruence was correlated with the satisfaction of travel information search; the degree of their congruence was also correlated with their travel experience and travel information search experience.

Detailed protocols regarding travel information search include web pages visited, computer programs used, time spent on a web page, verbalization of subjects (information processing and evaluation of the web page) and interpretation. The data were used to test the model of navigation and reading process during information search and different episodes in information search were identified.

4 Preliminary Results

4.1 Basic Subject Profile

The 15 subjects included 10 undergraduate students, 1 graduate student, and 4 researchers in the university. Their ages ranged from 19 to 45 years old with an average age of 25 years; 6 subjects were men and 9 were women. Four of the subjects have been to San Diego. The 15 subjects took an average of 36 minutes to plan their trips to San Diego, with minimum time of 20 minutes and a maximum time of 55 minutes. The subjects visited from 7 to 26 web sites with an average of 15 web sites. They visited from 60 to 312 web pages, averaging 124 web pages from different web sites. About half of the subjects (7 of 15) used a printer to print out information and organize information. Most subjects (12 of 15) used pencils and paper to organize information.

4.2 Tourism Information Space

The web sites the travel information searchers visited constitute the travel information space of one specific destination (San Diego). From the preliminary analysis, we can see this space is large and diversified. In total 145 web sites were visited by the subjects, which include not only general search engines (e.g. http://www.google.com, http://www.yahoo.com/), general tourism reservation systems (e.g. http://www.expedia.com/, http://www.orbitz.com/, http://www.hotwire.com/), local official tourism offices (e.g. http://www.sandiego.org/), web sites of local attractions (e.g. http://www.cafesevilla.com/, http://www.balboapark.org/), but also information portals provided by different commercial parties (e.g. http://www.sandiego.cc/, http://www.a-zsandiego.com/), and general recreational and educational web sites (e.g. http://www.nfl.com/, http://www.nps.gov/, http://www.sdsu.edu/) (see Table 1 for the list of web sites that at least two subjects visited).

A common theme among all the subjects was that they were all looking for an "information portal" or a "hub" page, which is an "authoritative" web page (in the travel information searcher's view) with many links to famous attractions or accommodations in San Diego. For example, the San Diego Convention and Visitors

Bureau web site (http://www.sandiego.org/), the official web site for the municipal government of San Diego (http://www.sannet.gov/), or a commercial information portal for San Diego (http://www.sandiego.cc/) were chosen by most subjects as the starting points for exploration of San Diego on the Internet.

4.3 Travel Information Searcher's Semantic Mental Models

The pre-experiment interview was semi-structured in order to elicit the subject's background knowledge of San Diego. The texts of the transcribed responses were aggregated together and analyzed using semantic network analysis (Doerfel, 1998). Table 2 shows top 38 top words the travel information searchers used to describe their knowledge about San Diego and their general travel preference.

Table 1. Tourism Information Space for San Diego, California

Web Site	Number of subjects	Web Site	Number of Subjects
www.google.com	13	www.sannet.gov	2
www.sandiegozoo.org	8	www.sandiego-online.com	2
www.sandiego.org	7	www.sandiego.cc	2
www.sdcommute.com	5	www.revup.biz	2
www.mapquest.com	5	www.reservetravel.com	2
www.trafficmp.com	4	www.portofsandiego.org	2
www.seaworld.com	3	www.orbitz.com	2
www.sandiego.com	3	www.netster.com	2
www.expedia.com	3	www.infosandiego.com	2
www.blueescape.com	3	www.hotwire.com	2
www.a-zsandiegoattractions.com	3	www.fodors.com	2
www.yahoo.com	2	www.citysearch.com	2
www.thebigbay.com	2	www.cafesevilla.com	2
www.sdsu.edu	2	www.balboapark.org	2
www.sdro.com	2	www.arestravel.com	2
www.sdnhm.org	2	www.4adventure.com	2
www.sdinsider.com	2		

4.4 Conceptual Model of Online Tourism Information

The San Diego Convention and Visitors Bureau web site (http://www.sandiego.org/) was the most importation information "hub" page for providing comprehensive travel information. We captured the text from the first page of their homepage and analyzed the text using semantic analysis. Table 3 shows the top 38 most frequently appeared keywords. Comparing this list of concepts with the one used by the travelers, we can see only 10 concepts (San Diego, Hotel, Visitor/visit, Art, Information, Activity, Attraction, Beach, California, Zoo) appear in both lists. However, the 10 concepts have different weights. "Beach" is rated as the most important attraction in travel information searcher's mind, but it only slightly important in the first page of San Diego Convention and Visitors Bureau web site. In the following protocol analysis, we did find some subjects had problems finding "Beach" information in that web site and also other tourism web sites regarding San Diego. Comparing the differences of the lists of concepts, the conceptual model of online tourism information has a clear marketing focus with concepts like Book, Member, Reservation, Order, Save Casino, and Free. On the other hand, the traveler's mental model is more focusing on activities (Food, Walk, music, Club, and Water) and subjective feelings (Fun, Pretty, and interested). These concepts were buried in different categories (What to Do or Getting Around) of the first page of San Diego Convention and Visitors Bureau.

Table 2. Concepts in Traveler's Semantic Mental Model

Words	Frequency	Words	Frequency
San Diego	40	Attractions	10
Beach	35	Nice	10
Museum	30	Club	9
Food	27	Interesting	9
California	25	Ocean	9
Restaurant	23	Shopping	9
Hotel	22	Check	8
Information	21	Fun	8
City	17	Night	8
People	17	Pretty	8
Walk	16	Special	8
Tourist	14	Time	8
Zoo	14	Activities	7
Big	12	Area	7
Different	12	Interested	7
Good	12	Love	7
Visit	12	Transportation	7
Music	11	Trip	7
Art	10	Water	7

4.5 Protocol Analysis

Using the clickstream data from the Internet monitor software, the movie file from the online camcorder along with the subject's verbalization, a detailed protocol can be obtained including behavior (Click Link, Type In, Print, or Write on paper), time spent on the page, starting time, the subject's verbalization and interpretation (see Table 4 for a part of the protocol from one of the subjects). An initial analysis shows that the search process can be broken up into different episodes where each episode targeted a sub-problem. The switch of episodes can be identified from the subject's verbalization, for example, "… Now hotel has been taken care of… next I'm going to see what to do over there…" Interestingly, most subjects started with hotel choices, and then switched to transportation and different activities. It was also found out that geographic information was an important constraint affecting choice of hotel, transportation, and activities. That is, the subjects always had trouble locating various hotels and attractions whereby they were forced to switch between the hotel/attraction web sites and online map web sites such as MapQuest to find geographical locations.

Table 3. Concepts in the Conceptual Model of Online Tourism Information

Words	Frequency	Words	Frequency
San Diego	33	Beaches	3
Hotel	10	Benefit	3
Book	7	California	3
Member	7	Card	3
New	7	Casino	3
Visitor	7	Family	3
Art	6	Famous	3
Event	6	Fast	3
Golf	6	Free	3
Information	6	Military	3
Reservations	6	Sea World	3
Save	6	Services	3
World	6	Tours	3
Activity	5	Weather	3
Attractions	4	Zoo	3
Dining	4	Adventure	2
Order	4	Animal	2
Park	4	Announces	2
Vacation	4	Apple	2

5   Conclusions and Implications

The initial results confirmed that the overall tourism information space is huge. When searching information on the Internet, each subject navigated through the space in his/her idiosyncratic way and accordingly, encountered totally different information space. Additionally, travelers appear to have a different mind sets compared with the information on the Internet and the discrepancy contributed to the usability problem of the Internet. However, more in-depth analysis needs to be carried out on the tremendous amount of qualitative data obtained in this study.

Table 4. Travel Information Search Protocol

Time Started	Time Spent	Behavior	Content	Verbalization	Interpretation
17:18:41	7	Type In	Sand Diego Vacation	You know what, that's bad type. I am gonna type something else. "San Diego Vacation".	Search for "San Diego Vacation"
17:18:48	4	Click Link	http://www. San+Diego+Va cation	...	Result page
17:18:52	7	Click Link	http://www.sand iego.cc/sandieg o/	Here we go, "San Diego hotels and resorts online". It looks like it has everything I need.	Click on the first result.
17:18:59	6	Write	Saturday, October 19th Sunday October 20th	If I plan getting there Saturday morning, I have only a few short days, two days to be exact.	Write time frame on the paper.

* Time Spent is measured in seconds.

A useful result of this study was that the web interface can be improved using the language and vocabulary of the information searcher instead of the designer (Furnas, Landauer, Gomez and Dumais, 1987). The link anchors can be improved according to the information searcher's semantic mental model to provide more meaningful proximal cues. Semantic network analysis can also be used to explore tourism web sites in order to assess the appropriateness of its content. In usability testing, the interaction process can be captured in order to investigate if the travel information searchers can arrive at appropriate information space which contains marketing information. Furthermore, this study demonstrated that the congruence between semantic mental models and the semantic models of travel information is a useful predictor of satisfactory information search.

The semantic web represents the next generation of Web technology and has been a heated topic in computer science research area. However, the generation of ontologies of tourism should consider the travel information searcher's semantic mental model as well as semantic model of the tourism industry. This research demonstrated that semantic network analysis is a useful tool in exploring semantic mental models of travel information searchers and can be used to generate tourism ontologies or at least provide useful validations.

References

Boechler, P.M. (2001). How Spatial is Hyperspace/ Interacting with Hypertext Documents: Cognitive Processes and Concepts, *CyberPsychology & Behaivor*, 4(1): 23-46.

Card, S.K., Pirolli, P., Van Der Wege, M., Morrison, J.B., Reeder, R. W., Schraedley, P., and Boshart, J. (2001) Information Scent as a Driver of Web Behavior Graphs: Results of a Protocol Analysis Method for Web Usability. *Proceedings of the ACM Conference on Human Factors in Computing Systems* (J. Jacko, A. Sears, M. Beaudouin-Lafon, and R.J.K. Jacob, eds.), pp. 498-505. ACM, Seattle, WA.

Carley, K., & Palmquist, M. (1992). Extracting, Representing, and Analyzing Mental Models. *Social Forces*, 70(3): 601-636.

Doerfel, M. (1998). What Constitutes Semantic Network Analysis? A Comparison of Research Methodologies. *Connections*, 21(2): 16-26.

Ericsson, K, & Simon, H.A. (1993). *Protocol Analysis: Verbal Reports as Data*, Cambridge: The MIT press.

Fesenmaier, D. R. and J. Jeng (2000). Assessing Structure in the Pleasure Trip Planning Process, *Tourism Analysis,* 5(3), pp. 13-28.

Furnas, G.W., Landauer, T.K., Gomez, L.M., & Dumais, S.T. (1987). The Vocabulary Problem in Human-System Communication. *Communications of the ACM*. 30(11): 964—971.

Hsieh-Yee, I. (2001). Research on Web Search Behavior. *Library and Information Science Research*. 23: 167-185.

Ingwersen, P. (1996). Cognitive Perspectives of Information Retrieval Interaction: Elements of a Cognitive Theory, *Journal of Documentation*, 52(1): 3-50.

Jeng, J. (1999). *Exploring the Travel Planning Hierarchy: An Interactive Web Experiment*. Unpublished Dissertation. University of Illinois at Urbana-Champaign.

Kim, H., & Hirtle, S.C. (1995). Spatial Metaphors and Disorientation in Hypertext Browsing, *Behavior and Information Technology*, 45(2): 93-102.

Krackhardt, D. (1987). QAP Partialling as a Test of Spuriousness. *Social Networks*, 9: 171-186.

Krackhardt, D. (1988). Predicting with Networks: Nonparametric Multiple Regression Analysis of Dyadic Data. *Social Networks*, 10: 359-381.

Nakayama, T., Kato, H., & Yamane, Y. (2000). Discovering the Gap between Web Site Designers' Expectations and Users' Behavior, *Computer Networks*, 33: 811-822.

Nielsen, J. (1993). *Usability Engineering*. Boston: AP Professional.

Norman, D.A. (1988). *The Psychology of Everyday Things*. New York: Basic Books.

Pan, B., & Fesenmaier, D.R. (2000). A Typology of Tourism-Related Web Sites: Its Theoretical Background and Implications. *Information Technology and Tourism*, 3(3/4).

Pirolli, P., & Card, S.K. (1999). Information Foraging. *Psychological Review*. 106(4): 643-675.

Holiday Scheduling for City Visitors

Beatriz López

Institut d'Informàtica i Aplicacions
Universitat de Girona, Spain
blopez@eia.udg.es

Abstract

This paper presents an approach to holiday scheduling which schedules the activities a tourist can enjoy in a city through an Internet service available 24 hours a day, 7 days a week (at information desks in airports, train stations, etc.). Holiday scheduling turns out to yield constraint satisfaction problems, with the particularity that the amount of constraints in this problem is not critical, but the number of activities is huge. Hence, the methodology proposed is a combination of case-based reasoning and constraint solving algorithms. The benefits of this combination are particularly useful in scheduling real life problems, and is suitable in terms of the particularities of the domain. A case study at the city of Girona is provided.

Keywords: City tourist, museums, case-based reasoning, constraint satisfaction, scheduling.

1 Introduction

Holiday scheduling consists of organizing, for a given period of time, a set of activities that a tourist wants to do at a given destination. Holiday scheduling can be looked at with different degrees of detail. Some researchers concentrate on the elaboration of packages in which the main idea is to allocate different tourist spots and attractions to a set of days according to pre-established catalogues (Soo and Liang, 2001; Ricci and Werthner, 2001; Huang and Miles, 1995). Others concentrate on the selection of a given hotel for a given destination (http://www.cbr-web.org/ [July, 2002]). While others look at particular tourist services in a given region (Blanzieri and Ebranati, 2000). This work is in line with this latter approach.

The aim of this project is to develop a 24-hour Internet service (at information desks in airports, train stations, etc.) that will enable a tourist arriving at anytime in a city to schedule his/her stay. The user selects, from a list, the set of activities he/she wants to enjoy. There are several kinds of activities: visiting museums, monuments, art exhibitions, temporary cultural events (such as festivals, markets and performances), sightseeing, parks and sports facilities among others). Some of the activities are constrained by timetables such as museums. Furthermore, some also require an entrance fee, so the user should have a budget for them. Each place to visit requires a minimal period of stay, estimated by those responsible for the activities. For example, a visit to the City Museum is estimated as lasting 2 hours. According to (Soo and Liang, 2001), every city has several "mandatory" spots: places that every tourist should visit. However, it could be the case that the user is coming to the city for a second, third, or even a tenth time. So in our approach, only a minimal set of spots is required: lunch, dinner and hotel (with the corresponding minimal staying period related to resting hours), depending on the duration of the visit.

Once it is checked that the user has enough time and a big enough budget to do the selected activities, the system schedules them, based on the experiences of past users using a combination of Case-Based Reasoning and Constraint Satisfaction Problem Solving techniques. In addition, temporary events in the city, as well as other activities not selected by the user, are proposed to the user in order to fill in time gaps.

This paper is organized as follows: Section 2 deals with the formulation of the holiday scheduling problem in terms of constraint satisfaction. Section 3 provides the methodology. Since this paper is exploratory, the results given in section 4 are case studies regarding the scheduling of activities for a visitor to the city of Girona (Spain). Finally, section 5 provides some conclusions and the details of related work.

2   Problem formulation

Holiday scheduling consists of organizing, for a given period of time, a set of activities that a tourist wants to do at a given destination. Every activity has a duration d_i which expresses the minimal staying period. For the sake of simplicity, the time required to get to the location of the activity is included in the minimal staying period.

Every activity has an associated cost c_i in Euros (resource). The cost can be 0 € for free entrance activities (for example, a sightseeing activity). No precedence relations between activities exist. However, some of the activities are only allowed at a given set of times (timetables) that may depend on the day of the week or a particular date, as well as on the season. For example, the Cathedral Museum is closed on Monday, and is open from Tuesday to Saturday from 10:00 to 14:00, and from 16:00 to 19:00. On Sundays and holidays it opens in the morning: 10:00-14:00. In addition, the Museum has three different timetables: one from October to February, another from March to June, and yet another from July to September. Hence, to handle all these temporal variables, we associated each activity with a timetable array, T_i, of month per day dimension (365 cells). Each cell of the array, $L_{j,i}$, is a list of available timetables for the corresponding month (j) and day (i) of the activity, $L_{j,i} = ((init_1, end_1), ... (init_n, end_n))$. Some $L_{j,i}$ values may be empty such as, for example, the ones corresponding to closing days as wells as non-existing days (i.e. February 30th). Thus, for the Cathedral Museum, we have $L_{Jan,1} = ((10,14))$ and $L_{Apr,9} = ((10,14),(16,19))$, both dates, January 1st and April 9th, being a Tuesday.

In addition to the activities, the user also specifies constraints: the duration of his/her stay and his/her budget. Duration and budget establish the following constraints:

- The sum of the minimal staying period of all activities must not exceed the holiday period of the user. That is: *duration* $\geq \sum_i d_i$.
- The sum of the cost of all activities must not exceed the user's budget. That is: *budget* $\geq \sum_i c_i$.

If the above constraints are not satisfied, there is no schedule possible. So the user is forced either to remove activities or to change the constraints before starting the problem solving.

3 Methodology

To solve the holiday scheduling problem the methodology proposed is based on the combination of Constraint Satisfaction Problem solving techniques and Case-Based Reasoning (CBR). On the one hand, holiday scheduling can be formulated as a Constraint Satisfaction Problem (CSP), in which we need to assign time slots to a set of activities without violating any tourist constraints (budget, trip duration). Constraint Satisfaction Problems are hard to solve, however. On the other hand, Case-Based Reasoning is a particular example of a Machine Learning Technique in which past experiences (problem solutions) can be used to solve new problems, making CSP solving efficient by providing solutions at one shot.

With regard to the problem at hand, i.e., holiday scheduling, the advantage of following this integrated approach is twofold. First, CBR provides complete past schedules in which other activities, in addition to the ones selected by the user, can be suggested. The number of activities in a city is huge. Thus, a similar past case can provide alternatives to the user who may not be aware of them. And second, CSP is not started from scratch, but from a past solution. Since the domain is not critical (not too constrained), the effort required for adapting a past solution to the current problem following a CSP is feasible.

3.1 The CBR component

According to the standard methodology, any CBR has four main steps: retrieval, reuse, revise and retain. In the retrieval phase, old solutions from the case base are recovered in order to solve a new problem. In the reuse phase, the solution of a past problem is used to solve a new problem. It is precisely during this phase when CSP techniques are used in order to adapt the old solution to the new problem. Then in the revise phase, the solution for the case is checked, and finally, it is learned in the retain phase.

Case Base. The case base is a collection of solved CSPs. A case is a CSP in which the following parts are distinguished (see Figure 1 for an example):

- Identification: a numerical value
- Constraints to be provided by the user: duration of the stay, budget, and temporal constraints (season, starting date, starting weekly day).

- Selected activities: activities the user wants to do.
- User profile: age, sex, marital status, profession and hobbies.
- Solution of the case: the final price, the starting time, the final time, and the scheduling. The latter consists of an ordered list of tuples representing the sequence of activities. Each tuple (A, S, D, C) contains the information of the activity name (A), the starting time (S), the duration (D), and the cost (C).

Note that we assume to have 5 constraints for every case that are used for case retrieval. Other constraints related to activity timetables are handled by the CSP component during case reuse.

Case Retrieval. Case retrieval is based mainly on a three-step process: activity matching, constraint matching and selection. First of all, past cases containing either the whole set of activities or part of them are recovered, $\{r_i\}$. Then, a constraint matching process is applied in which a similarity degree for each case r_i is computed, based on the constraints. No case is rejected. The constraints must be satisfied in the final scheduling, but previous approximations to these constraints may be useful when producing the new one. Finally, the one that best matches the probe (highest similarity degree) is selected for adaptation in the reuse phase.

Activity matching is based on the number of activities of the probe P present in the past experience C_i:

$$Sim_a(P, C_i) = \frac{|Act(P) \cap Act(Ci)|}{|Act(P)|} \tag{1}$$

where Act (X) are the activities of a case. Constraint matching is based on a similarity function that follows an Ordered Weighted Average (OWA) (Yager, 1988):

$$Sim_C(P, C_i) = \sum_{j=1}^{n} w_j * sim_{\sigma(j)}(cons_j(P), cons_j(C_i)) \tag{2}$$

where: n is the number of constraints; $cons_j(X)$ is the j-constraint of case X; $sim_{\sigma(j)}(x_j, y_j)$ is the similarity between the two j-constraints (see (López, 2002)); $\sigma(j)$ is a permutation of the values 1...n so that $sim_{\sigma(j-1)}(x_{j-1}, y_{j-1}) \geq sim_{\sigma(j)}(x_j, y_j)$ $\forall i = 2,...,n$; and w_i is the relevance of each constraint in the similarity. The OWA is suitable for computing similarity, because we can order the different constraint similarities based on the degree of their similarity. In our model we have n=5, and we set $w_1=0.35$, $w_2=0.25$, $w_3=0.17$, $w_4=0.13$ and $w_5=0.1$.

```
Case 1
Duration: 3
Budget:  200
Temporal: (Season June), (Date 22-June-2002), (Week day Saturday)
Activities: Jewish Museum, Cathedral Museum, City Museum
Profile: 23, Female, Single, Computer Science, (Traveling, Sports, Photography)
Solution: (Price 190.85€),(Starts at10h),(Ends at 20h)
          ((Jewish Museum, 10h, 3h, 1.8€),(Lunch, 13h, 1h, 9€),(Beach, 14h, 6h, 0€),
          (Dinner, 20h, 2h, 15€),(Pub, 22h, 2h, 20€),(Disco, 24h, 3h, 20€),
          (Hotel, 3h, 7h, 45€),(Cathedral Museum, 10h, 2h, 4.25€),
          (Sightseeing, 12h, 1h, 0€),(Lunch, 13h, 1h, 7€),(Beach, 14h, 6h, 0€),
          (Dinner, 20h, 2h, 15€),(Pub, 22h, 2h, 20€),
          (Disco, 24h, 3h, 20€),(Hotel, 03h,7h, 45€),(City Museum, 10h, 2h, 1.80€),
          (Sightseeing, 12h, 1h, 0€),(Lunch, 13h, 1h, 7€),(Beach, 14h, 6h, 0€)
Case 2
Duration:2
Budget:  100
Temporal: (Season February),(Date 2-Feb-2002),(Week day Saturday)
Activities: Art Museum, Cathedral Museum, Cinema Museum, City Museum
Profile: 23, Female, Single, Computer Science, (Travel, Sports)
Solution: (Price 90.05€),(Starts at 10h),(Ends at 19h),((Art Museum, 10h, 3h, 3€),
          (Lunch, 13h, 1h, 7€),(Sightseeing, 14h, 2h, 0€),
          (Cathedral Museum, 16h, 2h, 4.25€), (Devessa, 18h, 2h, 0€),
          (Dinner, 20h, 2h, 12€),(Pub, 22h, 2h, 8€),(Hotel, 24h, 10h, 45€),
          (Cinema Museum, 10h, 3h, 2€),(Lunch, 13h, 1h, 7€),
          (Sightseeing, 14h, 2h, 0€), (City Museum, 16h, 2h, 1.80€)
```

Fig.1. Case base used in the example.

Case Reuse. In the reuse phase, we adapt the past case solution to the new problem. On a few occasions we will have an identical case, so the solutions are carefully elaborated to guarantee that constraints are satisfied. The procedure for adapting the past solution is based on CSP techniques and is explained in the next section.

Case Revise. Once the solution has been produced, it is given to the user. All solutions are valid ones from the point of view of a CSP, that is, all constraints are satisfied. However, the solution might not be the best one. Moreover, the user may not like the activities suggested for filling in time, or the solution in general. Thus, he/she can accept it or modify it. That is to say, it is the user who evaluates the outcomes.

Case Retain. The new solution is not always retained in memory. The storage of new solutions depends on the adaptation effort. If the solution comes from an adaptation process in which more than one access to memory has been performed (i.e. more than one case has been used to solve the case), then the new case is stored. It will probably be useful in other situations, and avoid wasting time in accessing the case base. Otherwise, the case will not be stored, since the adaptation process is simple and not time consuming. Avoiding the storage of all cases, we keep the case base to a reasonable size.

3. 2 The CSP component

The procedure for adapting the retrieved solution to the current problem is as follows:

1. Copy the past solution to the new case
2. Remove activities not selected by the user and those that do not take place in the corresponding season
3. Check for the following inconsistencies:
   3.a. Duration exceeded
   3.b. Budget exceeded
   3.c. Timetable violation according to calendar holidays and day of the week
   3.d. Overlapping activities
4. Solve one inconsistency and go to 3 until no more inconsistencies hold
5. Complete the set of selected activities
6. Add activities for filling up time according to user profile

Solving duration inconsistencies. This situation may occur when, for example, the user has specified a 1-day duration and the retrieved case is a 2-day duration. In such a situation, only the scheduling of the activities requested by the user are transferred, together with the components that make up the minimal set (lunch, dinner, hotel). Remember that at the beginning of the problem solving, it is guaranteed that the user will have enough time to perform all the activities he/she has selected.

Solving budget inconsistencies. Budget violation may occur as a consequence of the constraint matching process, since recovering cases with similar budgets is allowed. However, it does not mean that the case has no solution. Added activities, such as lunch and dinner, are revised in order to suit the budget.

Solving timetable inconsistencies. Timetable violation may occur when the scheduling retrieved does not exactly match either the date or the week day. Hence, certain activities planned for one day cannot be performed. To solve this inconsistency, a new case that provides an alternative scheduling for the inconsistent activities is recovered.

Solving overlapping activities. To solve this inconsistency the following procedure is applied (following the Dynamic Minimum Conflicts Algorithm developed by Lisa Purvis (Purvis, 1997)):

1. Choose, from among the conflicting activities, the one that is least constrained.
2. Search the case-base for alternative schedulings for that activity.
3. If one is found, return the new scheduling.
4. Otherwise, search for available time gaps in the current schedule compatible with the activity. If one is found, then re-schedule it.
5. Otherwise, leave the conflicting activity chosen in step 1 unchanged and proceed by selecting the next least conflicting activity for re-scheduling.

6. Continue step 2 until all overlapping activities are solved, or no solution is found (that is, all conflicting activities have been analyzed without success).

Complete the set of selected activities. If an activity selected by the user was missing (it was not among past cases), it is added to the scheduling either by exchanging it with activities to fill up time or by placing it in empty slots.

Adding activities to fill up time. The addition of activities to fill up time is performed according to the user profile, following recommender system approaches, such as the ones in (González et al. 2002).

4   Case study

In this section, the methodology presented in this paper is illustrated by means of an example in the city of Girona. For this purpose, consider the cases that make up the case base, shown in Figure 1. Assume that a visitor to the city selects the following activities (new problem P_1): Art Museum, Cinema Museum and City Museum. At the first step of the retrieval phase, the activity matching formula is applied for every case in memory (equation 1) and gives the following results: $Sim_a(P_1,C_1) = 1/3$, $Sim_a(P_1,C_2) = 1$. Case C_2 is therefore retrieved as the most similar case. No further similarity computations are performed. The next step is the adaptation process.

The past solution is transferred to the new case. Since Season is the same in both cases, all activities scheduled in C_2 are still valid for P_1. So, no activities are removed. Next, the inconsistencies are checked: the current case is for a duration of 1 day whereas the retrieved case is for 2 days, so, the activities selected by the user are only transferred as a provisional solution:

(Art Museum, 10h, 3h, 3€), (Lunch, 13h, 1h, 7€),
(Cinema Museum, 10h, 3h, 2€), (City Museum, 16h, 2h, 1.80€)

The budget is OK. According to the calendar, the starting day for C_2 is Saturday, while for P_1 it is Sunday. Such a difference is ignored since the museums have the same timetable on Saturday and Sunday.

At this point, the following activities are found to overlap: (Art Museum, 10h, 3h, 3€) and (Cinema Museum, 10h, 3h, 2€). According to the procedure for adaptation (see section 3.1), the Art Museum is more constrained than the Cinema Museum (because of a more restrictive timetable). Hence, the Cinema Museum activity is selected for re-scheduling. First, an alternative past case is looked for, where the Cinema Museum was planned for a different time slot. But no case is found in the current case base. Second, the available empty slots in the current scheduling are 14-16h and 18-24h. The minimal staying period for the Cinema Museum is 3h, so it is not possible to place it between 14 and 16h. The museum closes at 18h, so it cannot be placed after 18h, either. So no solution is found for changing the Cinema Museum. As a consequence, the scheduling for the Cinema Museum is left unchanged and the Art

Museum is re-scheduled. This time, there is a previous case, C_1, where the Art Museum was scheduled at 14h. Minimal staying period for the Art Museum is 3h, so the end of the activity is 17h. This new scheduling for the Art Museum does not violate any constraint. At this point the solution is: (Cinema Museum, 10h, 3h, 2€), (Lunch, 13h, 1h, 7€), (Art Museum, 14h, 3h, 3€), (City Museum, 16h, 2h, 1.80€).

In this solution, a new inconsistency arises between (Art Museum, 14h, 3h, 3€) and (City Museum, 16h, 2h, 1.80€). The visit to the Art Museum finishes at 17h, while the scheduled time for the City Museum is 16h. Since no other alternative has been found for the Art Museum, the next step is to re-schedule the City Museum.

There is no previous case in the case base regarding an alternative schedule for the City Museum. The available empty slots in the current schedule are from 17-24h. Hence, the City Museum can be scheduled from 17h to 19h, two hours duration, leaving at closing time. The final schedule is as follows: (Cinema Museum, 10h, 3h, 2€), (Lunch, 13h, 1h, 7€), (Art Museum, 14h, 3h, 3€), (City Museum, 17h, 2h, 1.80€).

5 Conclusions

In this paper, the problem of holiday scheduling for city visitors has been formulated as a CSP and then a methodology to solve it with the integration of Case-Based Reasoning and Constraint Satisfaction techniques has been provided. The approach needs to be tested with experimental work. It would be especially interesting to evaluate the gain in critical situations, that is, to measure the trade-off between an exhaustive exploration of the CBR and the use of a CSP mechanism, in a quite unconstrained domain such as the one studied in this paper.

There are previous works in which the integration of CBR and CSP has been tackled. Purvis (Purvis,1997) has developed COMPOSER, a system that combines both techniques in order to solve configuration problems. Her practical experiments showed that the approach is efficient. Sqalli and Freuder also combine CBR and CSP in the problem of testing protocol interoperability (Squalli and Freuder, 1998). In their system, CBR is used when CSP fails. Scott and Simpson propose using CBR to solve a nurse-rostering problem (Scott and Simpson,1994). Cases are then used when hard constraints are satisfied. The work presented here is in line with the work of Scot and Simpson since the amount of constraints considered is not critical. Avesani et al. propose integrating CBR and CSP to build plans for fire fighting (Avesani et al. 1993). Bilgic and Fox use a case library to store design problems indexed by constraints (Bilgic and Fox, 1996). However, no adaptation process is performed on the solutions.

With regard to holiday planning, Huang and Miles also propose the use of CBR and CSP techniques, to configure packages for users according to a pre-established catalogue (Huang and Miles, 1995). His main goal is to help users making reservations. The approach presented in this paper focuses on scheduling at a different level. The two approaches may complement one another. Other approaches to holiday planning use CBR alone, such as (Blanzieri and Ebranati, 2000), or (Haig et al. 1997).

The latter uses several cases to build a whole route, and then an A* algorithm to combine the results. Soo and Liang (1997) and Ricci and Werthner (2000) also work at the level of travel packages configured from catalogues. The innovation of the work presented in this paper, is that it is designed to provide a service at the basic level of holiday planning, that is, when visiting a city.

In future work, capacity constraints on resources should be considered. Particular museums accept a maximum number of visitors per day, theatres have a limited forum, hotels may be full in festival seasons, etc. A good scheduling system should be able to cope with these problems.

References

Avesani, P., Perini, A. & Ricci, F. (1993). Combining CBR and Constraint Reasoning in Planning Forest Fire Fighting. *Proc. EWCBR, Kaiserslautern,* 235-239.

Bilgic, T. & Fox, M.S. (1996). Constraint-Based Retrieval of Engineering Design Cases: Context as constraints. J. Gero and F. Sudweeks (eds.), *Artificial Intelligence in Design,* Kluwer Academic Publishers, 269-288.

Blanzieri, E. & Ebranati, A. (2000). COOL-TOUR: A Case Based Reasoning System for Tourism Culture Support. *Technical Report #0002-05,* Instituto per la Ricerca Scientifica e Tecnologica, Italy.

González, G., López, B. & de la Rosa, J.L. (2002). The Emotional Factor. An Innovative Approach to User Modelling for Recommender Systems. *RPeC02.*

Haigh, K.Z., Shewchuk, J.R., & Veloso, M.M. (1997). Exploiting Domain Geometry in Analogical Route Planning. *Journal of Experimental and Theoretical Artificial Intelligence,* 9:509-541.

Huang, Y. & Miles, R. (1995). A Case Based Method for Solving Relatively Stable Dynamic Constraing Satisfaction Problems. *ICCBR-95.* LNAI, 1010, Springer.

López, B. (2002). Combining CBR and CSP: A case study on holiday scheduling. Technical Report, University of Girona, Spain.

Purvis, L. (1997). Dynamic Constraing Satisfaction using Case-Based Reasoning Techniques. *Constraint Programming'97, Workshop on Dynamic Constraint Satisfaction.*

Ricci, F. & Werthner, H. (2001). Case-based destination recommendations over and XML data repository. *Proc. Enter 2001.*

Scott, S. & Simpson, S. (1998). Case-Bases Incorporating Scheduling Constraint Dimensions. Experiences in Nurse Rostering. *EWCBR'98,* 392-401

Soo, V.W. & Liang, S.H. (2001). Recommending a trip plan by negotiation with a software travel agent. *Proc. Cooperative Information Agents.*

Sqalli, M.H. & Freuder, E.C. (1998). Diagnosing InterOperability Problems by Enhancing Constraint Satisfaction with Case-Based Reasoning. *Ninth International Workshop on Principles of Diagnosis,* Sea Creast Resort, Massachussetts, 266-273.

Yager, R.R. (1988). On Ordered Weighted Averaging Aggregation Operators in Multi-Criteria Decision Making. *IEEE Transactions on SMC,* vol. 18, 183-190.

Usability Evaluation of Hong Kong Hotel Websites

Tom Au Yeung[a]
Rob Law [b]

[a] Grand Hotel Group Limited, Hong Kong
tomauyeung@grandhotel.com.hk

[b] School of Hotel & Tourism Management,
The Hong Kong Polytechnic University, Hong Kong
hmroblaw@polyu.edu.hk

Abstract

Usability and functionality are two major elements which can significantly affect to the quality of a website. Although a number of prior studies have investigated the functionality of hotel websites, the usability of hotel websites has largely been overlooked in the existing hospitality and tourism literature. To make a contribution in this important area, the present study proposes a model which uses a modified heuristic evaluation technique to evaluate the usability of hotel websites. Research findings are expected to provide a benchmark for hotels to rectify website usability flaws, and to make improvements.

Keywords: website, functionality, usability, heuristic evaluation, hotel.

1   Introduction

The Internet, especially the World Wide Web (WWW or Web), has had a large impact on the all industries, in particular on the hotel industry, in recent years. Many hotels have been, and will be, making their largest efforts to develop their own websites. The resulting websites, therefore, play an important role for mediating between hotels and customers for information acquisition and transactions.

Making a profit from the Internet is a major goal of many businesses. Whether or not a customer makes a purchase on a website depends greatly on the level of trust towards that particular website. A recent study revealed that the quality or usability of the interface design is a crucial determining factor for customers to establish trust of a website (Roy, Dewit, & Aubert, 2001). Loosely and intuitively defined, usability is the ease with which users can use a software product to perform its designated task. Souza, Manning, Goldman, and Tong (2000) reported that 65 percent of visitors to retail websites leave due to usability barriers. Hence, usability problems are a major issue facing Web customers, which ultimately affect the success of a commercial website.

According to the framework proposed by Lu and Yeung (1998), usability and functionality are the two major elements contributing to the usefulness of a website. Hotel practitioners should, therefore, attempt to continuously improve the usefulness of their websites. This, in turn, lead the websites generate more business transactions and revenues. While various studies have been conducted to examine the functionality of hotel websites (Chung & Law, 2002; Law, Leung, & Au, 2002; O'Connor & Horan, 1999; Procaccino & Miller, 1999; Murphy, Forrest, Wotring, & Brymer, 1996), the usability of hotel websites has been largely overlooked in the existing hospitality and tourism literature. To bridge the gap of the dearth of published articles in this particular area, this study presents a model which evaluates the usability of hotel websites, with an application to hotel websites in Hong Kong.

2 The Theoretical Background

A hotel website is not only a place for displacing information about its products and services. Law and Leung (2000) argued that websites should have a commercial value in terms of profitability. In order to achieve the business purpose of building a website, every important design consideration should be viewed in a proper perspective. A comprehensive framework for developing effective commercial Web applications is proposed by Lu and Yeung (1998) as an extension of prior researches. This framework regards website as a special type of software system with specific target user populations. The acceptability of this system is viewed as a combination of social acceptability, system feasibility and usefulness. The framework, as shown in Figure 1, states that the usefulness of a website design includes two major components, namely: functionality and usability.

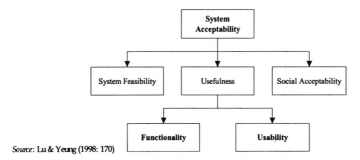

Source: Lu & Yeung (1998: 170)

Fig. 1: A framework for effective Web application development

Performing well in functionality does not necessarily mean that the website is useful to users. To a large extent, a website's usefulness also depends on users' views of the site's usability. The ultimate goal of usability is to make a website more efficient and enjoyable to use. According the definition given by Preece (1993:10), "The goals of Human Computer Interaction (HCI) are to develop and improve systems that include computers so that users can carry out their tasks safely, effectively, efficiently, and enjoyably. These aspects are collectively known as usability".

Several approaches have been proposed to evaluate Web usability (Cunliffe, 2000; Faulkner, 2000; Lin, Choong, & Salvendy, 1997; Nielsen, 1994). These approaches vary from complex to simple. The approach that is mostly widely adopted is the generic heuristic evaluation originated by Nielsen (1994).

A heuristic evaluation involves the study of a user interface by a small group of evaluators who look for violations of common usability criteria (heuristic). A comprehensive evaluation checklist for identifying usability hazards was developed by Abeleto (2002) based on the criteria proposed by Nielson (1994, 2000) and Rosenfeld and Morville (1998). In this checklist, the usability criteria were divided into five dimensions, each representing an area where usability hazards are common. These dimensions include: 1. Language: This dimension refers to the choice of words used to present information on the Web page. 2. Layout & graphics: This dimension pertains to how elements are visually rendered on the Web page. 3. Information architecture: This dimension pertains to how a website's content and features can be arranged. 4. User interface: The user interface of a site determines the ease of navigating through its content. 5. General: This dimension refers to the general practice of design and maintenance.

3 Methodology

This study investigates the usability of hotel websites in Hong Kong using a modified heuristic evaluation approach, which was introduced by Nielsen (1994, 2000). This study is exploratory in nature and the research is based on gathering primary data and secondary data via the research instrument after comprehensive literature review. The research process is shown in Figure 2, which is basically divided into the following four stages:

1. identifying the chosen usability criteria,
2. gathering opinions about the usability of hotel websites in Hong Kong,
3. analyzing and comparing the usability of hotel websites in Hong Kong,
4. providing recommendations for improvement.

264

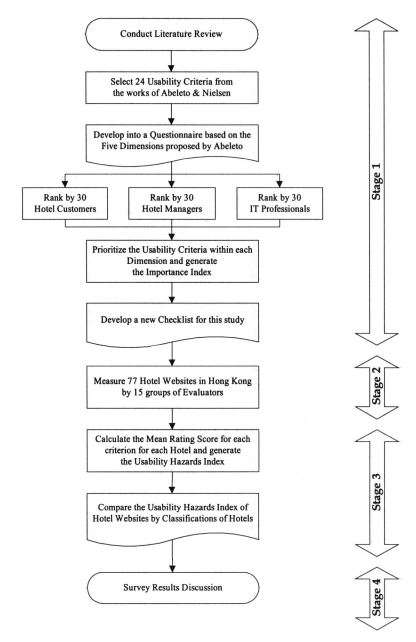

Fig 2: Research process

3.1 Identifying the Chosen Usability Criteria

To establish the attributes for evaluating hotel websites, twenty-four usability criteria were selected from the studies of Abeleto (2002) and Nielsen (1996, 1999). It should be noted that, while most of the criteria in Abeleto and Nielsen's studies were applicable to evaluating various websites, only those criteria suitable for the hotel industry were selected in this study. For instance, while recognizing that the criterion "insulting, derisory or offensive language" is a major hazard to a website, it was abandoned since this is unlikely to happen in a hotel website.

The selected criteria were then developed into a questionnaire based on the five dimensions proposed by Abeleto (2002). This questionnaire was then completed by ninety respondents from three different groups, namely hotel customers, hotel managers, and information technology professionals. To meet the normal distribution requirement, thirty respondents participated in each group, and a convenience sampling method was adopted. Qualified respondents should have previously visited hotel website(s).

In the questionnaire, respondents were asked to rank the usability hazards according to their perceived damage to a hotel website. These criteria were ranked from the most offensive practice to the least offensive practice within each dimension. Respondents were also asked to weigh the relative importance of responses from each group so that an Importance Index can be calculated for each usability hazard within the dimension using the following derived mathematical formula. A new checklist for this study was established based on this final rating.

x = Number of respondents in the hotel customers group

y = Number of respondents in the hotel managers group

z = Number of respondents in the IT professionals group

m = Total number of respondents; where $m = x + y + z$

n = Number of criteria within a dimension

$\overline{C_r}$ = Mean score for the rth criterion given by hotel customers; $r = 1, 2, \ldots n$

$\overline{H_r}$ = Mean score for the rth criterion given by hotel managers; $r = 1, 2, \ldots n$

$\overline{T_r}$ = Mean score for the rth criterion given by IT professionals; $r = 1, 2, \ldots n$

P_i = Relative importance of responses from the hotel customers group given by the ith respondent; $i = 1, 2, \ldots m$; where $P_i + Q_i + R_i = 100\%$

Q_i = Relative importance of responses from the hotel managers group given by the ith respondent; $i = 1, 2, \ldots m$; where $P_i + Q_i + R_i = 100\%$

R_i = Relative importance of responses from the IT professionals group given by the ith respondent; $i = 1, 2, \ldots m$; where $P_i + Q_i + R_i = 100\%$

I_r = Importance Index of criterion r; $r = 1, 2, \ldots n$

$$I_r = \left(\dfrac{\overline{C_r}}{\sum\limits_{r=1}^{n} \overline{C_r}} \dfrac{\sum\limits_{i=1}^{m} P_i}{m} \right) + \left(\dfrac{\overline{H_r}}{\sum\limits_{r=1}^{n} \overline{H_r}} \dfrac{\sum\limits_{i=1}^{m} Q_i}{m} \right) + \left(\dfrac{\overline{T_r}}{\sum\limits_{r=1}^{n} \overline{T_r}} \dfrac{\sum\limits_{i=1}^{m} R_i}{m} \right) \quad (1)$$

3.2 Gathering Opinions about the Usability of Hotel Websites in Hong Kong

In this research, all members of the Hong Kong Hotels Association (HKHA) with developed hotel websites were selected for analysis. A total of seventy-seven websites were finally included. A panel of thirty evaluators was formed which composed of hospitality practitioners to gather opinions about the usability of hotel websites. These evaluators were divided into fifteen groups of two. Each group evaluated five to six randomly selected hotel websites. The evaluators rated the hotel websites using a five-point Likert scale, with 1 being "no problem on this criterion", 2 being "minor problem", 3 being "medium problem," 4 being "major problem", and 5 being "destructive problem identified". Prior to the evaluation, evaluators were briefed about the criteria with examples. The evaluation was then performed with the researchers' presence to clarify queries and to answer questions. To ensure unbiased reliability is maintained, each of the websites was evaluated in a double-blind fashion by two evaluators in the group. If the two ratings for each website differed by one point, the average of the two ratings was taken as the final rating. If the two ratings differed by more than one point, the two evaluators discussed the reasons of their choices, and then to make a collective decision on their ratings.

3.3 Analyzing and Comparing the Usability among Hotel Websites in Hong Kong

For each hotel and in each criterion, a mean rating score was calculated based on the result of group evaluation. This score was then transformed to a Usability Hazards Index derived from the following formula:

n = Number of criteria within a dimension i, i = 1, 2, ...5

$\overline{S_r}$ = The mean rating score for the rth criterion; r = 1, 2, ...n

I_r = Importance Index of criterion r; r = 1, 2, ...n

U_r = Usability Hazards Index of criterion r; r = 1, 2, ...n

D_i = Usability Hazards Index of dimension i, i = 1, 2, ...5

H = Usability Hazards Index of a Hotel Website

$$U_r = (\overline{S_r} - 1) * I_r * 5 \quad (2)$$

$$D_i = \sum_{r=1}^{n} U_r \quad (3)$$

$$H = \sum_{i=1}^{5} D_i \quad (4)$$

It should be noted that the Usability Hazard Index of each dimension was running from the minimum "0" representing "no problem found on this dimension" to the maximum "20" representing "destructive problems found on this dimension". When adding up the scores from all five dimensions, the Usability Hazards Index of a Hotel Website was generated with minimum score "0" to maximum score "100".

4 Results

4.1 Importance Index of the Usability Criteria

From ninety successfully completed questionnaires, the Importance Indexes of all Usability Criteria were calculated, and the findings are presented in Table 1.

Table 1. Importance Index of the Usability Criteria

Dimension 1: Language	Importance Index
1. Headlines that make no sense out of context	0.402
2. Spelling and grammatical errors	0.300
3. Internet jargon/popular buzzwords	0.298
Dimension 2: Layout & graphics	
1. Scrolling text, marquees, and constantly running animations	0.153
2. Illegible text colours	0.135
3. Large images being used solely for visual appeal	0.135
4. Functional design that looks like advertising	0.124
5. Horizontal scrolling	0.118
6. Graphical images used to deliver a critical message	0.118
7. Improper use of graphical bullets and graphical divider bars	0.113
8. Pages longer than 4 screens in length	0.103
Dimension 3: Information architecture	
1. Outdated information	0.389
2. Moving addresses	0.341
3. Poor labelling of navigation systems	0.271
Dimension 4: User interface	
1. Internal link that does not work	0.214
2. Breaking or slowing down the 'Back' button	0.168
3. Lack of navigation support	0.152
4. Opening new browser windows	0.140
5. Non-standard use of Graphic User Interface (GUI) widgets	0.116
6. Scrolling front pages	0.111
7. Non-standard link colours	0.099
Dimension 5: General	
1. Long download times over 10 seconds	0.431
2. Slow server response times over 1 second	0.335
3. Unnecessary use of latest technology	0.233

As shown in Table 1, 'Headlines that make no sense out of context', 'Scrolling text, marquees, and constantly running animations', 'Outdated information', 'Internal link that does not work' and 'Long download times over 10 seconds' are perceived to be the most important criterion within their respective dimensions.

4.2 Usability Hazards Index of Hotel Websites in Hong Kong

Based on the result of the group evaluation, the Usability Hazards Indexes of all hotel websites were calculated and a comparison of the indexes among the three hotel categories in Hong Kong was conducted, and results are presented in Table 2.

Table 2. Comparison of the Usability Hazards Index by Classification of Hotels

Dimension	Usability Hazards Index			
	High Tariff A Hotel (n=17)	High Tariff B Hotel (n=30)	Medium Tariff Hotel (n=30)	All Hotels (n=77)
1. Language	1.691	2.217	2.568	**2.238**
2. Layout & graphics	3.744	4.792	4.181	**4.323**
3. Information architecture	1.969	2.803	3.853	**3.028**
4. User interface	4.286	4.661	5.918	**5.068**
5. General	6.207	4.989	4.104	**4.913**
Overall Score	17.897	19.462	20.623	**19.569**

As shown in Table 2, the industry average of Usability Hazards Index of hotel websites in Hong Kong was 19.569 out of 100. Generally speaking, minor problems were found in Hong Kong hotels' websites. Among the five dimensions, 'User Interface' creates most of the problems and is followed by 'General', 'Layout & graphics' and 'Information architecture'. 'Language' creates least problems among all.

5   Conclusion

In this study, a technique was developed to evaluate the usability of hotel websites. It also provides a benchmark for hotels to rectify website usability flaws. The results show that the average usability performance of all hotels in Hong Kong is 19.6 point out of 100 (the lower the better). The result also shows that hotel managers in Hong Kong should pay more attention to the design of user interface, especially they have to ensure that all internal links and 'Back' button are functioning properly. Equally important, the download speed and the server response speed should be fast so that their hotel website can become more efficient and enjoyable to use.

References

Abeleto (2002). *Objective evaluation of likely usability hazards - preliminaries for user testing* [Online]. Available: http://www.abeleto.com/resources/articles/objective.html (April 14, 2002)

Chung, T. & Law, R. (2002). Success Factors for Hong Kong Hotel Websites. *Proceedings the Fifth Biennial Conference on Tourism in Asia: Development, Marketing and Sustainability,* May 2002, Hong Kong, pp. 96-104.

Cunliffe, D. (2000). Developing usable Web sites – a review and model. *Internet Research: Electronic Networking Applications and Policy,* Vol. 10, No. 4, pp. 295-307.

Faulkner, X. (2000). Usability evaluation. *Usability Engineering,* Macmillan Press, pp.137-175.

Law, R. & Leung, R. (2000). A Study of Airlines' Online Reservation Services on the Internet. *Journal of Travel Research,* Vol. 39, pp. 202-211.

Law, R., Leung, K. & Au, N. (2002). Evaluating Reservation Facilities for Hotels: A Study of Asian Based and North American Based Travel Web Sites. *Proceedings of Information and Communication Technologies in Tourism 2002 Conference,* Innsbruck, pp. 303-310.

Lin, X.L., Choong, Y.Y. & Salvendy, G. (1997). A proposed index of Usability: a method for comparing the relative usability of different software systems. *Behaviour & Information Technology,* Vol. 16, No. 4/5, pp. 267-278.

Lu, M.T. & Yeung, W.L. (1998), "A Framework for Effective Commercial Web Application Development", Internet Research: Electronic Networking Applications and Policy, Vol. 8, No. 2, pp. 166-173.

Murphy, J., Forrest, E.J., Wotring, C.E. & Brymer, R.A. (1996). Hotel Management and Marketing on the Internet. *The Cornell Hotel and Restaurant Administration Quarterly,* 37(5), pp. 70-82.

Nielsen, J. (1994). Usability Heuristics. *Usability Engineering,* Academic Press Professional, Cambridge, pp. 115-164.

Nielsen, J. (1996). Top ten mistakes in Web design. *Jakob Nielsen's Alertbox* [Online]. Available: http://www.useit.com/alertbox/9605.html (April 14, 2002)

Nielsen, J. (1999). Top ten new mistakes of Web design. *Jakob Nielsen's Alertbox* [Online]. Available: http://www.useit.com/alertbox/990530.html (April 14, 2002)

Nielsen, J. (2000). Introduction: Why Web Usability. *Designing Web Usability: The Practice of Simplicity,* New Riders Publishing, pp.8-15.

O'Connor, P. & Horan, P. (1999). An Analysis of Web Reservation Facilities in The Top 50 International Hotel Chains. *International Journal of Hospitality Information Technology,* Vol. 1, No. 1, pp. 77-85.

Preece, J. (1993). Introduction to HCI. *A Guide to Usability: Human Factors in Computing.* Addison-Wesley, pp.10-20.

Procaccino, J.D. & Miller, F.R. (1999). Tourism on The World Wide Web: A Comparison of Web Sites of United States – and French-based Business. *Information Technology & Tourism,* Vol. 2, pp. 173-183.

Rosenfeld, L. & Morville, P. (1998). What Makes a Web Site Work. *Information architecture for the World Wide Web,* California, O'Reilly & Associates, pp. 1-9.

Roy, M.C., Dewit, O. & Aubert, B.A. (2001). The Impact of Interface Usability on Trust in Web Retailers. *Internet Research: Electronic Networking Applications and Policy,* Vol. 11, No. 5, pp. 388-398.

Souza, R., Manning, H., Goldman, H. & Tong, J. (2000). The Best of Retail Site Design. *Forrester Research: Techstrategy* [Online]. Available: http://www.forrester.com/ER/Research/Report/Summary/0,1338,10003,FF.html (April 14, 2002)

Hotel (and) School:
Integrating Hospitality Information Technology into a Live Learning Environment

Robert Govers
Floor Bleeker

Emirates Academy of Hospitality Management, PO Box 11416, Dubai, United Arab Emirates,
{robert.govers; floor.bleeker} @emiratesacademy.edu

Abstract

Information technology in hospitality and tourism education has gained popularity as a research topic since the first few conceptual papers appeared at the end of the last century, particularly focussing on curriculum issues. This paper diverges slightly from the existing research stream and examines how industry specific hospitality information technology can be integrated into a hotel school's IT-infrastructure, creating a live learning environment where systems are not just used in "standalone" training sessions, but also in operations where students can get "real" hands-on experience with the systems themselves as well as how these systems interface. The authors developed a prototype based on the IT-infrastructure that they developed at their own institution, and tested this externally against the practice and opinion of their peers (other schools) and internally with students and colleagues, based on the experience that they had with the system infrastructure during its first year of operation. The results show that the developed prototype is quite advanced compared to other hotel schools, but that challenges still lie in the traditional problem of how to integrate the technology into the curriculum.

Keywords: Hospitality Information Technology, Hotel Schools, IT-curriculum, Prototyping.

1   Introduction

In his synoptic work to formulate a research agenda in the area of information technology in tourism, Frew (106) concluded in the year 2000, that research into the application of hospitality and tourism information technology in a training and education environment had largely been neglected until then. Since that time however, quite a few papers have been published in this area, and in its inaugural year 2001, the Journal of Teaching in Travel & Tourism devoted a whole issue (No. 2/3) to the topic of the Internet and Travel and Tourism Education. If it was Frew's work that sparked this renewed interest in the topic of IT in tourism and hospitality education, will be difficult to say, but nevertheless a change in emphasis has been observed by the authors.

Before the turn of the century few authors touched on the topic of information technology in hospitality and tourism education and then primarily through conceptual papers discussing the issue of how to integrate this subject into the curriculum (Buhalis 1998; Cooper et al. 1996; Daniele and Mistilis 1999; O'Connor and Buhalis 1999). However, empirical research on how to provide students with the skills and understanding to actually operate industry specific systems and what systems to focus on, had largely been ignored until then. That changed when several publications appeared, particularly in the Journal of Hospitality & Tourism Education (Breiter and Hoart 2000; Mandabach et al. 2002; Mandabach et al. 2001). Mandabach et al. (2001: 52) first measured the software skills among 358 hospitality students at different levels in nine 4-year university degree programmes. Also they identified to what industry specific systems (besides general software applications such as Office) these skills related to. These included Point of Sales Systems, Hotel Front Desk Systems, Recipe Management Systems, Inventory Control and Purchasing Systems, and Report Analysis and Report Generation. In a subsequent study Mandabach et al. (2002: 10) identified the rapid increase in the use of computer technology at degree granting culinary programmes between 1996 and 2001. What is common among all the referenced papers though, is the shared consensus that IT has now probably become one of the most important areas determining competitiveness in the hospitality and tourism industry and therefore higher education institutes serving these industries should focus not only on fully integrating IT into their curriculum, but probably also their operation and administration. Later studies, as mentioned, have provided some insight into the extent to which this has happened and on what systems educational institutions have focused.

2 Background

The information provided in the Introduction was used to develop the IT infrastructure at the hospitality management academy where the authors are employed (hereafter the Academy) as a prototype for an IT-integrated hotel school learning environment. Although integration of technology into instruction and curriculum (incl. online delivery) is the number one IT-related issue faced by higher education institutions in the US in general (Educause 2002: 20), this will not be the main focus of this paper. Alternatively, this paper will primarily look at what systems to incorporate in such a prototype and how they are integrated, in terms of interfacing, and also in terms of integration into the actual operation and administration of a hotel school, besides being used as training systems. All this in order to provide students with hand-on experience on the actual technologies used in industry, which are operational in the students' "real-world" living environment as well as in a non-threatening learning environment such as training labs. In order to test the prototype and its compatibility with other hotel schools, a global online survey was conducted among higher education institutions offering hotel/hospitality management programmes.

2.1 Brief History

The Academy is a new hospitality, travel and tourism management school, located in the Middle East and established in academic association with Ecole hôtelière de Lausanne. The academy is owned by the Jumeirah International, a locally based international hospitality management group, that focuses on, amongst other things, the complete application of the latest technologies in all guest amenities, and operational and management processes. With the backing of such a group with its obvious good relationships with hospitality industry technology suppliers, the Academy has been able to build a learning environment where the latest industry technologies have been fully integrated within the school and its operational and administrative processes and systems. The Academy opened its doors for the first intake of students on an International Hospitality Management degree in September 2001. 40% of the curriculum of these students in the first two semesters consisted of operational training in the areas of food & beverage (F&B) and front office, with another 20 % of training occurring on placements in the Jumeirah International properties. Therefore students have had the opportunity to get hands-on experience with the latest industry technologies in a learning environment as well as in a real life setting.

2.2 Integrated Systems

Figure 1 provides a schematic representation of the systems used and how they are integrated at the Academy. There are three main systems on campus; the Student Information System (SIS), the (Hospitality) Property Management System (PMS) and the Point of Sales System (POS). To meet the need of today's mobile teachers and learners all systems have been made available anytime from anywhere, over the Internet as well as over the campus LAN connection. The Student Information System handles admission, registration, scheduling, grading, student accounting and other administrative tasks related to students. Students have access to this system through the Intranet and the system is fully interfaced with; the Library System for student records and late return fees, the PMS for on-campus accommodation booking and charging (incl. telephone, meals, printing, etc.), the Internet Payment Gateway for online payments, and the online educational course management system for record transfer, payment of fees and transfer of grades. The Property Management System is used to manage the school and its campus, but is also accessible from the computer lab and on-campus accommodation office for front office training. Room bookings and charges are also managed through this system. All students are provided with a virtual wallet that they can use to pay for facilities all over campus. This hybrid ID card with a magnetic strip, transponder and chip serves as a student ID card, library card, door key and charge card for restaurants, vending machines and pay phones. The print/copy and call/fax accounting systems will charge expenditures to the PMS system. The student is identified by its active directory username for printing, copying and faxing and by its telephone number for call accounting. Every student has a dummy folio in the PMS system where charges are being collected. The POS system is the central system for the F&B operations. The POS system interfaces with a restaurant management system for restaurant reservations, CRM information and yield management; with an F&B control and purchasing system for automatic stock deduction; and with the electronic purse system and PMS for charging to the student's folio. The recipe management system is interfaced with the purchasing system to get the latest market prices in recipe and menu analysis. All applications used in the F&B operations are being used in a teaching situation as well as in the live operations and therefore run a separate database for training purposes. All interfaces are online real time except for the recipe management interface. A batch runs every night to update the prices in the system.

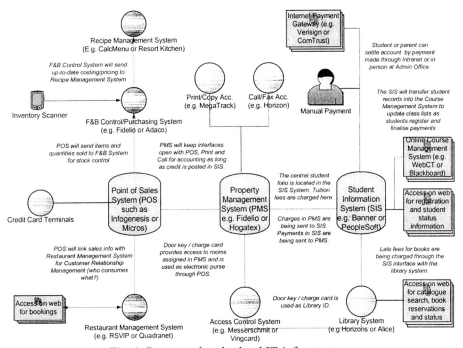

Fig. 1. Prototype hotel school IT-infrastructure

3 Purpose

With this situational outline in mind, the authors wanted to perform an internal and external testing of the prototype. Internally, the performance of the IT infrastructure of the Academy and its application within the learning environment was assessed in two ways; a review of operational and management issues with the IT-department and faculty, and, an analysis of student evaluations of the IT-integrated learning environment, compared to the placement situation in the "real world". Externally, the authors wanted to test the suitability of the prototype in two ways; comparing the prototype to the IT infrastructure of other hotel schools in order test its compatibility, and, to solicit colleagues' opinions in qualitative form, in order to acquire expert guidance on how to improve the prototype.

4 Methodology

Three types of research approaches were used to uncover the answers to the research questions listed in the "Purpose" section above. For the internal analysis the following two studies were conducted:

- In-depth interviews were held with the IT-department and Faculty in order to identify operational and management issues related to the implementation and use of the described technologies in the learning environment;
- A focus group with students took place in order to analyse student evaluations of the IT-integrated learning environment, compared to the placement situation in the Jumeirah International properties.

The external analysis was performed through an online Survey as discussed in the next section.

4.1 Survey

An online survey was conducted among hospitality/hotel management schools/ departments. The survey was divided into three "click-through" sections:

- Section 1 asked respondents to identify the various systems available on campus and what they are used for, and how, if at all, they are interfaced with each other;
- Section 2 presented the prototype as displayed in Figure 1 including a brief description, and solicited open responses in terms of similarities and differences between the prototype and the respondent's own systems, strengths and weaknesses of the prototype as perceived by the respondent, and additional comments;
- Section 3 finally collected some statistical details regarding the respondents' institution, such as contact details, location and size.

The survey was piloted internally with the help of six faculty members, and externally by two IT-managers of the hotel management group that the Academy belongs to.

4.2 Sampling and Profile

Electronic mail was used to send letters of invitation to IT-managers of hospitality management / hotel schools and departments around the world, asking them to participate in the survey, with an explanation of the research objective and a link to the online survey, which was located on the Academy's Internet server. Various channels were used to target appropriate educational institutions:

- The first approach was to try to access hotel school websites through links found on Yahoo.com, eHotelier.com and Hospitality-1st.com. On the hotel school websites the contact-us or e-mail links were used to send the invitation to the general info@ or admission@ e-mail addresses;
- A second approach was to use the personal e-mail addresses of the contact persons listed on membership pages of for instance the Euhofa International Association of Hotel Schools' website;
- The third and last approach was the use of CHRIE and EuroCHRIE ListServ mailing lists.

In all these cases the e-mail was headed with a request to forward the e-mail to the IT-manager of the hotel school. The latter obviously did not work as intended, because of the ten responses received, not one was completed by an IT-manager. In six cases the respondent was someone in a management position within the hotel school / hospitality management department. Three questionnaires were completed by professors/instructors (one of which specialised in IT though) and one response was anonymous (but including the institution's details). Responses that failed to complete the last section of the survey and therefore did not forward any institutional details, were ignored. The whole approach produced disappointing results in terms of response rate, considering that only ten usable responses were received. It is difficult to estimate the actual response rate, as some mailing lists do not indicate to how many members of the list the message is sent, but since at least a hundred individual e-mail addresses were contacted, the response rate is definitely far below 10%. It is widely known that in general, response rates to online surveys are rather low and therefore it is not surprising that data collection in this case was troublesome, taking into account the constraints and the sampling method used. Nevertheless, although only ten responses were received, a rather good spread in terms of demographics was achieved. The breakdown in terms of the location of the schools is U.S. (3), Australia (2), Switzerland (2), Canada (1), Ireland (1) and the Netherlands (1). Six schools indicated to be public, and also in terms of size there was a wide range of responses ranging from small to very large:

- 3 small schools with 200-300 students, 20-28 faculty and 10-14 admin staff
- 3 medium sized schools with 1000-5000 students and total staffing of 120-450
- 2 large schools with 12,000-17,000 students and around 1000 staff members
- 1 very large university of 45,000 students, 3000 faculty and 5000 admin staff

As a result of this, IT budgets obviously also varied largely, ranging from less than 250,000 US$ to over 1 million US$, both in capital expenses as well as the operational budget.

5 Results

Figure 2 lists the education management and administrative systems as included in Figure 1 and the number of responding institutions that have adopted such systems. The most widely used type of education management systems are course management systems such as WebCT (in 4 cases) and Blackboard (in 3 cases). It is also encouraging to see that all eight institutions that have indicated to be using some sort of course management system, are not just using it purely to offer e-learning courses, but also to support traditional classroom based courses as well as to replace some of the traditional methods of delivery, combining online and offline learning. Student ID-card systems are generally just used for student id, particularly for the library and in some cases also for copying (3 cases).

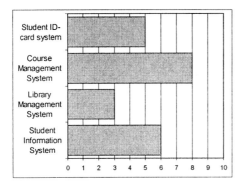

Fig. 2. Adoption of education systems (above) and administrative systems (right, n=10)

In terms of administrative systems, Call Accounting Systems and Intranets are the most adopted applications (but still only four out of ten). Some of these systems are also commonly used by the hospitality industry and could therefore enhance student training in order to familiarise students with the wide variety of technological solutions used in hotels. Instead of only focusing on the major industry systems as listed in Figure 3, it might be useful if students come in contact with auxiliary systems such as Fax Accounting Systems, Access Control Systems, Online Payment Systems and Call Accounting Systems. Only in the latter case, two respondents indicated that they not only use the Call Accounting System to manage administrative and students telephone extensions, but also to teach students the skills of how to operate it. It is encouraging to see though that already three out of the ten institutions offer online payment facilities. On the other hand it is unfortunate that automatic access control systems, that are now very common in industry, are not used by hotel schools at all. Figure 3 displays the number of responding institutions that indicated to have any of the five major industry specific systems on campus. Least common are Recipe and Restaurant Management Systems.

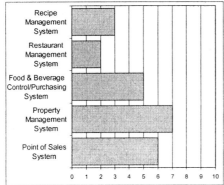

Fig. 3. Adoption of industry specific systems

Table 1 shows the various applications used and how they are utilised. In almost all cases, when systems are available on campus, they are not just used in a training environment, but also in the live operations, giving students the opportunity to get hands-on experience with these systems in situations more closely resembling the reality of an industry setting.

Table 1. Industry specific systems' usage

	Recipe Mngt. Sys. (3)	Restaurant Mngt. Sys. (2)	F&B Control Purch. Sys.(5)	Property Mngt. Sys. (7)	Point of Sales Sys. (6)
Systems used	CalcMenu (1) On Cooking (1) Unknown (1)	Respak (1) Unknown (1)	Fidelio (1) De Haan (1) CBorf (1) Unknown (2)	Fidelio (3) LANmark (2) De Haan (1) Unknown (1)	PASHA (1) Squirrel (1) Silverware (1) De Haan (1) Unknown (2)
Used for:					
Teaching in training lab	3	2	3	6	5
Running live operations	3	1	4	5	6
Teaching in live operations	3	1	4	5	6

Table 2. System interfaces reported by responding institutions

Between system A	and system B	Batch	On-line
Student Information System e.g. to exchange charges and payments	Property Management System		2
Student Information System e.g. for accounting of students' F&B consumptions	Point of Sales System	1	3
Student Information System e.g. for students and parents to track academic performance and financial info.	Intranet / Internet		3
Property Management System e.g. to make room bookings and track financial information online	Intranet / Internet	1	2
Library System e.g. for use of smart student ID-card as Library card	Smart student ID-card system		2
Library System e.g. for online payment of book late return fees	Internet Payment system		4
Point of Sales System e.g. to share items and quantities sold for stock control	Food & Beverage Control System		2
Point of Sales System e.g. for automatic charging of F&B consumptions to student account	Smart student ID-card system		2

In terms of the interfacing of all these systems, Table 2 lists those interfaces that were reported by at least two or more responding institutions. The most common interfaces reported are between the Point of Sales System and the Students Information System, for instance to account for students' food and beverage consumptions, and between the Library System and the Internet. Based on the above information, the IT infrastructure of the best equipped responding institution can be depicted as shown in Figure 4. Although there is quite some resemblance between the prototype and what has been reported through the survey, the prototype has several additional systems incorporated and particularly the interfacing is more complex. Respondents raised issues in terms of user friendliness, cost, and what would happen if one of the systems breaks down in terms of the effect on the total infrastructure. However, respondents did identify that if such a prototype was fully up and running it would cover most processes of a hotel schools' operation and increase efficiency, particularly in automatically keeping all data up-to-date. What was surprising is that none of the respondents commented on any issues regarding integration of the systems in the learning environment, particularly in terms of the training of skills needed to operate the systems and problems associated with integrating those into the curriculum. This was the primary issue identified in the internal analysis.

278

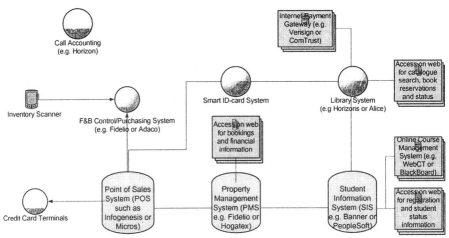

Fig. 4. Best reported hotel school IT-infrastructure

Although the prototype is almost fully operational, the students had little knowledge of the unique resources that are at their disposal. Therefore also during the two one-month internships, students gained limited experience on the systems, as supervisors felt uncomfortable with the insufficient skills of students. Even faculty recognised this issue and attribute it to the constraints of curriculum content, credit hour load and available time for student contact. Instructors still have to cover the normal content of classes, and teaching students the related IT skills on the appropriate systems comes on top. The biggest problem therefore is to find a balance.

6   Conclusions

Although the prototype as displayed in Figure 1 seems to be very advanced compared to other hospitality management / hotel schools / departments, some institutions come a long way, particularly when concentrating on the major educational and industry specific systems. Especially the fact that the latter systems are not just used in pure training environments, but also in live operations, giving the students the opportunity to learn in a live environment which closely represents industry reality, is encouraging. What is often missing is the interfacing, prohibiting full use of the potential of the systems, in terms of covering all operational processes and resembling industry reality as well as keeping records up-to-date. The main conclusion however has to be that regardless of the level of innovation and efforts made to completely incorporate IT, it is the USE of the technology that determines its success. In this case, to the original question raised in most of the literature, in terms of how to integrate hospitality IT in the curriculum, would have to be revisited. A model already published in 1998 by O'Connor and Buhalis (probably adapted from (Buhalis 1998)) might provide the key to the solution. The model is displayed in figure 5.

279

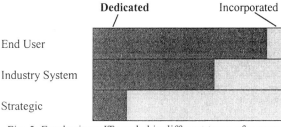

Dedicated   Incorporated

End User

Industry System

Strategic

Fig. 5: Emphasis on IT needed in different types of courses
(O'Connor and Buhalis 1999)

End user courses are IT-dedicated courses providing students with general IT-skills such as on Microsoft Windows, Office, Internet and e-mail. The Academy provides students with two such courses, and in general they have a well established curriculum that does not create many challenges, particularly as it involves low level skills and such courses are offered early in the curriculum. Strategic courses are those courses that address IT from a strategic perspective, examining how it can and should be used to gain and sustain competitive advantage. Such courses are generally not IT-skill focused but emphasise strategy and incorporate IT related issues from a conceptual perspective into the content of the course. It is the training of skills on industry specific systems that forms the main challenge. Dedicated hospitality IT courses are not a solution, because it is during the operations that students get the best learning experience. The Academy has tried to fully incorporate industry specific systems into the teaching of relevant hospitality operations courses. This proves difficult as the curriculum of such courses is already loaded with content. Particularly as the emphasis on IT increases when more systems are incorporated in the total infrastructure and sophistication is expanded dramatically through complex interfacing, the burden of an infrastructure such as presented in the prototype in this paper, might become too heavy if not very carefully balanced within the curriculum – a major research challenge.

References

Breiter, D. and H. Hoart (2000). Competencies in Foodservice Technology Expected by Foodservice Industry of Graduates of Bachelor Degree Hospitality Programs in the United States. *Journal of Hospitality & Tourism Education*, 12 (2).

Buhalis, D. (1998). Information Technology in Tourism: Implications for the tourism curriculum. in: D. Buhalis, A.M. Tjoa, and J. Jafari (eds.) *Proceedings of the International Conference on Information and Communication Technologies in Tourism.* (ENTER 1998), Vienna, Austria: January, Wien-New York: Springer Verlag, 289-297.

Daniele, R. and N. Mistilis (1999). Information Technology and Tourism Education in Australia: An industry view of skills and qualities required in graduates. in: D. Buhalis and W. Schertler (eds.) *Proceedings of the International Conference on Information and Communication Technologies in Tourism.* (ENTER 1999), Vienna, Austria: January, Wien-New York: Springer Verlag, 140-150.

Educause (2002). *The Pocket Guide: to U.S. Higher Education.* Washington: Educause.

Frew, A.J. (2000). Information Technology and Tourism: A research agenda. *Information Technology & Tourism*, 3 (2), 99-110.

Mandabach, K.H., R. Harrington, D. VanLeeuwen, and D. Revelas (2002). Culinary Education and Computer Technology: A Longitudinal Study. *Journal of Hospitality & Tourism Education*, 14 (2), 9-15.

Mandabach, K.H., D. VanLeeuwen, and P. Bloomquist (2001). Hospitality Technology Education: Students Successes in Mastering the Hidden Curriculum. *Journal of Hospitality & Tourism Education*, 13 (1), 49-56.

O'Connor, P. and D. Buhalis (1999). IT in the Curriculum. in: A.J. Frew (ed.) *Hospitality Information Technology.* (HITA Conference), Edinburgh, HITA, 1-8.

Technology Enabled Distribution of Hotels: An Investigation of the Hotel Sector in Athens, Greece.

Konstantina Kaldis,
Raymond Boccorh,
Dimitrios Buhalis

Centre for eTourism Research, School of Management, University of Surrey, UK
kkaldis@hotmail.com, {r.boccorh; d.buhalis} @surrey.ac.uk

Abstract

Recent major advances in information and communications technology have enabled provision of multiple hotel distribution channels to the customer. The advent of the Internet has revolutionised the traditional distribution model and has consequently led to a reconfiguration of distribution channels, enabling simultaneous competition and co-operation between principals and intermediaries. The objective of the study is to examine existing practices in the area of hotel distribution, highlight important trends and explore their potential evolution in Athens. The analysis illustrates the electronic means for the distribution of Athens hotels.

Keywords: Hotel; Reservations; Distribution channels; The Internet; Athens.

1   ICT enabled hospitality distribution

The hotel industry is characterised as information-intensive, rather than a technology-intensive industry. This is partly attributed to the perishable nature of the hotel product. As the functionality of the entire industry depends on accurate and reliable information, the mechanism used for dissemination is considered central to its overall success. Use of Information Communication Technologies (ICTs) in hotels improves quality and productivity, and consequently confers competitive advantage. A number of studies (Cho and Connolly, 1996; Namasivayam, et al., 2000) have highlighted the benefits attainable through adoption of such technologies. Innovations in ICTs will thus be expected to increase customer satisfaction and long-term profitability (Dubé et al., 2000). ICTs can also transform the strategic position of organisations by altering efficiency, differentiation, operational cost and response time (Buhalis, 1998). Van Hoof et al. (1995) specified improved service quality, enhanced profitability and efficiency, integration of departments, speedier communications and reduced costs as beneficial outcomes of increased use of technology in the hotel industry. Distribution and procurement have also been cited as suitable areas for innovations in ICTs (Dubé et al., 2000; Buhalis, 1998; Leong, 2001).

The distribution channel system is a critical link in the marketing mix between demand and supply, and consumer and producer. In the hotel sector, electronic distribution systems are those, which use electronic media to provide relevant information to the customer to facilitate a purchase decision. Such systems enable transactions to be completed by facilitating ordering and product purchasing of the product (O'Connor, 2000 and 2001). Several studies (Buhalis and Laws, 2001; Richer and O'Neill-Dunne, 1999; Waller, 1999; Go, 1995; PricewaterhouseCoopers, 2001) have suggested that changes occurring within traditional channels in the hotel sector were mainly due to economic pressures and transformations in the industry's environment. In 1994, the World Wide Web was quickly recognised as an ideal tool for a new form of distribution (O'Connor and Frew, 2000).

Travel-related products, including hotels, will be among the fastest growing to utilise the Internet as a route for purchases. Whilst sales of airline tickets are presently the fastest growing segment within the travel industry, rates of expansion in hotel reservations are expected to increases over the next few years, and to approach 25% of total online travel sales by 2003 (Sangster, 2001). The Internet offers hotels an opportunity to sell direct to consumers without having to pay sales commission to intermediaries. From the hotel's perspective, this approach is expected to bypass legacy systems such as Global Distribution Systems, and to result in faster and more cost efficient bookings by the direct-to-consumer channel (Connolly et al., 1998). Organisations can also use the infrastructure and standards of this medium to build *intranets* and *extranets* (Collins and Malik, 1999) allowing access to the network to internal users and trusted partners respectively (Buhalis, 1998). Other benefits are related to enhanced data capture about customers (Alford, 1999) and improved customer relationships through effective use of electronic mail strategies (Price and Starkov, 2002). O'Connor and Frew (2001) described the impact of the Internet as a catalyst. The majority of hotel chains, the global distribution systems, third party reservation systems providers and Switch companies have introduced websites, all with the objective of transacting business directly with the customer. The main effect of this phenomenon has been a decrease in dependency between different parties within the distribution chain, resulting in increased levels of competition (O'Connor and Frew, 200; Dombey, 1998; Richer and O'Neill-Dunne, 1999).

2 Methodology

The primary objective of this study was to survey the application of electronic distribution channels of hotels in Athens. A further aim was to examine how hotels perceived the Internet as a channel for distribution. Future trends within the industry were also investigated. A thorough literature review was carried out. Issues related to both advantages and disadvantages of the Internet as a distribution channel, and predicted aspects of future channels, were adapted from Licata et al. (2001). Primary data collection was based on questionnaire development for use during structured interviews, and also for distribution through electronic mail to hotels that preferred this way of responding. Feedback received from pilot interviews revealed a number of minor flaws which were duly rectified.

The scope of this research was restricted to the examination of electronic distribution systems in a sample set of hotels classified within the three higher hotel categories (Lux, A and B), in Athens. Hotels of these categories normally operate more than 70 rooms. A judgmental sample set was drawn from the total number of hotels located in Athens. Hotels used in the study fulfilled a number of criteria including *location:* Athens and city suburbs; *category:* Lux, A and B; *capacity:* More than 70 rooms; *connectivity:* E-mail addresses available in official guides; *availability:* Operational all year round and not currently closed for renovation; *affiliation:* One hotel selected in a group of hotels operated by the same company. Table 1 shows the sampling process and actual response rate. Although this process creates several limitations for the study, it still provides a good indication of the hospitality industry in Athens. A total of 70 hotels fulfilled the selection criteria and were targeted, resulting to 22 responses or 27.8% response rate. The major limiting factor within the above sample was reluctance of some accommodation establishments to either participate in, or respond to the survey.

Table 1. Sampling Process And Response Rate

Category	Total Hotels in Athens	More than 70 rooms	Open 12 Months Currently Operational FINAL TARGET	Responses	Response Rate
Lux	21	20	18	8	44.4%
A	48	26	22	10	45.5%
B	94	45	39	5	12.8%
Total	163	91	79	22	27.8%

3   Results

Whilst the majority of respondents worked in the sales departments within participating hotels, a significant proportion of the respondent pool were owners, General Managers or Assistant General Managers.

3.1   Customer mix and Source of reservations

Athenian hotels provide hospitality to a wide range of customers including business and leisure customers. Travel agencies remain as important players for reservations in all hotel types in Athens, with Internet reservations providing, on average, only 8% of total bookings to hotels. Connectivity was a central selection criteria, and participant hotels had web pages, with respective e-mail addresses available as avenues for reservations. An average of 63% offered real-time booking access. An interesting observation was that, category 'A' hotels were, on average, more advanced than 'Lux' hotels for this attribute. A similar pattern was observed between these two groups for use of Internet technology for communication purposes within the company (Intranets). The majority of category 'B' hotels relied on an electronic form to generate reservation requests from their web-page. Category 'A' hotels offered both e-mail and real-time booking options, while 'Lux' hotels reported making available all options to their customers. Hotels practise a variety of Internet promotional activities. Category 'B' hotels tend to rely mainly on forwarding their web-pages on search engines and placing e-banners on host sites. Category 'A' hotels score higher than other categories on hosting e-promotions, namely offering a better price for customers who book through their web-page.

3.2 Distribution channels

The most important channel cited by all hotels was direct communication with each property. Travel agents also remain the most important partner for hotels. Figure 1 shows overall responses related to distribution channels used by each hotel category. 'Lux' and category 'A' relied heavily on GDS, CRS and representative companies where such systems are available. This is explained as a number of these hotels are affiliated to international hotel chains.

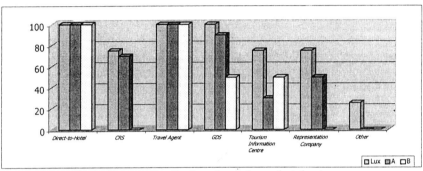

Fig 1: Current distribution channels per hotel category

On-line presence is considered quite important for all participant hotels. They adopted other electronic routes to the customer, most frequently through travel agents and GDS, in addition to having their own web-sites. Influence of travel portals were found to become increasingly important, especially among category 'B' hotels, with discounts increasingly being used in selling of distressed inventory (Figure 2).

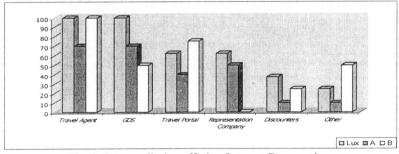

Fig. 2. Intermediaries offering Internet Reservations

The most frequently cited route to be adopted in the short-term was *a discount web channel*, independently from hotel category. Hitherto the most important channels for addressing business customers have been travel agencies, direct-to-hotel communication, and GDSs. The development of electronic distribution channels significant potential for this market segment. Travel agencies and tour operators are more important as intermediaries than GDS for the leisure market. Web channels, were however more frequently cited as important when addressing individual leisure customers. Travel portals are also considered of strategic importance.

The pilot-testing phase revealed a number of concerns among hoteliers. Overall, transaction costs were revealed to be a central issue, as it was felt this rendered some of the electronic distribution channels unapproachable. Also, all hotel groups reported the regular updating of on-line records was a crucial, but tedious task. In most other issues, however, each hotel type assigned different priorities to different concerns (Figure 3). For 'Lux' hotels the most significant concern was ability to formulate a coherent and complete overall strategy. Availability of multiple routes to the end-customer, as well as evaluating different web channels were also causes of concern.

3.3   Internet as a distribution channel.

Overall, hoteliers in Athens have high regard for the Internet as a communications and reservations medium, with most respondents in this study having a thorough understanding of the potential of this medium, as a means of gaining competitive advantage and promoting themselves effectively. This medium is also viewed as an effective tool for optimising customer interaction, and minimising distribution costs, and dependency on third parties (Figure 4).

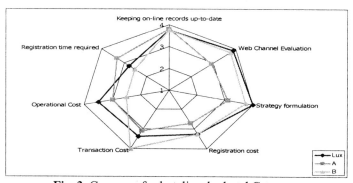

Fig. 3. Concerns for hoteliers by hotel Category

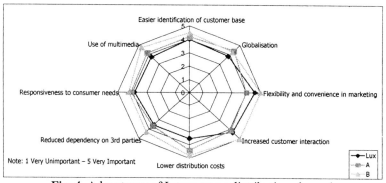

Fig. 4. Advantages of Internet as a distribution channel

Table 2 shows the comparisons of results from the present study, with that of Licata et al. (2001). The latter study, however, addressed a different audience, i.e., tourism industry experts. The Internet is also effective in providing greater flexibility and convenience. Athenian hoteliers do not consider minimising distribution costs as important, as respondents in the study by Licata et al. (2001). Considering concerns expressed for effective distribution through electronic channels, it was apparent from the present study that the Internet is mainly regarded as a marketing and promotional tool, rather than one for gaining competitive advantage through reduction of distribution costs. Overall, hotels in Athens were more interested in the advantages gained from using the Internet as an electronic distribution channel.

Table 2. Advantages of Internet as a distribution channel

	Mean obtained by Licata	Results obtained in Athens	
		Mean	Rank
Greater flexibility and convenience	4.20	4.41	1
Lower distribution costs	4.00	4.13	6
Increased customer interaction	3.87	4.31	2
Easier to identify and target customer base	3.83	4.18	3
Globalisation of product	3.77	4.18	3
Use of multimedia	3.48	4.14	5
Average	3.86	4.23	

Note.:A 5-point Likert scale - Attributes from 1= very unimportant, to 5= very important.

Overall, however, hotels in Athens feel more strongly about disadvantages brought about from the Internet as a distribution channel (Table 3). Respondents in this study placed greater importance to security issues compared to those used in Licata et al. (2001). For both studies, however, lack of human contact remained a central issue. Volatility of customer base and the threat of alienation of potential customer groups, rank lower in importance for the Athenian sample. Overall though, hotels in Athens feel more strongly about disadvantages brought about from the Internet as a distribution channel.

Table 3 Disadvantages of Internet as a distribution channel

		Results obtained in Athens	
	Mean by Licata	Mean	Rank
Lack of human contact during the transaction	3.07	3.86	2
Volatile customer base	2.53	3.09	7
Security issues	2.40	3.95	1
Danger of standardisation of product	2.40	3.59	3
Alienation of potential customers	2.20	3.36	6
Difficult to market effectively	1.90	3.45	4
Difficulty in keeping records up-to-date	1.87	3.45	4
Average	2.34	3.54	

3.4   Future distribution channels

Hoteliers in Athens are still unsure about the future of electronic distribution channels as illustrated in Table 4. The Internet was, however, cited as the medium with the highest potential, relative to mobile phones and other portable devices. It is predicted that with such tools, hotels will be in a position to practise location-based marketing. While devices such as self-service kiosks were not viewed as having significant electronic devices that will have a significant impact, it is envisaged GDS will continue to play an vital role in the short-term. Interactive TV is not considered as having a future potential in Athens, principally because it is largely since this medium is still largely unknown in Greece, with a penetration of less than 10% (Nua Internet Surveys, 2002). Hotels have also recognised gradual reductions in the roles of travel agents and conventional telephones; and penetration of mobile phones, and closely-related technologies on the increase: 50% of the Greek population now make use of a mobile phone; 20% use a personal computer; 10% use the Internet (Community Research and Development Information Service, 2002). These media have high potential of further increases in momentum as electronic distribution channels.

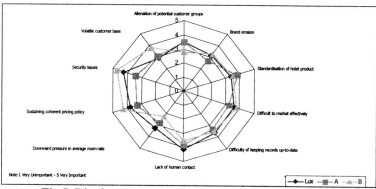

Fig 5. Disadvantages of Internet as a distribution channel

Table 4. Future distribution channels per hotel category

	Mean by Licata	Results obtained in Athens	
		Mean	Rank
Internet	4.70	4.45	1
Interactive TV	4.10	3.50	5
Mobile Phones / Portable Devices	4.03	3.82	3
Conventional Telephone	3.86	2.86	6
GDS	3.61	4.14	2
Travel Agencies	3.41	3.55	4
Self Service kiosks	3.04	2.81	7
Average	3.82	3.59	

4 Conclusions and Recommendations

This study has revealed that although hotels in Athens tend to, and still rely on traditional distribution channels, executives and management have realised potential benefits attainable through adoption of web-based channels. This realisation has shifted the trend towards adoption of a range of electronic distribution channels. Hotels are, however, reluctant towards immediate adoption of such new technologies and strategies, for fear of respective disadvantages (known and unknown), and also for avoiding risks associated with disrupting well established relationships with intermediaries. Another major concern is the confusion caused by the vast range of channels currently available.

Although the Internet has been described previously as having several significant advantages, it is still not a major avenue for travel-related reservations. There are, however, numerous potentials for this medium in Greece, and ultimate adoption for the reservation function can only be expected to be a matter of time. The primary study research has confirmed the growing influence of the Internet as a marketing tool. Increasingly, hotels have realised the opportunities emerging through developments of this medium, and have utilised new media to support distribution of their product. Findings in this study have revealed that among different distribution channels available to customers today, the most common still remains the simplest of all: the walk-in customer. In addition, there is reliance on traditional channels, such as GDS and travel agents, indicating existing potential for new entrants, and thus major new developments can be expected. Some hotels use the Internet primarily as an electronic directory to disseminate information regarding hotel rooms, and other facilities and services, that are typically accessible in their brochures. Athenian hoteliers are often concerned about the vast range of options on offer, and have complained about not receiving objective information from reliable sources, to determine criteria for effective assessment of such options. The effect is reluctance to adopt such technologies, despite initial enthusiasm about prospective benefits.

References

Alford, P. (2000). E-business models in the travel, *Travel and Tourism Analyst*, 3, 67-86.

Buhalis, D. (1998). Strategic use of information technologies in the tourism industry, *Tourism Management*, 19(5), 409-421.

Buhalis, D. (2001). Tourism in Greece: strategic analysis and challenges, *Current Issues in Tourism*, 4(5), 440-480.

Buhalis, D. and Laws, E. (eds.) (2001) *Tourism Distribution Channels – Practices, Issues and Transformations*. London: Continuum.

Cho, W. and Connolly, D. J. (1996). The impact of information technology as an enabler on the hospitality industry, *International Journal of Contemporary Hospitality Management*, 8(1), 33-36.

Collins, G. R. and Malik, T. (1999). *Hospitality Information Technology: learning how to use it*, 4th ed. Iowa: Kendall/Hunt Publishing Company.

Community Research and Development Information Service (CORDIS) (2002). *The results of panhellenic research into the use of computers, the Internet and mobile telephones*. Accessed on-line: http://www.cordis.lu/greece/news_rd8.htm

Connolly, D. J., Olsen, M.D. and Moore, R. G. (1998). The internet as a distribution channel, *Cornell Hotel and Restaurant Administration Quarterly,* August, 42-54.

Dombey, A. (1998). Separating the emotion from the fact – the effects of new intermediaries on electronic travel distribution. Buhalis, D., Tjoa, A. M. and Jafari (eds.), *Information Communication Technologies in Tourism,* Vienna: Springer-Verlag.

Dubé, L., Enz, C. A., Renaghan, L. M. and Siguaw, J. A. (2000). Managing for excellence: conclusions and challenges from a study of best practices in the US lodging industry, *Cornell Hotel & Restaurant Administration Quarterly,* October, 30-39.

Go, F. and Pine, R. (1995). *Globalisation Strategy in the Hotel Industry.* New York: Routledge.

Leong, C. (2001). Marketing practices and internet marketing: a study of hotels in Singapore, *Journal of Vacation Marketing,* 7(2), 179-187.

Licata, M., Buhalis, D. and Richer, P. (2001). The future of the travel e-mediaries (CRS, GDSs, Switch Companies, Videotext). Sheldon, P. J., Wober, K. W. and Fesenmaier, D. R. (eds.), *Information and Communication Technologies in Tourism,* Vienna: Springer-Verlag.

Namasivayam, K., Enz, C. A. and Siguaw, J. A. (2000). How wired are we?, *Cornell Hotel & Restaurant Administration Quarterly,* December, 40-48.

Nua Internet Surveys (2002). *Strategy analytics: Britain leads the way in digital TV,* January 16. http://www.nua.com/surveys/index.cgi?f=VS&art_id=905357558&rel=true

O'Connor, P. (2000). *Using Computers in Hospitality,* 2nd ed. London: Cassell.

O'Connor, P. (2001). The changing face of hotel electronic distribution, *EIU Travel and Tourism Analyst,* 5, 61-78.

O'Connor, P. and Frew, A. J. (2000). Evaluating electronic channels of distribution in the hotel sector. Fesenmaier, D. R., Klein, S. and Buhalis, D. (eds.), *Information and Communication Technologies in Tourism 2000,* Vienna: Springer-Verlag.

O'Connor, P. and Frew, A. (2001). Expert Perceptions on the Future of Hotel Electronic Distribution Channels. Sheldon, P. J., Wober, K. W. and Fesenmaier, D. R. (eds.), *Information and Communication Technologies in Tourism,* Vienna: Springer-Verlag.

Price, J. and Starkov, M. (2002). *Developing a total email marketing strategy,* June 21. Accessed on-line: www.eyefortravel.com

PricewaterhouseCoopers (2001). *New Europe and the Hotel Industry,* PricewaterhouseCoopers.

Research Institute for Tourism (RIT) (2000). *Greek Economy and Tourism (in Greek).* Athens: Research Institute for Tourism.

Richer, P. and O'Neill-Dunne, T. (1999). *Distribution technology in the travel industry strategies for marketing success,* London: Financial Times-Retail & Consumer.

Sangster, A. (2001). The importance of technology in the hotel industry, *EIU Travel and Tourism Analyst,* 3, 43-56.

Van Hoof, H. B., Collins, G. R., Combrick, T. E and Verbeeten, M. J. (1995). Technology needs and perceptions: an assessment of the US lodging industry, *Cornell Hotel & Restaurant Administration Quarterly,* October 64-69.

Adapting to Cognitive Styles to improve the Usability of Travel Recommendation Systems

Andreas H. Zins

Institute for Tourism and Leisure Studies
Vienna University of Economics and Business Administration, Austria
zins@wu-wien.ac.at

Abstract

Testing user models empirically in a rigorous way does not seem to be a very common and wide-spread activity (Chin, 2001). This study is intended to contribute to this field of research by investigating the effect of adapting the graphical user interface to the user's cognitive style. 176 test persons were assigned to four groups of a 2-by-2 experimental design and inspected the demonstrator of a Web-based travel recommendation system. The basic premise that this type of a user adaptive counseling system is more appreciated by the user than a conventional rigid user interface could be affirmed by this study.

Keywords: travel recommendation; usability; human-computer interaction, experimental design.

1   Introduction

Serving and supporting the prospective traveller in an efficient yet convenient way when browsing and planning for the next trip on the Internet is a common goal for many service providers on the Web. More often than not the user of a particular web site is unknown to the system since she has not registered so far or she is a first-time visitor. If the information and maybe booking services are offered to a broad, hence, inhomogeneous audience the interface to navigate through the digital information space is either designed to be unique to every user or is adaptive and sensitive to certain user characteristics (Stephanidis, 2001). For unidentified clients the navigation system or software agent can provide for an implicit adaptation of the ongoing dialogue or may offer explicit options how to be guided through the web pages (particular references for this kind of knowledge-based human-computer interaction in: Fischer, 2001). To find a favourable solution a compromise has to be found among various usability requirements: e.g. clarity and structure of information, avoiding information overload, sensually appealing presentation format, overview, time efficiency.

A recent study which investigated ongoing travel planning behaviour from an observational perspective came up with six different decision styles (Grabler and Zins, 2001). Travellers belonging to a particular decision style followed a more or less distinct way how the information gathering, evaluation, and selection process for a vacation trip planning was shaped. It was argued that a good recommendation system should cope with these different cognitive patterns. In developing such an improved travel recommendation system the flexibility and adaptability claimed to meet the requirements for five of the six decision styles could be built in one comprehensive interface. Yet, one style with an outstanding holistic characteristic expecting the system to actively drive the dialogue instead of the user is hypothesised to require a completely different treatment and consequently a different graphical user interface. It is the purpose of this study to investigate the potential advantages of two effects of the human-computer interaction: 1. giving the user the choice to select among two alternative navigational options A (more sequential) and B (more pictorial-compound), and 2. classifying the user in advance into one of two broad categories of cognitive style (A' and B') to direct her to the potentially more suitable navigational option. For the empirical test a demonstrator (i.e. a not yet fully operable system of related web pages to collect or identify the user's travel preferences and wishes) had been developed and presented to 176 test persons (Internet users only). This demonstrator acts as a preliminary graphical user interface for a comprehensive web-based travel recommendation system covering information and booking facilities for travel elements such as destination units, accommodation, dining, cultural and other leisure facilities.

2 Methodology

2.1 Study Design

Decision making for travel purposes are more often than not related to an extensive problem solving task. From this perspective it is evident that after recognizing the problem a sequence of information retrieval, evaluation, and filtering steps have to be run through in order to solve the decision problem. As similar cognitive capabilities are required for learning problems it seems appropriate to make use of the framework of learning style dimensions. Related research questions, however not in the tourism domain, are currently investigated within the HomeNetToo project (see Biocca et al. 2001 and Jackson et al. 2002 for referring publications). A comprehensive definition of "learning style" reads as the composition of cognitive, affective, and physiological learning preferences that serve as relatively stable indicators of how learners perceive, interact with, and respond to the learning environment (Keefe, 1979). Various other definitions and conceptualisations (e.g. in socio-psychology and neuro-psychology) can be found in the literature which are often addressed as hemisphericity (e.g. Springer and Deutsch, 1998). However, it does not depend on the particular model chosen but on the adaptation and personalization of the information delivery (Felder, 1986). Taking into account the recommendations on the selection of an instrument for measuring learning styles (Mamchur 1996:2) the following reduced set of nine test items have been chosen (see Table 1, initially 11 items from Hilliard; cited in Gill and Dietrich, 2001; further reference Hilliard, 1992). These statements represent one major dimension of cognitive or learning styles: either being more analytical or being more holistic which covers several aspects of the proposed Felder-Silverman Learning Style model (Felder, 1996; sensing vs. intuitive, visual vs. verbal, inductive vs. deductive, active vs. reflective, and sequential vs. global).

Table 1. Cognitive Style Measurement Instrument

Item	Responses	
To which type would you assign yourself?	*verbal*	visual
Do you plan more?	impulsive	*ahead*
Do you recall more easily?	*people's names*	people's faces
Do you speak with?	gestures	*with few gestures*
Do you respond more?	*to logic*	to emotions
In general, are you?	less punctual	*punctual*
Do you prefer?	*formal study design*	sound/music background while studying
Do you stick more to?	the tone of voice	*word meaning*
Do you process information more?	*linearly*	in varied order

Note: italic responses represent indicators for the analytical style.

As an appropriate counterpart for these two fundamental cognitive styles the demonstrator of the web-based travel recommendation system was designed in two variants: GUI option A incorporates a more traditional way of collecting travel wishes and preferences by asking questions and/or filling in forms or checking boxes in several item lists. GUI option B, in contrast, comes up immediately with a mixed set of six prototypes of trip plans represented by two pictures (landscape and accommodation) and two keywords describing the offer.

Respondents were randomly assigned to one of the four groups of the following experimental design (see Figure 1). The respondents were asked to answer the nine cognitive style statements: 50% in advance before browsing through the demonstrator, 50% follow-up. For those respondents asked in advance it was possible to identify the preferred cognitive style. Only for Group 2 the result of this quick test was immediately used to assign the respondents to the 'correct' option of the interface demonstrator. This resulted in two sub-groups: A'-A whose members were identified as being more analytical (A') and therefore shown the more analytical option of the interface (A), and B'-B whose members were identified as being more holistic (B') and therefore shown the more holistic-pictorial option of the interface (B). Members of experimental Group 4 were free to choose their preferred GUI option despite the fact that they had to reveal their learning style in advance. As a consequence they selected the appropriate (A'-A, B'-B) or the inappropriate interface option (A'-B, B'-A). For those experimental groups (1 and 3) whose members were asked for their dominant learning style after having inspected the demonstrator the same correct and incorrect combinations of GUI option and style could occur (either A-A', B-B' or A-B', B-A'). To test the first effect, 50% of the respondents were assigned to Groups 3 or 4 who were free to choose among the two options. They used that starting page which attracted them more at first glance without having inspected the remaining option.

	Knowledge about Learning Style (A' or B')	
Interface Options (GUI A or B)	Asked afterwards	Asked in advance
No choice	**Group 1** n = 44 A-A', A-B', B-A', B-B' n = 7 / 15 / 10 / 12	**Group 2** n = 44 A'-A, B'-B n = 20 / 24
Free of choice: A or B	**Group 3** n = 44 A-A', A-B', B-A', B-B' n = 12 / 15 / 5 / 12	**Group 4** n = 44 A'-A, A'-B, B'-A, B'-B n = 8 / 14 / 11 / 11

Fig. 1. Experimental design for the demonstrator evaluation

This balanced or so-called crossed design enables a check for unintended influences while investigating the central research question.

2.2 Evaluation Concept

After allocating each respondent to one of the four experimental design conditions a brief oral introduction about the objective of the study and the sequence of tasks was given by pre-trained experimental help staff. After the browsing task for familiarizing with the interface features a paper-and-pencil questionnaire had to be filled out. Socio-demographic characteristics (e.g. age, gender, Internet usage intensity) were collected using this fully structured self-administering questionnaire. The main section of this instrument was used to evaluate the overall utility, functional aspects, the attractiveness and other relevant facets of the usability of the recommendation demonstrator. The following 11 statements (Table 2) were selected to evaluate both variants (A and B) of the demonstrator using seven-point Likert-type scales (ranging from 1 'poorest' to 7 'best' rating).

Table 2. Evaluation criteria for the usability of the demonstrator

Statements

How easy did you find to use the counseling system?
Did you comprehend how to get to travel recommendations?
Did you find the input possibilities as systematically structured?
Did you feel personally overloaded with information?
Did you appreciate the design of the Web pages?
Was the list of travel recommendations clear?
How apt do you find the system to support the initial phase of your trip planning?
Can you imagine to plan a whole trip with this kind of system?
Can you imagine to end up with an interesting result by using this system?
Did you appreciate the details of the particular travel recommendations?
What was your overall perception of the system?

3   Results

For a period of two days four notebook PCs were set up in a typical fair booth in front of a household electronics shop in a high frequency shopping mall. Respondents were selected randomly on the premise that they are Internet users and interested in travel and tourism. In total, 176 experiments could be conducted allocating an equal number of respondents to the four experimental conditions (see Figure 3). From the next Table 3 some socio-demographic characteristics can be reported. The gender distribution is with a share of 61% male respondents somehow biased. A significantly higher share of 60% females was allocated to Group 3 or 4 (unconditional choice among interface option A and B) compared to Group 1 or 2. The gender distribution in the second experimental condition (prior versus afterward activation of someone's cognitive style) is unbiased. No bias can be found with respect to the age distribution within the two design factors. The majority of respondents are at the age below 30. One quarter is older.

The Internet usage was monitored for several aspects. The overall 'connect time' was roughly estimated with a 58% share using the Internet more than 10 hours a month. Only 16% has a very low usage rat of less than 5 hours a month. No significant deviation from an equal distribution of these general usage patterns can be observed within the 2-by-2 experimental design. General information search on the Internet was reported to apply for almost 100% of the respondents. 62% of the sample use the Internet as a source for travel planning purposes Specific activities over the Internet were used as a measure of trust in the medium. Almost 50% of the respondents use the Internet for their banking transactions. One third has already run through some purchase activities while about one quarter has used the Internet for booking some travel arrangements. Again, the information behaviour as well as the buying behaviour on the Internet is equally distributed in each of the four experimental groups.

Table 3. Socio-demographic characteristics of the respondents

Criteria	Share	Criteria	Share
Male	61%	Internet usage:	
Female	39%	for information search	99%
		for banking	48%
Age:		for shopping	36%
< 20 years	19%	for travel planning	62%
20 – 29 years	53%	for travel booking	26%
30 – 39 years	17%		
40 – 49 years	8%	< 5 hours a month	16%
50 – 59 years	2%	5 – 10 hours a month	26%
60+ years	1%	10+ hours a month	58%

The responses to the nine cognitive style items were summed and immediately used for classifying the respondents. The distribution of analytical items chosen does not follow a normal distribution. The share of 57% of more holistically inclined people in the sample is significant. In addition, female respondents can be more often found (71% share) among holistic participants of this study than male. If the classification would have been based on only the first item (self-assignment to either verbal or visual type) the distribution had been even more skewed: 74% visual types. Interestingly, this – maybe lead – variable does not correlate significantly with the overall classification based on the summed index over nine variables.

A first attempt of investigation looks at the distribution of respondents within the four experimental groups (see Figure 3). Without asking the respondents in advance about their cognitive style the arbitrary assignment/selection of one of the two GUI options was 50:50. About 50% got in touch with the GUI option which did not correspond to their cognitive style. Option B (the more visual-compound version) was correctly selected in 27% of the cases (Option A: 22%). However, the difference is not sufficient to be significant. On average, 60% of the respondents got their 'correct' GUI option. A significant higher share of 72% could be observed among those experimental groups (2 and 4) which had been asked in advance about their cognitive style. No significant advantage could be detected when comparing the two cognitive styles and GUI options respectively.

So, giving the user, generally, the choice to select among two different navigational options yields a 50% chance to hit the appropriate user interface. Activating in advance something like the user's cognitive style does neither disturb nor improve this choice behaviour. In addition, no advantage turned out to be associated with one of the addressed cognitive style. If, however, the recommendation systems asks explicitly for the cognitive style and leads the user immediately to the corresponding interface option a significant gain may be the result. Whether this gain is an improvement or not cannot be determined from outside. To answer this question perceptual measures should be compared – especially – between correct and incorrect GUI option assignments/selections.

This leads to the second attempt of investigation which is based on evaluative criteria (Table 4). On the 7-point Likert-scale the ease-of-use aspect was rated best (6.03 on average) whereas the design of the Web pages had a significant worse appreciation (5.31 on average). The least variation could be observed in judging the details of the (simulated) recommendations. Much more differences are related to the perception of the information overload (standard deviation: 1.74) and assistance for trip planning (SD: 1.66).

A principal axis factor analysis was applied to the 11 criteria (KMO = .883) in order to reduce the semantic evaluation space. After a non-rectangular rotation a two-dimensional factor solution could be identified (see Table 4 for factor loadings above .4). The first factor can be addressed as usability, the second as decision support. Further analyses are conducted on factor score level.

In addition to several t-tests with each single experimental factor a comprehensive model of variance decomposition was specified:

$$Total\ variance = Constant + EF\text{-}1 + EF\text{-}2 + (EF\text{-}1\ x\ EF\text{-}2) + (EF\text{-}1\ x\ Corr\text{-}GUI) + (EF\text{-}2\ x\ Corr\text{-}GUI) + Corr\text{-}GUI + Error$$

The treatment factors EF-1 (prior activation of cognitive style) and EF-2 (no choice vs. freedom of choice between the GUI options) did not show a significant difference of the evaluation: neither in the t-test nor in the compound variance analyses as single factor nor as an interaction effect. In contrast, it was hypothesized that when someone's cognitive style is matched with a corresponding GUI this advantage should be reflected in a better perception and evaluation from the user's point of view. While the single undesired (EF-1, EF-2) and desired (Corr-GUI) influences did not show any potential significance level for both evaluation dimensions the interaction between using the correct GUI option and giving the user a choice passes with a p-value of .086 the 10%-threshold for exploratory purposes for the usability dimension. The parameters of this interaction effect show clearly that a better evaluation of the demonstrator system was achieved with respect to the usability criteria if the interface implicitly directs the dialogue to the 'correct' interface option.

Table 4. Evaluation criteria for the usability of the demonstrator

Statements	Mean	SD	Usability Factor 1	Decision Support Factor 2
How easy did you find to use the counseling system?	6.03	1.34	.86	
Did you comprehend how to get to travel recommendations?	5.73	1.37	.81	
Did you find the input possibilities as systematically structured?	5.54	1.46	.71	
Did you feel personally overloaded with information?	5.51	1.74	.69	
Did you appreciate the design of the Web pages?	5.31	1.44	.51	
Was the list of travel recommendations clear?	5.51	1.37	.44	
How apt do you find the system to support the initial phase of your trip planning?	5.56	1.28		.88
Can you imagine to plan a whole trip with this kind of system?	5.63	1.66		.82
Can you imagine to end up with an interesting result by using this system?	5.88	1.23		.70
Did you appreciate the details of the particular travel recommendations?	5.60	1.13		.49
What was your overall perception of the system?	5.79	1.12	.41	.43

4   Conclusions and Recommendations

This study was initiated in order to validate one fundamental assumption of a user model for developing an intelligent Web-based travel recommendation system: adapting to the user's cognitive style (analytical vs. holistic) improves the quality of the human-computer interaction. For knowledge-based HCI systems it is necessary to exploit a minimum of information about the user characteristics in order to adapt the user interface accordingly. In this experiment a rather short instrument (a nine item-questionnaire) was used to identify the respondent's cognitive style. From the rich theoretical framework of learning styles it can be derived that more than one single dimension (e.g. verbal versus visual) plays a substantial role in explaining differences in information processing. Hence, the proposition of testing the contrast between more analytical and more holistic users of a travel recommendation system should be extended to competing and related assumptions: e.g. alternative measurement instruments for identifying cognitive styles, other predominant paradigms for information processing such as sensing vs. intuitive, inductive vs. deductive, or different involvement conditions (Strebinger et al. 2000).

When reflecting the experimental design it is worth mentioning that neither the activation of the cognitive style nor the choice condition imposed an undesirable bias on the preference for the two alternative user interfaces. However, it is important to note that classifying the respondents on the basis of only one item instead of nine would have resulted in very different segmentation. As one part of the sample was forced to follow an ad hoc-determined GUI alternative the consequences of a different classification cannot be analyzed with the available data.

Despite the somehow weak indicators for corroborating the hypothesized desirable effects one major interaction effect is almost significant at the conventional 5% error level: If the users are asked in advance about their cognitive style it is important to direct them to the appropriate user interface. In this case the usability dimension achieved a better evaluation comparing all other experimental conditions.

From other indicators – not reported extensively in this paper – it can be concluded that better and more rigorous tests are expected to be feasible with a prototype of such a travel recommendation system covering much more functions than the demonstrator of this study was able to perform. However, the request for more and more systematic tests of user models as stated by Chin (2001) is supported by this empirical study. Its basic premise that Web-based travel recommendation systems staffed with interface options related to different cognitive styles are to the benefit of the user (= customer) could not be disproved completely.

References

Biocca, F.A., L.A. Jackson, H. Dobbins, G. Barbatsis, A. von Eye, H.E. Fitzgerald & Y. Zhao (2001). Adapting computer interfaces to differences in cognitive and cultural style. Paper presentation in the symposium on the *Digital Divide, International Communications Association (ICA) and the International Association of Media and Communication Research (IAMCR)*, University of Texas, Austin, Texas, November 16-18.

Chin, D.N. (2001). Empirical evaluation of user models and user-adapted systems. *User Modeling and User-Adapted Interaction* 11(1-2): 181-194.

Felder, R.M. (1996). Matters of Style. *ASEE Prism* 6(4): 18-23.

Fischer, G. (2001). User modeling in human-computer interaction. *User Modeling and User-Adapted Interaction* 11(1-2): 65-86.

Gill, J. & Dietrich, U. (2001). Creating an oasis from a resource desert: integrating right and left brain learning approaches in developing health promotion resources for rural and remote areas. Paper presentation at the 6*th* *National Rural Health Conference*, Canberra, Australia, March 4-7.

Grabler, K. & A. Zins (2002). Vacation trip decision styles as basis for an automated recommendation system: lessons from observational studies. In: Wöber, K., A. Frew & M. Hitz, (eds.), *ENTER 2002, Information and Communication Technologies in Tourism 2002*, Springer, Wien, 458-469.

Hilliard, A. (1992). *Behavioral style, culture, and teaching and learning*. New York, Free Press.

Jackson, L.A., A. von Eye, F.A. Biocca, G. Barbatsis, H.E. Fitzgerald, & Y. Zhao (2002). Personality, cognitive style and Internet use. *Swiss Journal of Psychology*, "Special Issue, Studying the Internet: A challenge for modern psychology" (in press).

Keefe, J.W. (1987). *Learning styles theory and practice*. Reston, VA, National Association of Secondary School Principals.

Mamchur, C. (1996). *A teacher's guide to cognitive type theory & learning style*. Alexandria, VA: Association for Supervision and Curriculum Development.

Stephanidis, C. (2001). Adaptive techniques for universal access. *User Modeling and User-Adapted Interaction* 11(1-2): 159-179.

Springer, S.P. & G. Deutsch (1998). *Left brain, right brain. Perspectives from cognitive neuroscience. Freeman*, New York.

Strebinger, A., S. Hoffmann, G. Schweiger, & T. Otter (2000). Verbal versus pictorial stimuli in conjoint analysis: The moderating effect of involvement and hemisphericity. In: Gundlach, G.T. & P.E. Murphy (eds.), *Enhancing Knowledge Development in Marketing, 2000 AMA Educators' Proceedings, Vol. 11*, American Marketing Association, Chicago, 182-188.

3D Maps for Boat Tourists

Volker Coors [a]
Ove Gjesdal [b]
Jan Rasmus Sulebak [b]
Katri Lakso [c]

[a] Fraunhofer Institut for Computer Graphics, Germany
HfT Stuttgart, University of Applied Sciences, Germany
Volker.Coors@igd.fhg.de

[b] SINTEF Institute of Applied Mathematics, Norway
{Ove.Gjesdal; Jan.R.Sulebak} @sintef.no

[c] Nokia Research Center (NRC), Finland
katri.laakso@nokia.com

Abstract

This paper presents the results from a study where 3D city maps is used on mobile computers. The aim of this study is to get feedback from the users on how they experience the use of 3D maps as an interface in search for tourist information. To do this an application is developed that use a 3D map for routing and way finding. The application is tested on a group of boat tourists in the harbour of Tønsberg (Norway). The results showed that a rather large group of people found the use of 3D maps valuable both as a navigational tool and as a tool for geographic orientation. In the paper, also some technical issues that enable the transmission of 3D maps in a wireless network and its use on mobile devices, is also discussed. The research is carried out as part of the EU-Project TellMaris (IST 2000-28249)

Keywords: 3D city maps, mobile computers, location-based services, tourist information

1   Introduction

An increasing number of service providers within the business of information and communication technology (ICT), are presently developing so-called Location Based Services (LBS). Such services are usually supported by existing GIS technology extended with functionality for interfacing with a wireless environment. Location Based Services provides application developers with a new set of tools to build new software solutions for the mass marked. Among the fastest growing application areas for Location Based Services are within tourism and travelling. The key issue here is to develop software solutions that provide the traveller or tourist with relevant and updated information on mobile computers that is both clearly understandable and readable.

1.1 TellMaris project

The objective of TellMaris (IST 2000-28249, www.tellmaris.com) is the development of a generic 3D‑ map interface to tourist information on mobile computers. The interface provides a new concept for creating value added information services on mobile computers for the European citizens related to geographical information. In TellMaris it is used for accessing tourist information relevant for boat tourism in the Baltic Sea area. According to Gjesdal (2002) a relative large number of boat tourists has already a Laptop computer or PC on board.

What differentiates TellMaris from other projects is the strong emphasise on 3D map models. Other projects focusing on maps and mobile computing still keep on the 2D map metaphor by providing only simple 2D maps as background information for placement and not for spatial query formulation. In TellMaris the 3D map is considered as the main communication interface for location based information retrieval.

1.2 TellMaris architecture

This section presents a brief overview of the TellMaris architecture. As reference, the TellMaris architecture use an architecture that follows the OpenLS initiative (OpenLS 2000) as shown in Fig. 1. In TellMaris two different applications are developed. The TellMarisOnboard application running on a laptop computer and the TellMarisGuide application running on a mobile computers or PDA. In this paper only 3D maps in the TellMarisGuide application will be discussed. In the TellMarisGuide application the Location Services concept is used to be conform to upcoming standardisation activities.

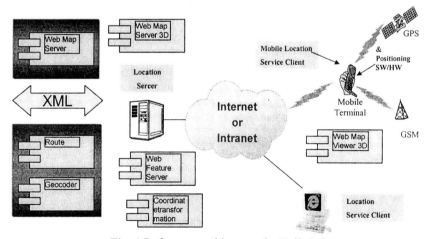

Fig. 1 Reference architecture for TellMaris

2 Concept of 3D maps for Location Based Services

2.1 Three-dimensional maps

A lot of cartographers and computer scientists have utilised Virtual Reality in a variety of ways to create three-dimensional maps with varying degrees of success. Robinson et al. (1995) defines a map as a visual representation of a spatial relationship in order to communicate environmental information. The primary theme that ties the material together is map effectiveness in thought and communication. The task of the map designer is to enhance the map user's ability to retrieve information. Dykes et al. (1999) point out that representing 'reality' authentically is not a necessary objective of the map though clear and believable representations are usually desirable. The map should rather represent an abstraction of the reality and emphasise on a selected set of features and their spatial relationships of importance for a specific audience. Whilst some map products provide a high fidelity replica of the vista at a location others aim to extract the information pertinent to the use to which the map will be put from the confusion of the surrounding clutter. Indeed a whole array of map products exists ranging from highly abstract, like an interactive three-dimensional Chernoff maps (O'Malley 1998), to extremely realistic ones. However, virtual and augmented reality offers a variety of new visualisation and interaction techniques in order to make maps not 'more real than real' (Dykes et al. 1999) but also 'more useful than before'.

2.2 Transfer rate and rendering of 3D map data

In recent years, research has focused on resource adaptive rendering (Martin 2000). Currently, small 3D graphic engines available on PDAs (Personal Data Assistants) can render up to 100.000 triangles. It is expected that the rendering capability will grow very fast in the future as it did on PCs and Laptops in the past. With UMTS, a reasonable amount of bandwidth (384 Kbit/s in urban areas) will be available in a mobile environment. Combined with 3D compression techniques (Coors and Rossignac 2002) it will be possible to transmit 3D maps in a wireless network and use it on a mobile device. However, the high data volume of a 3D map will be critical even in high bandwidth wireless networks. A three-dimensional model with 100.000 triangles will lead to a 3 MB VRML file. Transmitting this data via UMTS will last about 60 seconds. A standard compression algorithm like GZIP reduces the data volume to 1 MB and transmission time to 20s. Still, 20s is a long time to wait for a normal user. Algorithms that are specialised on 3D meshes achieve a compression rate about 95%. A 100.000 triangle model can be compressed down to 180 KB and will take less than 4 seconds for transmission.

2.3 Relevance of 3D features in a 3D map

Besides rendering and data transmission, one remaining challenge is to decide how to use 3D objects to create a 'more informative' map. The tourist poses a query to get information. So we have an idea of his or her actual interests. Based on this users' inquiry, we derive some knowledge of the importance of features in the database. We propose a dominance function

$$dom: F \times Query \rightarrow \Re$$

that assigns a dominance value to each feature f in the database indicating the relevance of f according to a given query (Coors 2001).

In case of routing the starting point and the destination are of special importance. Eye-catching buildings are also helpful as visual landmarks, especially at points where the routing direction is changing, and should be accentuated in the representation. Further on, additional buildings can sometimes deliver a helpful context, but they can also overload the representation, so that the navigation support is ruined.

In our work we specify the degree of abstraction of a feature by a dominance value, which reflects the feature ranking in the request. One influence factor of the dominance value is the relevance of a feature concerning a user specific query. A formal description of distance is specified by a distance function between an attribute of the feature and the corresponding query value.

However, not only the query dependent relevance factor influences the dominance of a feature. Actually, the dominance is composed by three components, the relevance factor R, the use of a feature as a reference object O while posing a query, and the general use of this feature as a landmark L.

2.4 Example

Fig. 2 illustrates the use of dominance values by showing a sequence of 3D routing maps. The sequence represents the route from the railway station to the Fraunhofer Institute of Computer Graphics in Darmstadt. All buildings along that route are derived from the database by a query and are relevance. All other features, which do not match that query, are not relevant. Start and endpoint of the route are reference objects O, and all prominent features like the castle of Darmstadt and conspicuous building along the route are landmarks L.

The dominance of each feature is mapped on the transparency value for visualisation purposes. Features with a no dominance are not shown at all. Features with a small dominance value are show in a semi-transparent style and use one grey colour only. Features with high dominance will attract the user's focus by using a textured model.

Fig. 2. Sequences of a 3D route visualisation to Fraunhofer IGD Darmstadt. Navigation landmarks are shown in detail while less important building, transparency is used to give a context but not distracting the focus.

3   Methods and procedures

3.1   Evaluation of 3D–maps on mobile computers

Empirical results were collected during the evaluation of the first prototype of the TellMarisGuide application. TellMarisGuide is a mobile city guide developed for the Nokia Communicator. Its first prototype was evaluated in the August 2002 in the Tønsberg, Norway with a group of ten boat tourists. The application showed the presentation of route instructions in connection with a combination of 2D and 3D maps. The main purpose of this the study was to collect some initial feedback from the users about using mobile 3D maps in the city environment. More tests with later prototypes will be conducted next summer.

3.2   Participants

There were in total ten users and one pilot user participating in the tests. All users were selected at Tønsberg harbour without strict criteria. Nine of the actual users were males and one was female. All but one of them were Norwegians, one was from

Canada. The ages of the actual users varied from 33 to 63 years, half of them being between 50 and 60 years. Average age was 51,6 years. All users had visited Tønsberg before. About one third had been there less than five times before, second third from 10 to 20 times and the last third over 30 times. Despite of this majority of the users had not seen the map of Tønsberg before the tests. All users were also quite experienced with maps. Almost all of them use maps, especially sea charts, often, two people chose the alternative "occasionally". Users thoughts about their map using skills varied from professional to novice, but most of the participants rated themselves as being professional or skillful map users.

3.3   Test procedure

Each test session consisted of three parts: introduction, test tasks and interview. In the first part the project and the application were introduced to the user. The user filled out a questionnaire, with some basic data about age, sex, education and their prior knowledge about maps and the area they visited. Then they get a quick introduction to the test procedure and how the test application was running.

The test part included six similar tasks. In all of them the user was asked to go from one place to another. In the first four tasks the participants were asked to use the application with the 2D and 3D map. Starting and target locations were marked in the maps, but no GPS was available to keep the user's current position automatically up-to-date. In the last two tasks the test users were asked to use a traditional paper map. Each test session ended with an interview, where the user's opinions about the TellMaris Guide application, the 2D and 3D maps, and the test session were asked. The tests were performed with a laptop running a mobile phone emulator. The laptop was an IBM ThinkPad 240 with an Intel Celeron 366 MHz processor, 192 MB RAM memory and 10.4 inch LCD Active Matrix display with 800x600 pixel resolution and 16-bit colours.

4   Results

4.1   General results

Users' attitudes towards the prototype were in general positive and 75% of them would like to use this kind of service rather than 2D paper maps and guidebooks. The 3D map itself was found to be a good idea, although many experienced map users thought that an electronic 2D map would be sufficient for them. There were two major problems during the tests: The laptop screen was difficult to look at in sunlight and secondly, the users had to navigate in the model by themselves because no positioning system was available.

Both of these issues influenced the users' satisfaction ratings. Most of the users (80 %) tried to use the 3D model as a navigational aid while all of them used it to recognise buildings. Some users claimed that non-textured buildings were hard to distinguish from each other, but textured buildings, especially one white building were considered easy to recognise (Fig. 4). Some users complained that comparison between the 2D and 3D maps was difficult, because there was no clear correspondence between them.

4.2   User navigation in 3D maps

Apart from matching buildings the most common navigation strategy of a user was to follow the direction arrow in the 3D view, and the target location and current location being displayed in the 2D map. The users also had the possibility to choose the viewing height in the 3D view to switch between walking level (pedestrian view, 1.8m altitude) and flying level (bird's-eye-view, 25m altitude). Interestingly, the flying mode was found to be much easier for navigational purposes. 3D maps were found to be slower to use both in initial orientation and route finding compared to 2D maps. We defined the orientation interval to begin at the moment the user was shown the target location and to end when she starts to walk towards it. Then the route finding interval began which lasted until the user reached the target. An average orientation interval lasted 42 seconds when the users used the 3D map, and 10 seconds when the paper map was in use. Route finding interval times also depend on the lengths of the routes, so a proper measurement is needed. Therefore, we rely on the average speed of the user (optimal route length divided by the average time), which was about 1,1 m/s for the 3D map and 1,4 m/s for the paper map.

When the users were asked how they would like to improve the application four things were mentioned frequently. According to the users the 3D model should be more detailed and realistic and the target should be highlighted. Street names should be visible and a zoom function should be included in the 2D map.

Fig. 3. Evaluating 3D maps: The white house in the middle was found easy to recognise.

5 Discussion and Conclusion

The purpose of this pilot study was to collect experience on how to improve future prototypes of our navigation system. The small amount of test users and the fact that the choice of participants was not random, suggest that the results of this study cannot be generalised. However, all users performed the same tasks in different order using four times a 3D map and two times a 2D map. Therefore we collected 40 samples for 3D maps and 20 samples for 2D maps, which makes us confident that the results are relevant for our prototype in spite of the small number of participants.

It should also be noted that the majority of the users were males and all of them were experienced with 2D paper maps. It has been shown that males and females use different strategies in navigation (Hunt and Waller, 1999). For example, it is possible that females, who are not accustomed to using maps, would have found 3D maps with landmarks more useful.

Despite the observation that experienced male users preferred the familiar 2D maps to the new 3D maps the results were promising. Users were able to recognize real world objects from the 3D model and use these landmarks as navigational aids. Many users also said that even though a 3D map would not give them much additional value, it was more fun to use.

Another interesting result was that users generally preferred the flying mode to the walking mode. The flying mode gave them a better overview of the surroundings and helped them in building recognition.

References

Coors, V. (2001). 3D-GIS in Networking Environments, International Workshop on 3D Cadastres, Delft, NL

Coors, V. & J. Rossignac (2002) Guess Connectivity: Delphi Encoding in Edgebreaker, *Technical Report*, Georgia University of Technology

Dykes, J. A., Moore, K.E. & D. Fairbairn (1999). From Chernoff to Imhof and Beyond: VRML and Cartography, *Fourth Symposium on the Virtual Reality Modeling Language*, ACM Press

Gjesdal, O. (2002). Market research in the boat tourism segment, ENTER conference

Hunt, E., & D. Waller (1999). Orientation and wayfinding: A review. (ONR technical report N00014-96-0380). Arlington, VA: Office of Naval Research.

Martin, I.M. (2000). ARTE - An Adaptive Rendering and Transmission Environment for 3D Graphics. *Proceedings of ACM Multimedia Conference*, Los Angeles, USA,

O'Malley, J. (1998). Visualising Multivariate Social Data using VRML, Master Thesis, Birkbeck College, London,

Robinson, A.H., Morrison, J.L., Muehrcke, P.C., Kimerling, A.J. & S.C. Guptill (1995) Elements of Cartography: Sixth Edition, Wiley

Assessing the Determinants of the Success of Web-based Marketing Strategies by Destination Marketing Organizations in the United States

Youcheng Wang
Daniel R. Fesenmaier

National Laboratory for Tourism and eCommerce
University of Illinois at Urbana-Champaign
{ywang13, drfez}@uiuc.edu

Abstract

Internet marketing is becoming an inseparable, oftentimes determining part of destination marketing organizations' (DMO) overall marketing efforts. This study attempts to contribute to the understanding of DMOs' Internet marketing strategies by examining the Web marketing practices of the DMOs in the United States. The results of a path analysis indicate that both hierarchical and nested relationships exist between the three groups of variables included in the study: 1) organizational characteristics (i.e., innovativeness and organizational size) and organization technology climate (i.e., management support and organization Internet literacy); 2) Web marketing strategies (including Web technology application, website marketing strategy, and website promotion strategy); and 3) the overall successfulness of Web marketing strategy. Discussion and implications are provided based on the study results.

Keywords: Internet marketing; destination marketing organization; tourism marketing

1   Introduction

The importance of information technology in tourism, especially the World Wide Web, has increased tremendously in the past decades, and more and more companies have realized that Internet marketing is becoming an inseparable, oftentimes a determining part of their overall marketing endeavor. As a result, marketing and advertising on the Internet is increasing in importance and impact. For example, Internet advertising spending in 1999 was $2 billion to $3 billion in the US alone (eMarketer, 1999), and this figure is expected to rise to $22 billion by 2004, with the Internet surpassing most of the traditional advertising media (Rothenberg, 1999).

The travel industry is one of the first areas doing business electronically (Copeland and Mckenny, 1998; Schultz, 1996), and significant transformation has taken place in the travel industry. Today, tourism is among the most important application domains in the World Wide Web (WWW) where estimates of between 33% and 50% of Internet transactions are tourism-based (Strassel, 1997). In responding to this new media and purchasing patterns, destination marketing organizations (DMOs) are trying to adjust their marketing strategies by integrating the Internet for destination marketing and promotion (Schmid, 1994; Werthner, 1996). However, many questions remain unanswered regarding DMOs' efforts in launching and executing digital marketing strategies.

2   Study Purpose

The purpose of this study was to examine the factors affecting the successfulness of web marketing strategies of tourism marketing organizations in the United States. Specifically, the success of their web marketing strategies were examined from the perspectives of: 1) Web technology applications, 2) website marketing strategies, 3) website promotion strategies, 4) organizational characteristics such as size and innovativeness, and, 5) organization IT climate such as management support to Internet operations and Internet literacy of the organization.

3   Theoretical Framework

The effectiveness/successfulness of web marketing strategy for destination marketing organizations (DMOs) depends on, among other things, the implementation and coordination of three interrelated aspects: an efficient use of the "right" combination of applications of the web technology, an effective marketing strategy for the website, and an effective promotion strategy for the website.   Owing to different technical backgrounds, financial resources, and marketing objectives, destination marketing organizations use web-based technologies in different ways and with varying intensity (Yuan and Fesenmaier, 2000). Some DMOs are at a preliminary stage of utilizing web technologies for marketing activities where the web site is used only to broadcast information by providing brochure-like information. Others may be more advanced and sophisticated in this regard, taking advantage of web technologies to make their business activities more effective and efficient, or even to reengineer the whole business practices. This pattern of web technology application is supported by recent research examining the alternative usage patterns as organizations adopt technology. For example, Contractor, Wasserman, and Faust's (1999) findings suggest that organizations use technology at three levels/stages: substitution, enhancement, and transformation. Substitution involves simple replacement of existing technology with new ones to accomplish the very same organizational tasks. Enhancement involves redesigning an existing process to make the best use of the new technology, improve product quality and provide additional and related services. Transformation involves taking a systems perspective on the role of the process within the organization as a whole and changing/reconceptualizing these processes in order to more effectively reach the consumer.

An important aspect of any Web marketing program relates to the strategies used to promote the web site. Like any business endeavor, a web site must attract customers by launching a good promotional plan targeted to the sites' audience. That is, the best site is a wasted resource if it is not used (Hanson, 2000). A site promotion plan should identify a set of specific actions that will inform a majority of the target audience that the website exists and is worth their time to visit. Typically, the plan will include both online and offline promotional activities that cover both the existing customer base and potential new customers. Several basic promotional tactics that cover these objectives may be: 1) Portal presence, including registration with online directory providers and search engines (Hanson, 2000); 2) Traded links, which is a basic activity wherein one web page references or points to other pages of associated information; 3) Paid links, which is actually an extension of online directory registration that provides the company with banners, basically paid-for advertisement, through the directory service or at related industry sites (Sweeney, 2000); and, 4) Placements in traditional media and editorial coverage, for example, adding the web page address and a small description of the service to printed collateral, advertisements, videos, first-response packages, and other marketing material.

Letting people know how to find your site is only the first part of a successful Web marketing story; the other part lies in retaining customers that visit the site and converting visitors into repeat customers (Hanson, 2000). By maintaining loyal customers tourism marketing organizations can capitalize on other benefits such as having more time to communicate the message of the site, additional opportunities to engage in commerce, more chances to build commitment and loyalty, and exposure to more advertising or alliance partner impression. These marketing techniques may include value-adding offerings to customers such as free news, email newsletters, personalization/customization, and special offers/best buys; security functions such as secure transaction through SSL, privacy policy, web seal certificate so that customers are assured that any activity on the site is safe and secure; site stickiness programs such as community functions and customer loyalty programs, etc.

Given the growing importance of web technology in organizational functions and areas of destination marketing organizations (DMOs), these organizations have to adapt to technological changes in order to succeed. Productivity in the information era requires the "right" combination of organization structure and culture, creative thinking, flexibility, and the ability to change and adapt quickly (Koch and Steinhauser, 1983). In addition, the technology climate of the organization and the availability of financial resources dedicated to the project have proved to be important factors for the successful implementation of new technologies. Thus, in this study it is argued that effective Web technology application, website marketing strategy and promotion strategy, and ultimately the overall success of the web marketing efforts will be determined by two groups of organization related variables: 1) Organization characteristics, which include size and innovativeness of the organization; and, 2) Organization technology climate such as management team's support to the web marketing efforts, and Internet literacy of the organization.

DMOs are primarily marketing organizations and providers of visitor services with little actual power to plan and develop. Their rationale is that by pooling the marketing resources of the public and private sectors together, a greater marketing impact can be achieved than in the case of dispersed efforts. However, since DMOs are usually small in size, lack of funding is oftentimes a constraining factor for them to develop and launch effective web marketing programs. This situation can be worse in the US where the hotel tax is the main source of income for many DMOs. Thus, larger DMOs that possess more human and financial resources tend to be more successful in their Web marketing efforts. Besides the organizational characteristics discussed above, the technology climate of these DMOs can be another important determining factor for the success of their web marketing strategy (Wang, Hila and Qualls, 2002). Technological adoption and utilization are motivated by attempting to achieve a competitive advantage (Utterback, 1982). Yet in the attempt to attain a competitive advantage through technology, all organizations are not necessarily created equal and their technology climate might be quite different since not all organizations start with the same foundation of technological understanding, expertise, or resources. In this study organization technology climate refers to the organization's degree of technocratization and environment in implementing technology-related projects such as technology expertise, management support, and the availability of technology training programs.

The relationships between the constructs and variables can be represented through a hierarchical structure of three layers. In this study, the overall successfulness of the DMOs' web marketing strategies is considered as the endogenous variable, organizational characteristics and technology climate are exogenous variables, and Web technology applications, site marketing strategy, and site promotion strategy are mediating variables; these relationships are represented in Fig. 1.

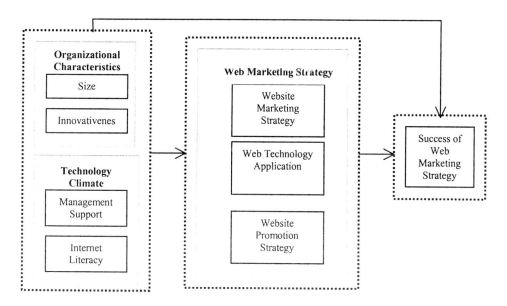

Fig. 1. Conceptual Model of Overall Web Marketing Success

4 Method

4.1 Sampling Frame and Data Collection

All the tourism offices at different levels (i.e., regional, county, and city levels) in the United States were used as the sampling frame of this study. A survey questionnaire was constructed and mailed to the CEO/Director of a sample of 600 convention and visitor bureaus. A cover letter was attached to the survey questionnaire explaining the purpose of the survey and seeking assistance and support from the tourism organizations. Two follow up mailings at an interval of two weeks were sent to those who did not respond. A total of 268 organizations returned the survey; responses from 260 were deemed usable, representing a 43 percent response rate.

4.2 Measures and Data Analysis

Multiple measures were used for the measurement of the exogenous constructs. Four items were used to define innovativeness by asking the respondents their organization's attitude to risks, reactions to outside changes, proactiveness in comparison to their competitors, and creativity. Organizational size was measured through the number of full time employees and yearly budget. Management technology support was measured using three items regarding management's support of and involvement in the Web technology implementation process, as well as the availability of technology training programs provided by the organizations. Organization's Internet literacy was measured by examining the technology expertise of both the management team and employees. Confirmatory factor analysis and Coefficient Alpha were used to assess the reliability of each construct; the results indicate that all the items loaded significantly on their respective constructs/factors, and Coefficient alpha values ranged from 0.73 to 0.90. The three mediating variables were measured using the summation of the number of web applications, website marketing or promotion strategies, respectively. Finally, the success of the overall web marketing strategy was measured using a single item (a 5 point bipolar scale) by asking the respondents (CEO/Director) how they would assess the successfulness of their organizations' web marketing strategies.

5 Results

5.1 Organization Profiles and Internet Capabilities

SPSS 10.0 and AMOS 4.0 were used to analyze the data. The profiles of the organizations and their Internet capabilities are reported in Table 1. Most of the tourism organizations surveyed are at county or city level, independent, and focus their marketing efforts on leisure travel (90.3%), meetings and conventions (63.7%), and festivals/special events (55.2%) being their major markets served. As expected, most of DMOs are small organizations with less than 10 full time employees (76.9%). In addition, though most of their computers are connected to the Internet and the majority of their employees have access to the Internet, their computing power is very limited.

Table 1 Organization Profiles and Internet Capabilities

Characteristics	Count	Percent (%)
Level of tourism office		
Regional level	21	8.4
County level	115	46.0
City level	114	45.6
Type of tourism office		
Independent organization	129	51.6
Part of county government	29	11.6
Division of economic development	4	1.6
Part of city government	24	9.6
Division of Chamber of Commerce	55	22.0
Major markets served		
Leisure travel	234	90.3
Sports	111	42.9
Business travel	106	40.9
Festival/Special events	143	55.2
Meetings/Convention	165	63.7
Other	38	14.7
Yearly budget		
Under $100,000	33	12.7
$100,001 - $250,000	60	23.1
$250,001 - $500,000	58	22.3
$500,001 - $750,000	19	7.3
$750,001 - $1,000,000	15	5.8
$1,000,000 and above	75	28.8
# of full time employees		
1 to 9	200	76.9
10 to 19	35	13.5
20 to 49	16	6.2
50 to 99	7	2.7
100 and above	2	0. .8
# of computers in use		
1 to 5	136	52.7
6 to 10	57	22.1
11 to 20	38	14.7
21 to 50	18	7.0
51 to 100	6	2.3
100 and above	3	1.2
% of employees having Internet access		
5% or less	8	3.1
6 to 25%	6	2.3
26 to 50%	6	2.3
51 to 75%	7	2.7
76 to 95%	25	9.7
96% or more	207	79.9

5.2   Web Marketing Strategies Adopted by Tourism Organizations

The overall marketing strategies were comprised of three major components: Web applications, website marketing strategies, and website promotion strategies. As shown in Table 2 most of the tourism organizations use Web applications in order to provide core information about the destination such as information about attraction, accommodation, restaurant, shopping, local events, location and direction, and travel guides/brochures. These are followed by applications providing peripheral information such as tour operator information, industry news, FAQs, and education materials. Importantly, applications for e-commerce activities are rare for these organizations. The tourism organizations appear to use a variety of different Website marketing strategies (Table 3). However, email newsletter, special offers/best buys, free news, and direct email campaign stand out as the major tactics. In contrast, they seldom employ customer relationship management tactics such as customer loyalty programs and community functions in their Website marketing efforts. As far as website promotion strategies are concerned, it seems that printing materials and search engines are the major sources they reply on for Website promotions (Table 4).

Table 2   Web Applications used by Tourism Organizations in the USA

Web Technology Applications	% of Organizations Utilizing	Rank
Activities/Attraction information	99.2	1
Accommodation information	99.2	2
Events calendar	97.3	3
Restaurant information	94.2	4
Shopping information	89.6	5
Brochure request capabilities	89.2	6
Links to regional/city/area pages	88.1	7
Maps/Driving directions	81.2	8
Travel guides/Brochures	77.3	9
Search functions	51.5	10
Tour operator information	50.8	11
Trip/Vacation planner	23.5	12
Industry news	22.3	13
Frequently asked questions	21.5	14
Online reservation	20.8	15
Banner advertisements	19.2	16
Education materials	18.8	17
Themed products	18.8	18
Publications/Reports	16.5	19
Virtual tours	13.8	20
Research & travel related statistics	9.6	21
Events tickets	9.6	22
Attraction tickets	8.8	23
Shopping carts & payment systems	8.5	24
Classified ads	3.8	25

Table 3 Website Marketing Tactics of Tourism Organizations in the USA

Website Marketing Strategies	% of Organizations Utilizing	Rank
Email newsletters	34.6	1
Highlight special offers/best buys	33.1	2
News	31.2	3
Direct email campaign	30.0	4
Interactive tools	28.5	5
Personalization/Customization	23.8	6
Privacy policy	22.3	7
Incentive programs to attract new customers	18.5	8
Cross-selling/up-selling opportunities	15.8	9
Secure transactions through SSL	10.4	10
Customer loyalty programs	4.2	11
Web seal certification	2.3	12
Community functions like chat rooms	0.8	13

Table 4 Website Promotion Strategies of Tourism Organizations in the USA

Website Promotion Strategies	% of Organizations Utilizing	Rank
Include Web address in organization's print materials	98.5	1
Register with search engines/directories	88.8	2
Magazine ads	84.2	3
Keywords for search engines	78.8	4
Include Web address in email signature	77.7	5
Newspaper ads	74.2	6
Page listings in search engines/directories	56.9	7
Meta-tags for search engines	58.5	8
Radio commercials	38.8	9
Direct marketing through email list	29.6	10
TV commercials	29.6	11
Billboards	28.5	12
Banner ads exchange	19.3	13

5.3   Model Testing

A path analysis was used to examine the relationships of variables proposed in the conceptual model; the goodness-of-fit indexes indicate that the model is robust and reliable (χ^2 = 62.63, df = 8, GFI = .94, NFI = .90, CFI = .90, and MRSEA = .06). Most of the hypothesized relationships were supported and the results are presented in Fig. 2. Specifically, Web technology application and website promotion strategy were found to be positively related to the overall web marketing strategy successfulness. Similarly, positive relationships were found between organization innovativeness, management support, Internet literacy and overall web marketing strategy successfulness. In addition, the path coefficients between the following

314

constructs were also found to be significant (a = .05): management support → Web application, management support → Website promotion strategy, organization size → Web application, organization size → Website marketing strategy, organization size → Website promotion strategy, Internet literacy → Website marketing strategy, Internet literacy → Web application, and Internet literacy → Website promotion strategy.

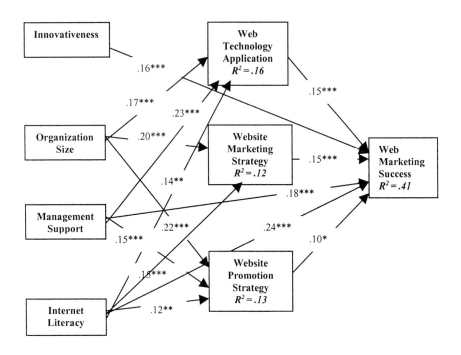

Fig. 2. Path Model and Coefficients of the Proposed Relationships

6   Discussion

The contribution of this study lies in providing empirical evidence assessing the determinants of successful web marketing strategies for these DMOs. The results of the study are consistent with previous research indicating that the integration of website marketing and promotion strategies and applications of web based technologies is a prerequisite for an effective overall marketing strategy. The study further indicates that successful integration requires a favorable organizational and technology environment and innovative approaches. The results provide further evidence indicating that successful implementation of web marketing strategy is more a management than a technical issue. That is, innovative capabilities need to be developed in order to be able to better respond to constant environmental change. In addition, creating a supportive organization technology culture is crucial in that a supportive organization technology culture reduces the resistance to effective technology implementation.

References

Contractor, N.S., Wasserman, S., & K. Faust (1999). Testing Multi-level, Multi-theoretical Hypotheses about Networks in 21st Century Organizational Forms: An Analytic Framework and Empirical Example, typewritten, University of Illinois.

Copeland, D., and J. L. McKenney (1998). Airline Reservation System: Lesson from History *MIS Quarterly*, 12 (3): 352.

Gretzel, U., and D. R. Fesenmaier (2001). Defining Internet Readiness for the Tourism Industry: Concepts and Case Study, in *Readings in E-Commerce,* Hannes Werthner (Editor*)*, Springer, 77 - 101.

Hanson, W. (2000). *Principles of Internet Marketing.* SouthWestern College Publishing.

Koch, D. L., and D. W. Steinhauser (1983). Changing Corporate Culture, *Datamatio*n, 29(10): 247-256.

Schmid, B. (1994). Electronic Markets in Tourism. In: Schertler, W., Schmid, B., Tjoa, A.M., Werthner, H. (eds): *Information and Communication Technologies in Tourism.* Springer, Vienna, 1-9, 1994.

Rothenberg, R. (1999). An Advertising Power, but Just What Does Doubleclick Do? *New York Times*, September 22. Available at:

http://www.nytimes.com/library/tech/99/09/biztech/technology/

Schultz, A. (1996). The Role of Computer Reservation Systems in the Travel Industry Today and in the Future. *Electronic Markets*, 6 (2). Available at:
http://www.businessmedia.net/netacademy/publications.nsf

Strassel, K.A. (1997). E-commerce Can Be Elusive. *Convergence*, 3/3, 1997.

Sweeney, S. (2000). *Internet Marketing for Your Tourism Business.* FL: Maximum Press.

Utterback, J. (1982). Innovation in Industry and the Diffusion of Technology. In M. Tushman and W. Moore (eds.), *Readings in the Management of Technology*, Marshfield, MA. Pitman, 102-124.

Yuan, Y. and D. R. Fesenmaier (2000). Preparing for the New Economy: The Use of the Internet and Intranet in American Convention and Visitor Bureaus, *Information Technology and Tourism*, 3(2): 71-86.

Werthner, H. (1996). Design Principles of Tourism Information Systems. In: Klein, S., Schmid, B., Tjoa, A.M., Werthner, H. (eds.), *Information and Communication Technologies in Tourism.* Wien, New York, 70-78, 1996.

Wang, Y., Hila, R., and W. Qualls (2002). Technology Adoption by Organization: The Effects of Organizational Capacity and Strategic Orientation. Proceedings, the Annual CBIM/ISBM Atlanta Conference, February 8-10, 2002, Atlanta, Georgia.

An Investigation into how Data collected by Destination Websites are Utilised as a Direct Marketing Tool

Hilary C. Murphy

Swansea Business School, Swansea Institute of Higher Education,
Wales, UK

hilary.murphy@sihe.ac.uk

Abstract

As buyers become more sophisticated in the use of the Internet when purchasing travel and tourism product, the wealth, breadth and depth of information collected as a result of site visits becomes massive. Therefore, the careful mining of this data becomes crucial to match customers and product offerings. Destination tourism planners must consider how many tourists are desired and which segments to attract and tourism marketers must know actual and potential customers and their needs and wants; determine which target markets to serve and decide on appropriate product offerings, services and marketing strategies. This new media offers the chance to improve the richness and reach of marketing communications with their target segments, particularly through e-mail marketing. The purpose of this paper is to examine how UK destinations collect, analyse and utilise their marketing data from their websites and transform them into meaningful marketing information that can be used in well-targeted direct marketing campaigns.

Keywords: direct marketing, destinations, database, e- mail marketing

1 Introduction

A variety of statistics show that the Internet and websites are becoming an increasingly important means of marketing and selling travel and tourism products, particularly flights and accommodation. Most major European destinations at city, regional and national level have an Internet presence, in various forms, ranging from a simple brochure style website to a fully developed portal for all tourist amenities within a destination.

A key way to focus on target markets and develop deeper customer relationships is by the use of *direct marketing*. The term direct marketing has taken on new meanings over the last few years and has broadened in focus from simply a form of marketing in which products or services move from producer to consumer without the use of an intermediary to a much broader definition that: "encompasses the use of one or more media to affect a measurable response and /or transaction at any location" (Thomas and Housden, 2002:4). Direct marketing allows precision targeting, personalisation, privacy and measurability. It usually means utilising a marketing database, incorporating some kind of offer and a specific call to action. Traditional direct marketing (i.e. via postal services) has already been used in the marketing of destinations (Kotler et al, 2000).

Many destinations have *traditional mailing lists* compiled as a result of fulfillment and also buy in mailing lists (Main, 2002). However, they fail to maximise the potential of the data collected internally from the websites and from direct e-mail enquiries. Additionally, there was a wide discrepancy on the integrity, integration and recency of these lists and the various lists were rarely stored on a central database where they could be mined and harvested for customer profiling. Some destinations actively engage in *e-mail marketing* activities e.g. to specialist travel and tour operators and to individual customers with specific campaigns. They find these campaigns difficult to track and the success of these campaigns was not always measured.

A review of both current trade publications and research literature into e-mail marketing and destinations indicates the growth of direct marketing as a marketing tool (Desai et al. 2001) and the decline of the mass market and growth of niche markets in tourism (Hsu and Kang, 2002). As markets have become more segmented we see the growth of, for example, city breaks which appeal most to families without children and the decline of the package holiday (http://www.scotexchange.net [March 10,2002]). Therefore, to target these niches with specific offerings is crucial to destination marketing.

This research looks at how UK destination websites collect data from site visitors and enquiries, both overtly via web-forms and covertly via "cookie files" and sophisticated tracking software. It then investigates the utilisation of that data, focusing on direct e-marketing activities, predominately in e-mail marketing and viral marketing.

2 Issues: Destinations and Marketing

2.1 The Internet and Marketing

The Internet is a growth channel for purchasing tourism products e.g. 13% of all Austrian travel bookings are through the Internet (European Internet Travel Monitor, 2002). A variety of statistics show that the Internet is becoming an increasingly important means of marketing and selling travel products, particularly flights and accommodation and provides extended choice and the means for consumers to choose and book without paying travel agent's commission. Most destinations have an Internet presence, in various forms, ranging from a simple website to a portal for cultural, travel, tourist amenities within a destination e.g. Tiscover, visitscotland. Most official destination websites are developed by National Tourism Offices and have public or quasi-public government funding. However this does not mean that the official tourist boards have exclusivity on the marketing of the destinations, by traditional means or electronic. Some regions divide responsibilities for promoting tourism and hospitality business between public and private sectors. In other situations there can be internal conflict within the destinations over promotion. The same can be said of their websites where destinations frequently have a variety of private and public agencies promoting their products and services. Therefore the control of the marketing activities for a destination, its image and communication with its stakeholders, is not solely in the hands of the official Tourist Boards.

A key way to focus on target markets is by the use of *direct marketing*. Direct marketing connects with individual customers to obtain an immediate response and generate lasting customer relationships. Direct marketing via websites is interactive and immediate. It reduces costs, can be timed accurately and offers flexibility (Kotler, 2000). Today's buyers are better informed and more selective (Thomas and Housden, 2002). There is more choice, more information available and greater pressure on consumer budgets. Whilst buying patterns and preferences have been changing, the major advances in technology mean that destinations can now identify the needs and motivations in diverse groups of customers and fulfil those needs cost efficiently. It is possible to identify "profiles" of customers from information collected from websites and use them to develop a whole series of selling propositions, which closely meet the real needs of our customer segments. This enables destinations to be more customer-focused in the truer sense. Direct marketing has already been extensively used in the marketing of cities. The city of Cleveland ran a direct marketing campaign in 1995 and received 70,000 visitor information requests in one quarter as a result, more than the whole total of inquiries for all of 1994, (Kotler, 2000).

2.2 Database Marketing and E-Mail Marketing

Database marketing has played an important role for many travel and tourism operators. Technological advances and the growth of data collection through Internet sites have fuelled the increase in database marketing. The travel and tourism sector has not been associated with sophisticated marketing practices (Alford, 1999). Recent evidence supports the theory that the travel and tourism industry is adopting more sophisticated techniques in database marketing. Alford would argue that there is a distinct difference between direct marketing and database marketing and that database marketing adopts a more holistic approach. Thomas and Housden would contend that a database is an essential component of direct marketing (Thomas and Housden, 2002). It is a the heart of market communications planning and, if correctly used, enables the destination to identify the right target for a particular communication, according to their potential value and their propensity to be interested in the offer. It tells us the right time to send the information and the right form of words to use. The database helps us to record and analyse the response in order to prioritise and target in the future. The scope, integrity of data that is fed into the database is crucial. Discussion seems to revolve, for many authors (Marinova et al. 2002; Mitchell, 2002) on the "intimacy" of the relationship between seller and consumer and the degree of information the customer is willing to impart and the level to which the seller can mine the data on the customer. Ultimately the level of the relationship will be limited by the customer, who perhaps does not want to be loyal or disclose data. Shaw proposes that to practise database marketing successfully it has to be part of a "virtuous planning" cycle where trust is established and built up over a lengthy period of time (Shaw, 1991).

A recent report has forecast that the market for *e-mail marketing* will be worth well over USD 1 billion by 2003 (http://www.nua.com/surveys/index.cgi[March 18, 2002]). The main benefits of e-mail marketing are simplicity, cost effectiveness and strength in customer retention. Companies which have tested e-mail marketing have predominately used their own lists. Increasingly, externally created e-mail lists are being used. The Henley Centre speculates that by 2009, 85% of UK consumers will be on-line, 2 hundred billion e-mails will be sent in 2004 and outsourced e-mail marketing activities will create a 3Billion pound industry (Gibson, 2000).

E-mails are cost effective, have global coverage, are fast and interactive. If they are personalised and use appropriate subject line headings then they are more likely to be read. On the negative side, there are no benchmarks established. There are, additionally, problems in areas such as: spamming, privacy and legislation. Ecommerce times reports that spam now represents a third of all e-mail sent, which marks a 15 per cent increase in the number of spam e-mails sent during 2001 (http://www.nua.com/surveys/index.cgi [Oct 1 2002]). The concerns about privacy are being addressed with the European parliament reviewing all aspects of the subject and have presented their findings in a document entitled " The Common Position- with a view to the adoption of the Directive of the European Parliament and of the Council concerning the processing of personal data and the protection of privacy in the electronic communications sector". It also proposes to ban e-mail "spam" throughout the EU. Consumers International's recent report highlights the fact that 58% of sites that collected information had a privacy policy but only a third highlighted any privacy policy when collecting data from sites. It also points out that very few of the privacy policies that did exist contained more than a bare minimum of information about the control consumers has over their own information. The cost of compliance with UK data protection laws and the directives emerging from the EU are a major concern for companies handling marketing data.
(http://www.which.net/whatsnew/pr/jan01,[Oct 1, 2002])

There is little available data on statistics on *destination's e-mail marketing activity*, however, there are indications that most cultural institutions within cities have a low e-mail usage rate varying from 6-11% of cultural establishments in European cities (Wober et al, 2000). He states that they have not fully realised the potential of e-mail and Internet for marketing activities and the possibility of reaching potential Internet customers, whose demographic profiles would seem to match the segment profiles that destinations are targeting. However, more generalised statistics indicate that consumers are receptive to e-mail marketing, particularly where permission has been sought, (http://www.e-marketer.com/articles.htm[March 23rd, 2001]). Predictions about behavioural profiles of travellers on the basis of Internet use segmentation suggest that destination marketers could benefit by knowledge of the determinants in Internet user characteristics (Bonn et al. 1999). This knowledge, Bonn suggests, permits consumers, service providers and marketing professionals to efficiently contact particular market segments through the Internet. Internet users are also more likely to be interactive with suppliers in terms of customer relationship management (Marketing Week, 14th March 2002). The potential is there for destination marketers to expand from a broadcast mass marketing medium to a tool that is useful for micro marketing and one- to- one marketing. For destinations this could mean the potential to reach visitor groups and niche segments that are not motivated by typical promotional activities. Customer and prospect databases are now an increasingly valuable asset. Consumers are demanding better targeting, so segmentation is going to have to work much harder and profiling must be accurate.

In addition there have been reviews of web activities in destinations in Germany (Dierich et al. 2002). 47% of these destinations "did not have knowledge about their target group" though 66% had stated that that they had built up information sources via incoming e-mails, on line bookings and through participation at online contest and games. This author also goes on to state out of those 66%, only 70% used the information gathered for promotional strategies, 92 % are not familiar with user profiles and 40% got no information about website promotion efficiency.

In an era where the public is more cynical about marketing activities the role of "peer to peer conversations" becomes a more important influence in the consumer buyer process. One of the most effective promotional activities is that of *word of mouth*, the process by which "an innovation is communicated through certain channels over a period of time among the members of a social system", (Fill, 1999:142). The Internet equivalent of word of mouth is "viral marketing" There are many terms that have been coined in the Internet era and the term "viral marketing" has also been referred to as "contagion marketing", "propagation marketing", "inertia marketing" and "multi-connected marketing". Nonetheless, many observers have described it as one of the most effective marketing methods on the Internet and is viewed as a virus to be admired (Wilson, 2001). Viral marketing is a strategy that encourages individuals to pass on marketing messages to others creating the potential for exponential growth in the messages exposure and influence. Viral marketing, in effect, encourages others to pass on the message with the consequent potential to multiply rapidly and communicate the message to thousands, or even millions mainly with the objective of driving business to your website. Can destinations target and nurture a key group of customer to be "champions" of their product and add creativity to their direct marketing campaigns? Limited research has been carried out on the use of viral marketing in the tourism sector, however, recent research reveals that , though virals are being considered by city destinations in Europe, they are not yet being utilised (Main , 2002).

2.3 Permission Marketing

Permission marketing is the idea where the marketer attempts to "cultivate a relationship with customers who have given him the go- ahead to send them information about a product, service, special offer or sale" (Godin, 1999). The main driver behind this concept is in the efficiency and efficacy of the marketing efforts of an organisation and, in direct marketing terms, this should decrease the mailing volume and raise the percentage for success. Permission marketing has been likened to "dating a customer" and that it requires a long-term process that requires an investment of time, information and resources by both parties (Godin, 1999). The result is an active, participatory and interactive relationship between customer and supplier. Customers may grant various levels of permission to engage with them and a key part is the level of trust established with the customer. They, the customers, may grant increasing levels of access from a limited access, one-off contact to the highest level of contact and trust where the customer trusts the marketer to make buying decisions for them. Permission marketing will improve the image of e-mail marketing where customers will choose to "opt-in" to receive marketing communications. Generally "opt-in" lists are targeted e-mail lists, which offer a "politically correct" way to reach the target audience on the net. Though targeted e-mail is more costly it has been proven to be more cost effective.

3 Methodology

The selection of the methods as well as the sample had to take into account some key considerations, destinations and their websites are marketed by a variety of agents via websites from brokers e.g. last minute.com to publicly funded sources, e.g. Wales Tourist Board. This study focuses on the publicly funded destination websites that are in the UK and contact details were retrieved from their websites. The destinations in this case, regional tourist boards, were contacted via semi-structured questionnaires by mail, at first, then followed up by fax and e-mail as the postal response rate was poor initially. Limitations must be acknowledged here in terms of response rate and bias introduced by the varied methods of data collection but this was minimised by using identical questionnaires and covering letters for all channels of communication with the tourist boards ensuring consistency in approach. Almost 42% of the regional tourist boards' marketing departments responded to the questionnaire with 26 being contacted and 11 completing the questionnaire. A semi- structured questionnaire was chosen as it was clear from the literature review that though the key issues and criteria were apparent that little published data was available in terms of the issues highlighted in this paper and therefore responses were difficult to anticipate. The results of the survey were analysed by inspection and the data is presented in both a quantitative manner and in a qualitative manner in terms of clusters of answers and generic themes in the responses to the open-ended questions. Respondents were guaranteed anonymity.

The questionnaire itself was divided into 4 key sections which were: 1: The traditional mailing list, direct marketing activities, sources, frequency of use, targets, 2: The database, sources, recency, maintenance and integrity 3: The destination's website, use of forms, cookie files, data collection opportunities and 4: The direct marketing activities, the campaigns, asking permission, intermediaries, measurement, incentives and creative use of e-mails and "virals".

Limitations are acknowledged in the scope of the survey in that it is UK based and therefore subject to similar public funding mechanisms and to the government policy in vogue. However by concentrating on a specific sector, contextual factors and cultural considerations may be minimised and a picture revealed of the utilisation of direct marketing activities of UK destinations.

4 Results

In terms of their *traditional direct marketing activities*, the destinations still used traditional mailing list, however only one respondent was still compiling mailing lists manually, all were catching data on the websites and only two responded that they bought their mail list in from a broker. These mailshots were conducted by 45% of respondents 1-3 times a year, the remaining number engaging in more frequent campaigns, 4-8 times a year. These were targeted at specific segments of the travel market, which were group travel organisations, travel trade and short break. Some traditional mailshot campaigns were targeted at individual customers but focused mainly on the trade. Direct mailshots were not the only form of traditional direct marketing these tourist boards engaged in, many also used inserts in magazines and direct response television advertisements as part of their direct marketing strategies.

The data on all destination *database* was compiled from similar sources i.e. existing and loyal customers but was only updated "continuously" on a real-time basis by 55% of respondents, the remaining were updated weekly or monthly. Only one respondent used various internal sources as a way of upgrading and supplementing the data and one used data from a "swap/ share" source. 36% of respondents maintained their database "in-house" with the remaining 73% outsourcing this activity with those outsourcing more likely to have their databases updated less frequently. Front-end access was via "Access" for all respondents. All but one respondent had their websites linked to their existing legacy database and this respondent stated that they maintained their websites database intact from data retrieved from other sources

Their *websites* claimed traffic of between 23,000 and 100,000 unique visitors per month, though consideration must be given to the fact that website traffic may be "counted" in a number of ways. All the websites used " forms" to collect data about visitors to their websites, though only 73% used " cookie" files to track site activity and behaviours on their websites itself. The Tourist Boards stated that there were, on average, between 1- 5 other opportunities to collect data from clients while they were on site. This included forms of various lengths, "opt-in" sections and joining interest groups and mailing lists. All respondents produced some form of reports or analysis from the data collected from their websites. When asked to specify which particular types of reports, they all fell into the category of web based statistics, web trends, page impressions, referral to member sites, number of hits etc. and only one stored web customer profiles.

All websites responded that they engaged in *direct marketing* via e-mail and these took the form of e-brochures, targeted e-shots and seasonal offers. Only two sought to personalise these campaigns. All specified that these campaigns were directed at specific market segments e.g. gardeners, attraction goers. All but a small percentage sought " permission" to contact their customer with direct e-mails, with 73% adopting an "opt-out" approach as opposed to the remainder who chose the "opt-in" approach. Two respondents used intermediaries to design or execute their direct marketing for them though no indication was given in terms of level of involvement between agency/ intermediary and Tourist Board. They all measured the success of direct e-mail campaigns via a selection of methods e.g hits on the "landing page", voucher redemption, response via advertisers. The incentives offered in their campaigns were limited to competitions and prize draws with no real innovative approach in terms of a "call to action". No real creative use of e-mail or virals was indicated in the responses.

In response to the open-ended questions, the issues that concerned respondents were those of privacy, "annoying customers with too much contact" "e-mail addresses changed frequently" and "compliance with legislation".

5 Discussion and Recommendations

A good clean well maintained customer database is arguably any company's best asset (Murphy, 2002). The problem arises here for destinations in that customer data seems to reside in different areas of the business and few seem to spend the time keeping the data up-to date. Once in place it opens up a wealth of possibilities for marketing purposes, enabling the segmentation of markets according to types of products of interest, time of year they are likely to buy, their purchasing history and demographic profile. The destination websites researched seem to focus on what they wished to offer as opposed to how customers choose to buy. Ideally the database should be spilt into a front end and back end. A back end where the data is continuously updated from websites and legacy systems and the front end, the revenue-earning end, with software tools that can help design effective campaigns.

There appears to be minimal activity on customer profiling. With the volume of traffic to websites and the variety of products and services marketed there, it may seem to be an insurmountable task. However, the profiling could be done incrementally as visitors return to a site or focused on a particular segment. Destinations seem to utilise direct marketing to fill up spare capacity, whereas a better approach might be to be customer focused and use direct marketing campaigns that match their customer needs (Mitchell, 2002). There are sophisticated tools available that design campaigns and sophisticated CRM solutions, the drawback being the expense, and the statistical ability of the marketer to mine the data. There is evidence to suggest that by using pro-active, consumer- alerting technology, such as e-mail, suppliers and providers can not only deepen customer relationships, but can also reduce service costs by 33 per cent, (http://www.jmmm.com/xp/jmm/press, [Oct 1,2002]).

Concern was expressed about "pestering" the customer too frequently, respondents did not want e-mails to be viewed by customers as an intrusion nor they did not want to be seen to be delivering junk mail. They also felt their customers suffered from information overload in general. This makes profiling more important and gaining permission to contact them crucial, via "opt-in" rather than the destination's choice of "opt-out". However, getting the timing and frequency of contact is essential. Recommendations are that contact is made every 20-30 days (http:://www.htmail.com/article6.html,[Oct 4, 2002]) and that keeping your name in front of the customer builds trust through familiarity. It is, therefore, vital that whatever message is used it is "something worth sharing" and relevant to the customer profile. Certainly, gaining permission and personalising the offer will increase the positive reception of an e-mailed marketing communication.

There were limited incentives used in their e-mail marketing activities and no real creative use was revealed in this survey on e-mail marketing. Customers are overwhelmed and, indeed, disinterested in traditional direct mass marketing techniques and they could utilise viral marketing, " peer to peer conversations", as a way of promoting the destination to target market segments. It also provides an opportunity to nurture product champions and build customer relationships. All destination-marketing departments in this research were in possession of the necessary databases and all they needed to do was design the necessary viral message for an initial, receptive target audience. Viral marketing also provides the opportunity to be creative about the type of viral message- this could take the form of a video clip about the destination itself.

This research indicated that the breadth and depth of data collected via the websites is not being exploited. Web statistics are limited to those of general trends, hits and page impressions. A more detailed analysis of logging files and customer tracking files would produce more meaningful statistics for marketing purposes e.g. exit information, time spent on page etc. would all assist on designing the website to meet customer criteria.

As competition between destinations increases, any competitive advantage becomes important and direct marketing has an important role to play in developing customer relationships. Viral marketing has further rewards in that it could be an effective means to harvest e-mail addresses for further marketing activities, it can help to build relationships with product champions and it can also be used to help build stronger links with tourism partners, internal markets and other stakeholders. Viral marketing may help make your website work harder for you and extend the depth and width of customer relationships.

If destination marketing fails to meet the challenge of direct marketing, perhaps they should consider outsourcing this function to agencies or intermediaries that could effectively manage and harvest the data and create successful direct marketing campaigns on their behalf.

References

Alford, P. (1999). Database Marketing in Travel and Tourism, *Travel and Tourism Analyst*, 1,87-105.

Bauer, H. Grether, M. and Leach M. (1999). Customer Relations through the Internet, *Working paper, Manneheim University*,1-18.

Bonn, M., Furr ,H. L., & Susskind, A., (1999). Predicting a Behavioural Profile for Pleasure Travellers on the Basis of Internet Use Segmentation, *Journal of Travel Research*, 37, May 1999, 333-340.

Bly, R., Felt, M., Roberts, S. (2000). *Internet Direct Mail*, New York, McGraw Hill.

Brady, M., Saren M., Tzokas, N. (2002). Integrating Information Technology into Marketing Practice- the IT Reality of Contemporary Marketing Practice, *Journal of Marketing Management*, 18,(5/6),555-578.

Coviello, N. Milley, R., Marcolin B. (2001). Understanding IT-enabled Interactivity in Contemporary Marketing, *Journal of Interactive Marketing*, 15(4), 18-33.

Desai, C., Fletcher, K., Wright, G. (2001). Drivers in the Adoption And Sophistication of Database Marketing in the Services Sector, *The Services Industry Journal*, 21(4), 17-32.

Dierich, J-C. et al. (2002). *The Use of Internet- Marketing by National Tourism Organisations in Germany- an Empirical Approach* in Information and Communications Technologies in Tourism 2002,K. Wober, A.Frew & M.Hitz(eds), NewYork, SpringerWein,115-225.

European Internet Travel Monitor, (2002), *Internet Booking Statistics*, March, 2002.

Fill, C. (1999). *Marketing Communications: Context, Contents and Strategies*, 2nd Ed., Europe, Prentice Hall.

Gibson, T. (2002). New Horizons in E-mail Marketing, *Direct Marketing Strategies*, 2(3) ,6-14.

Godin, S. (1999). *Permission Marketing: Turning Strangers into Friends and Friends into Customers*, New York, Simon and Schuster.

Hsu, C., Kang, S. (2002). Psychographic and Demographic Profiles of Niche Market Leisure Travellers, *Journal of Hospitality and Tourism Research*, 6, (1), 3-22.

Kotler, P., Bowen, J. & Makens, J. (2000). *Marketing for Hospitality and Tourism*, 2nd ed, New Jersey, Prentice Hall.

Main, H. (2002). *Developing the use of Viral Marketing in the Context of Web-based Marketing Communications Strategy for City Destinations* in City Tourism 2002, K.Wober(ed),286-295 New York, SpringWein..

Marketing Week (2002) *Feedback*-Taylor Nelson Sofres, 14th March 2002.

Marinova, A., Murphy, J., Massey, B. (2002) Permission E-mail Marketing, *Cornell Hotel and Restaurant Quarterly*, February, 61-70.

Mitchell, A. (2002), In One to One Marketing, Which One Comes First, *Interactive Marketing*, 1(4), 354-367.

Murphy, D. (2002), Data, Data Everywhere, *Marketing Business*, July/Aug 2002, 29-31.

Roberts, M-L. (2003). *Internet Marketing: Integrating On-line and Off-line Strategies*, New York, McGraw-Hill.

Shaw, R. (1991). *Computer Aided Marketing and Selling*, Oxford, Butterworth-Heinemann.

Solomon, M., Bamossy, G. & Askegaard, S. (2001). *Consumer Behaviour*, Harlow, FT Prentice Hall.

Thomas, B. and Housden, M. (2002). *Direct Marketing in Practice*, Oxford, Butterworth-Heinneman.

Wober, K., Grabler, K., & Jeng, J.M. (2000). Marketing Professionalism of Cultural Institutions in Europe, *Journal of Euromarketing*, 9(4), 33-55

Wilson, R. (2001). *Internet Archive Articles*, hhtp//www.wilsonweb.com{May 25, 2001]

http://www.wilsonweb.com/earticles/virals.htm[May 25, 2001]

http://www.euromonitor.com./gmidv1[March 9,2002]

http://www.scotexchange.net/knowyourmarket/kym-niche.htm[March 10,2002]

http://www.nua.com/surveys/index.cgi[March 18, 2002]

http://www.e-marketer.com/articles.htm[March 23rd, 2001]

http://www.nua.com/surveys/index.cgi [Oct 1 2002])

http://www.jmmm.com/xp/jmm/press, [Oct 1,2002

http://w.which.net/whatsnew/pr/jan01,[Oct 1, 2002])

http://www.htmail.com/article6.html,[Oct 4, 2002])

An Investigation into the Current Status of Corporate Travel Management in South Africa and the Use of Online and Wireless Technologies by Corporate Travellers

Berendien A Lubbe

Department of Tourism Management
University of Pretoria, South Africa
blubbe@orion.up.ac.za

Abstract

Centralising the corporate travel function in organisations with substantial travel expenditure is becoming an established practice worldwide. South African organisations follow this trend. However, corporations in South Africa, in contrast to the findings of studies done in the USA, has shown a minimal adoption of self-booking tools that allow corporate travellers to book their own travel by making direct contact with GDS or supplier inventory systems through corporate intranets accessible via personal computers and personal digital assistants and wireless application protocol. More importantly, it appears that corporations currently do not encourage corporate travellers to use self-booking tools. This research shows that the corporate travel agency is still in a strong position as the preferred distribution channel. Based on secondary research and supported by these results, this paper proposes a conceptual model for successful corporate travel management. This model requires ongoing research in the field as well as testing.

Keywords: Corporate travel, management model, corporate travellers, technology.

1 Introduction

Approximately 40% of the total travel market in South Africa can be classified as corporate or business-related. It is expected that by 2003 more than half of the travel business in South Africa will be driven by corporate travel (Palapies, 2001). Corporate travel spending for large organisations in South Africa range between $1 million and $10 million per annum and these costs are steadily on the increase. A contributing factor is that, in most cases, travel and entertainment expenditure is not incurred by purchasing professionals, but by the travellers themselves. The increasing complexity of available travel products and services adds to the dilemma of managing the travel account. According to Nigel Adams, previous Managing Director of American Express Travel South Africa, there are "as many as 27 different fee structures on a single flight where a difference of a flight connection after 23 hours and 55 minutes or 24 hours and 15 minutes can bring the price of a $3 000 flight down by as much as $700" (R/$ conversion added).

The pressure to implement changes in corporate travel management is mounting. This is due to the introduction and proliferation of online corporate booking tools; the increasing rate of adoption of corporate travel technology by companies (Chircu, Kauffman and Keskey, 2001); and the introduction of management fees by the travel management companies. These changes also mean that the role of the travel supplier and the travel management company is changing fundamentally with traditional travel servicing making way for partnership agreements with corporate clients.

Two main issues face the corporate sector, the management and control of travel expenditure and the management of the travel process to increase service levels to an optimum cost-benefit ratio. Evolving technology is central to both these issues. As a departure point this paper proposes a conceptual management model for corporate travel, which is loosely based on the so-called "soft value management model" developed by Liu and Leung (2002:341). The management model is conceptualised for application in a broader study on corporate travel in South Africa, of which this paper forms the first part. In this paper the initial results of the current status of corporate travel and, in particular, the use of technology is reviewed in the context of the proposed management model.

2 A conceptual corporate travel management model

2.1 The soft value management model

The soft value management model was specifically described in the context of project management in the engineering field but the underlying elements of this model are similar to those found in corporate travel management. Simply stated Liu and Leung (2002) say that value management is represented by the input of internal (personal/individual values) and external (environmental pressures) factors that triggers a decision-making process producing decision outcomes. This leads to the definition of project goals (figure 1). The transformation of values into goals – as a decision–making process – comprises objective setting, objective analysis and alternatives evaluation. The goals being set will initiate required action towards project realisation. Goals are underpinned by an individuals value system. What the person attaches value to affects the content specificity of the goals set. When more than one individual is involved in goal setting (as in corporate travel management), value conflict may lead to goal conflict. Agreement on value standards between individual participants, referred to as "value specificity", is a prerequisite to agreement of goals, also referred to as "goal specificity". The nature of the conflict (which may be inter- and intra-personal) arises from the expected probabilities of alternative actions and possible outcomes. This conflict may be latent or manifest and should be brought forth, supporting participative and creative problem-solving and stimulating the desire to transform value specificity to goal specificity. "People are more easily committed to performance with specific goals; vague goals have shown to

328

have a deterrent effect on high performance" (Liu and Leung 2002:345). Feedback is part of the goal-action-outcome cycle that guides the individuals' behaviour (through considerations of options/alternatives) towards goal attainment, and also provides a basis for assessing the project outcome. In determining the success of a project the individual considers the utility of its success/failure and the level of his/her satisfaction/dissatisfaction. Satisfaction is measured in terms of what the person wants to gain and or keep in the outcome and goal attainment.

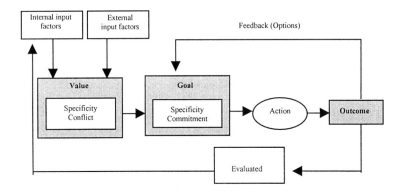

Fig 1. The soft value management model
(Adapted from Liu and Leung, 2002: 346)

2.2 Corporate Travel

Corporate travel can be described as travel undertaken by the employees of a particular organisation that has a substantial travel volume. Corporate travel management involves the centralised management of all travel and, in many instances, the establishment of a corporate travel department that oversees the entire travel programme. One of the primary functions of such a department is to consolidate the travel volume to obtain leverage from travel providers such as airlines, hotels and car rental firms in the negotiation of contracts at reduced prices (Bell and Morey, 1995). The overall goal of corporate travel departments is to control travel-related corporate spending, given the trips that firms take (Anderson, Lewis and Parker, 1999). In the pursuit of this goal, the corporate travel department is dependent on the support of individuals in senior management and the co-operation of corporate travellers (internal input), and on the support of travel providers and the travel management company (external input). Value conflicts can occur between corporate clients, suppliers and travel management companies because corporate clients typically want flexibility in their travel arrangements; airlines and other travel suppliers want to maximise yields by optimising revenue and travel management companies want to service client needs at the lowest cost (Roodt, 2001). Value conflicts also occur:

- Between management and travellers due to cost containment actions versus traveller comfort and "self-esteem" (Gilbert and Morris, 1995)
- Between suppliers, travel management companies and the corporate client in optimising the travel management process, particularly in the area of technology application, where the client often faces both a traditional channel partner and an electronic commerce channel partner. There is often confusion about which channel they should support because it is not clear whether the electronic commerce partner will be successful in the long term or if the traditional channel partner will be able to re-establish itself. According to Subramani and Walden (2000) having both a traditional and an electronic commerce partner allows the firm to play one off against the other, with the result that the new partner will not have the same level of incentive to invest in the relationship as he would otherwise.
- Between management and travellers in policy monitoring and compliance (Gilbert and Morris, 1995).

Value management through specification and agreement of values and goals for effective corporate travel management leads to commitment by all participants to appropriate action and goal attainment. This study covers the current status of corporate travel in South Africa, the relationship between certain participants and, in particular the use of technology in the corporate travel process. This review serves as a prelude to the future application and testing of the model.

3 Methodology

A quantitative ex post facto survey was undertaken. A questionnaire was sent to the individual concerned with managing the corporate travel function at each of 350 South African corporations registered on the Association of Corporate and Travel Executives (ACTE) South African database. These individuals were either officially designated as corporate travel managers or appointed to manage corporate travel as part of their designated function, which ranged from that of financial director to purchasing or procurement manager. The selected organisations have a substantial travel spend and are spread over diverse industries. This population was selected because membership of and participation in ACTE forums indicate awareness (and willingness) by these organisations to effectively manage corporate travel. Structured qualitative interviews with suppliers and travel companies were conducted, together with an assessment of current and available online booking tools and other electronic channels in preparation for the formulation of the questionnaire. The questionnaire was pre-tested on a small group of companies and then dispatched in electronic form. To ensure the highest possible response rate an incentive prize of two return airline tickets from South Africa to Switzerland, sponsored by Swiss Airlines, were given in a lucky draw. Follow-up of non-respondents was done telephonically with a second round of questionnaires being dispatched where necessary. For this paper, data-analysis primarily included the use of descriptive statistics of variables to assess the current state of travel-related activities in selected South African corporations; the organisational placement of corporate travel; current technology used; and the perceptions, concerns and expectations of corporations with regard to emerging

330

technologies. Cross-tabulation was used for comparing across size of organisations and current travel processes. Chi-square tests were done to establish statistical significance of results, where applicable. Results for this first part of the study are presented in this paper and cover mainly the use of technology in corporate travel in South Africa.

4 Results

The questionnaire was successfully sent to 350 electronic mail addresses and 109 responses were received, with 3 being unusable, this represents a response rate of 30%, which is considered good if the length of the questionnaire is taken into account. For the purposes of this paper, and mainly due to length limitations, only certain findings are presented.

4.1 Corporate profile and most-used distribution channel

For the purposes of cross tabulation, to establish whether the nature or size of organisations could be linked to the centralisation of the travel function and the use of technology, organisations were classified, amongst others, into the following categories: Ownership - mainly South African owned (57%) mainly international with offices in South Africa (43%); public company (47%), private company (42%), parastatal (9%), government organisation (2%); Industry - it was found that the organisations were spread over a diverse number of industries with the most being found in the manufacturing and financial services industries; Annual corporate travel expenditure - less than $1 million (53%), more than $1 million (47%). It was evident from the results of the cross tabulation that the nature and size of organisations did not affect the propensity of an organisation to centralise the corporate travel function, to use designated travel agents or to use technology for travel arrangements. Currently 89% of organisations use travel agencies (either in-house or outside) for travel arrangements and other travel-related services (figure 2). Of the remaining organisations 2% work through suppliers and 9% work through both travel agencies and suppliers.

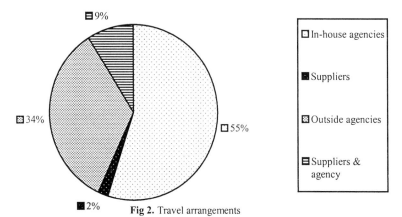

Fig 2. Travel arrangements

4.2 Service levels of travel agencies

Table 1 indicates how organisations perceive the level of service provided by their travel agencies. The service level of travel reservations is considered to be good (74%) or acceptable (26%) with other services such as travel advice, policy development and supplier negotiations showing more diversity of opinion. In a number of instances the travel agencies do not provide the service at all, for example, summary of car rental expenditure, policy development, compliance and monitoring and technology support and access. As far as technology support and access is concerned, only 20% of organisations rated the service as good with 46% rating it as acceptable and 12 % as unacceptable. In 21 % of the cases the service is not provided at all. Organisations were also asked what they believed the role of the travel agent should be in the future: that of travel advice and reservations (26%); that of travel management companies (63%); as irrelevant (3%); and other, such as business partners (8%). The vast majority of organisations see a role for travel agents in some form or other in the future.

Table 1. The level of service provided by designated travel agencies

	Good	Acceptable	Unacceptable	Does not provide the service
Travel reservations	74%	26%	-	-
Travel advice	40%	47%	11%	2%
Summary on airline expenditure	54%	35%	5%	6%
Summary on accommodation expenditure	41%	36%	7%	15%
Summary on car rental expenditure	35%	40%	3%	22%
Reconciliation reports matching reservations and actual billing	28%	42%	9%	20%
Policy development	21%	36%	5%	38%
Policy compliance monitoring	35%	29%	9%	26%
Supplier negotiations	33%	36%	11%	20%
Technology support and access	20%	46%	12%	21%
VIP services	28%	43%	9%	21%

4.3 Technology utilisation

Respondents were categorised into two groups with regard to their utilisation of technology in making travel arrangements: "users" and "non-users". Criteria used in classifying a respondent, as a 'user', were whether they use a computer based system to make reservations, where 'non-users' are respondents who either does not use a computer-based reservation system or who use the Internet for enquiries only. This resulted in the two groups being of similar size. In the "users" category, a number of organisations qualified on the strength of their use of hotel (21%) and airline web sites (20%), and are not necessarily users of Internet-based corporate self-booking systems (via the corporate Intranet). Had more stringent criteria been applied in classifying the "user" and "non-user" groups, with only those respondents who use Internet-based corporate self-booking systems being included, the ratio would have been substantially different as only 12% of respondents use corporate self-booking systems. Organisations also do not appear to encourage travellers to use the technology options themselves as the majority of respondents (76%) indicated that this is not encouraged by the organisation.

4.4 Technology as a critical success factor

One of the most interesting results found was on the question of whether organisations regard corporate self-booking systems that are linked to suppliers and travel agencies as a *critical success factor* in effective corporate travel management. While only 32% of all respondents regard this as a critical success factor and 13% indicated that they do not know, a substantial difference of opinion was found between the "user" and "non-user" categories.

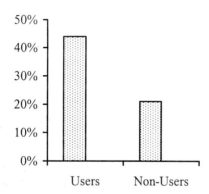

Fig 3. Is a corporate self-booking system a CSF?

As shown in figure 3, in the "user" category 44% believe it is a critical success factor while in the "non-user" category only 21% believe it to be so. This suggests a difference in expectations regarding the value and benefits of such a system and may be one of the reasons for "non-users" not actively pursuing changes in technology to become "users". In response to a question asked relating to whether the organisation is changing its current technology system 62% of respondents replied in the negative.

4.5 Traveller online bookings

Travel reservations are still mainly being done through the designated travel agency via the telephone, electronic mail, facsimile, online and personally, with the telephone being the primary mode of communication in most cases (72%). Reservations are done through the travel agency mainly by the corporate travellers themselves and the departmental secretary in 81% of cases. A comment that was repeatedly made was that travellers prefer personal contact when making their reservations, and this was supported by the response to a question on what hampers the use of technology for travel arrangements in the organisation. The option that was selected by 77% of respondents was that corporate travellers prefer personal contact. In trying to establish if there has been an increase in travellers doing their own bookings online over the past two years, respondents were requested to indicate the percentage of corporate travellers doing their own bookings online this year (2002) and last year (2001). The respondents depicted in figure 4 represent companies where some of their corporate travellers do their own bookings online. It must be mentioned that online bookings within these companies is not ubiquitous – in the companies with the highest percentage of corporate travellers doing their own bookings only 10% of travellers do so.

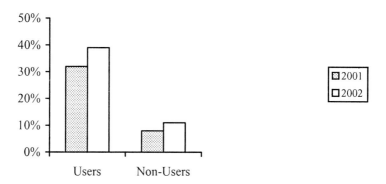

Fig 4. Companies where corporate travellers do their own bookings online

In cross-tabulating the "user" group of organisations (organisations where bookings are done through various websites and self-booking systems) with the increase in corporate travellers doing their own bookings online, a 7% increase was recorded from 2001 to 2002. In cross-tabulating the "non-user" group of organisations (organisations where technology is not used to the same extent) the increase in corporate travellers doing their own bookings online amounted to a 3% increase from 2001 to 2002. Therefore a small increase is indicated where companies have travellers doing their own bookings online, and this increase is shown to be greater in the "user" group. This indicates that adoption by travellers to do online bookings themselves is faster in the category of organisations where bookings are done through various websites and self-booking systems than that of travellers in organisations that do not use technology to the same extent. This result is based on what the corporate travel manager believes and it would be interesting to compare these perceptions with those of the travellers themselves. A further study on corporate travellers forms part of this research project but is not covered in this paper.

4.6 Perceived problems in technology utilisation

A final category of results that can be highlighted in this paper is that of respondent perceptions on what hampers the use of technology in the organisation. As already mentioned the most frequent response (77%) was that travellers prefer personal contact. Respondents also perceive complex travel reservations as not being suited to a system (60%) with 20% not knowing if this was the case. Half of the respondents (51%) indicated that the organisation would have less control over travel spending if a system is implemented with 17% not knowing if this was the case. The statement that elicited the least number of positive responses (26%) was that technical problems in terms of reliability is perceived to be a problem in technology use with 27% not knowing if this was the case.

5 Conclusions

In 1998 Anderson Consulting (Anderson, 1998) forecast that in the United States and Europe corporate travel markets, there would be a move towards the provision of self-booking tools, fundamentally changing the role of the travel intermediary. These tools would allow corporate travellers to book their own travel by making direct contact with GDS or supplier inventory systems through corporate intranets. Access would be obtained via personal computers, personal digital assistants or by telephone and using voice recognition systems. However, there would still be a need for more complex journeys and itineraries to be handled by highly qualified travel consultants. This view was echoed in South Africa in 1998 by Janet Aldworth, the vice-president of the Association of South African Travel Agents, who warned that about half of the travel agencies would be out of business within the next few years as a result of the impact of the Internet (Caras, 1998). With this survey in South Africa, five years on, it is evident that the adoption of corporate self-booking tools and mobile technologies has not yet reached a stage where it fundamentally changes the role of the travel agency in South Africa. One of the most important findings from this research is that, in South Africa, corporate travel agents are currently the most used distribution channel by the corporate client. There is minimal use of mobile technologies by corporate travellers to make reservations or changes while travelling, or even as a source of information. More importantly, this research showed that corporations do not generally encourage usage of this technology. This is also true for self-booking corporate tools. It would appear that corporations prefer to control the travel process through a centralised function and believe that some control may be lost if travellers do their own bookings online. Travellers and corporate travel secretaries generally do reservations through the designated travel agency. The most common modes of communication are the telephone, electronic mail, facsimile or personally. Corporate travel agencies are still perceived to be the most effective reservations channel. There is, however less satisfaction with their ability to offer travel advice, provide travel data, assist with policy development and compliance, conduct supplier negotiations and provide technology support. The research found that the majority of corporations believe that the role of corporate travel agencies should change to that of travel management companies.

Although only a small percentage of corporations are using corporate self-booking tools and supplier websites, an increase over the previous year was indicated. This, together with the perceived advantages of technology, and efforts to overcome current application and integration problems, should serve as warning signals to corporate travel agencies. Corporate travel agencies are expected to provide value-added corporate services with the necessary effective technology access and support in order to maintain their position as the preferred channel. With continuing technological advances and systems such as the Intranet and mobile technologies, travel management companies increasingly have to compete with suppliers who have far greater direct access to the client than ever before. This research found that, while the majority of corporations still do not access supplier websites, a number of major corporations use them for enquiries and also for bookings.

The future of travel management lies in a soft value management approach. The travel management company, the travel supplier, the corporate client and the traveller must specify their values, some of which have been identified in this research (in the model this represents input from the external and internal environments). They must together seek value for money and share in achieving a desired level of service (goal specification). Further research is required in travel policy development and compliance, preferred supplier agreements and technology (actions) as a tool to improved corporate travel management (outcome).

References

Andersen Consulting. (1998). The future of travel distribution. *Research Report Executive Summary*: 7.

Anderson, R.I., Lewis, D., and Parker, M.E. (1999). Another Look at the Efficiency of Corporate Travel Management Departments. *Journal of Travel Research,* 37(3): 267-272.

Bell, R.A., Morey, R.C. (1995). Increasing the efficiency of Corporate Travel Management through Macro Benchmarking. *Journal of Travel Research*, 33(3): 11-20.

Caras, D. (1998). Travel Agents. *Tourism Talk Southern Africa.* 5: 47.

Chircu, A. M., Kauffman, R.J. & Keskey, D. (2001). Maximizing the Value of Internet-based Corporate Travel Reservations Systems. *Communications of the ACM*, 44(11): 57-63.

Gilbert, D.C. & Morris, L. (1995). The relative importance of hotels and airlines to the business traveller. *International Journal of Contemporary Hospitality Management.* 7(6): 19-23.

Liu, A.M.M., Leung, M. (2002). Developing a soft value management model. *International Journal of Project Management,* 20: 341-349.

Roodt A. 2001. Flight plan. *Leadership.* September 2001: 61-64.

Palapies, F. 2001. Mixing business with pleasure is becoming more business than leisure. *TIR Southern Africa*, 75: 19.

Subramani, M.R., & Walden, E. Economic Returns to Firms From Business-To-Business Electronic Commerce Initiatives: An Empirical Examination, *Proceedings of the 21st International Conference on Information Systems*, Brisbane, 2000: 229-241.

E-ticketing Service Development in the Taiwanese Airline Market

Li-Jen Jessica Hwang
Jan Powell-Perry
Ching-Ying Crystal Lai

School of Management
University of Surrey, United Kingdom
{jessica.hwang, j.powell-perry}@surrey.ac.uk

Abstract

The benefits of E-ticketing have created new opportunities for relationship marketing in Asia, in particular in combination with the Chinese concept of Kuan-hsi, to understand the customer better and quicker. This empirical research examines the pre-purchase behaviour of Taiwanese travellers under the first three stages of Hoffman and Bateson decision making model. A total of 206 primary research data were collected inside the waiting lounge of Taiwan's CKS Taipei airport after passing through the pre-boarding security checks. The results showed that although airline passengers are more likely to use the Internet that the general Taiwanese populace, the vast majority (95%) of air tickets were purchased from travel agents. Safety and service were the highest priority for air ticket purchasers, which combined with the cultural relationship norm of Kuan-hsi form barriers to E-ticketing in Taiwan that must be overcome in order for it to be successful.

Keywords: Online sales; Taiwanese travel market; Airline e-ticketing; Pre-purchase Behaviour, Relationship Marketing; Kuan-hsi concept

1   Introduction

Since the introduction of e-tickets in the US airline industry has provided positive results, airlines in Europe and Asia have started to join this trend. In the Taiwan area, travel sites such as E-Z fly are moving towards domestic e-commerce and at the end of 1999 China Airlines, the biggest in the country, launched the first airline e-commerce site to sell international flight tickets via the Internet. These e-services are new to Taiwan and can effectively respond to consumer demand, yet there is little available research in understanding consumer behaviour in this cybermarket (Butler and Peppard, 1998). Using the tenets of relationship marketing, and in particular the framework of Hoffman and Bateson's purchase model (1999), this paper examines the pre-purchase behaviours of travellers toward airline e-tickets in the Taiwanese travel market and explores their validity in the new Internet context. By examining the motivations and decision making of the airline passengers in Taipei, this research can further assist cybermarketers in developing the strategic relationship marketing to anticipate Taiwanese airline marketspace.

2 Theory/Issues

2.1 Relationship marketing

Relationship Marketing (RM) has been recognised as a paradigm shift from focusing on product profit and customer orientation to developing mutually beneficial relationships. Researchers have provided various RM definitions, all of which emphasise the value of long term relationships and networking with all parties including consumers, distributors, dealers and suppliers (Grönroos, 1994; Sharma and Peterson, 1998; Gummesson, 1997, 1999). The Web environment, as part of a push marketing strategy, can provide a new level of interactivity and intimacy with customers and possibly form a non-threatening, reliable platform (Geissler, 2001). Significant shifts from the traditional *market place* to the current computer-mediated environment *market space* have enabled consumers to receive products and services without geographical limitations and establish faster and quicker relationships with providers (Butler, 1998). With a frequently updated web site, airlines can build up a stronger relationship with customers, encouraging repeat visits, establishing strong scale advantages for sale opportunities, and preventing them from switching to competitors through consistency of brand image.

The strategic positioning of the Web in a multi-channel market approach, however, becomes a critical issue as the Web enforces a fundamental restructuring of channel markets (Simons *et al.*, 2002). Schoenbachler and Gordon (2002) argue that the challenges of moving to a multi-channel strategy are magnified by the fact that little is known about what drives consumers to be single channel or multi-channel buyers. Marketers rush into the Web based environment on the assumption that "more is better", attracting new customers which drives growth and profits upward. The potential for information overload for consumers during the evaluation process may not support the argument of the pattern of decision making model. In addition, the e-commerce site cannot sustain business fraught with major outages, slow performance, content errors and broken transactions. Reardon and McCorkle (2002) discovered that consumer channel-switching behaviour is influenced by tradeoffs on the basis of the relative opportunity of cost of time, cost of goods, pleasure derived from shopping, perceived value of goods and relative risk of each channel. The confused and frustrated user may willingly pay for the services of travel agents as a familiar communication channel or seek a safe branded competitor. The argument for bypassing the role of intermediaries to save costs could be over-stated.

338

Marketing is recognised as an economic activity that can be affected by several factors. Kotler (1996) described the buyer's *black box* as different levels of cultural, social, personal and psychological factors that influence the outcome of product purchasing. This research considers the impact of Chinese culture on purchasing decisions. The differences in Eastern and Western methods of doing business can be striking. From the Western point of view, a relationship may follow after the first trading transaction, whereas the Chinese are more likely to do business after establishing the relationship. The concept of the Chinese term *Kuan-hsi* presents profound dynamic meanings regarding the establishment of a type of obligation in the form of a reciprocal relationship, which consists of *Mein-tsu* (showing one's respect) and *Jen-chin* (giving a favour with the unspoken rule of equity) (Gilbert and Tsao, 2000). When dealing with Taiwanese travellers raised under this philosophy, airline marketers should consider the concept of *Kuan-hsi* in their strategic planning.

2.2 Consumer's Pre-purchasing Behaviour

Traditionally, consumer behaviour is an extended and complex process involved in making decisions on product choices. A number of models of the buying process have evolved to help understand this complex procedure (Kotler, 1996; Butler and Peppard, 1998; Hoffman and Bateson, 1999). Much attention has focused on brand loyalty and post-purchase behaviours, with little research into understanding the phenomena of pre-purchase behaviour (Simintiras, 1997). Hoffman and Bateson (1999) deconstruct the decision-making process into three stages, the first of which is pre-purchase. Pre-purchase influences include purchasing stimulus, problem awareness, information research, and the evaluation of solution alternatives. **Figure 1** illustrates their linear relationship and the possible cues and factors which may have an influence on the buyer's *black box*.

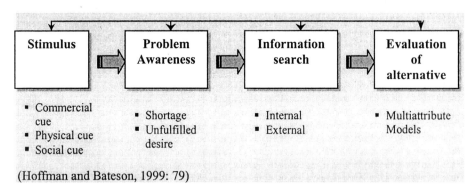

(Hoffman and Bateson, 1999: 79)

Fig. 1. Pre-purchase Stage of Consumer Decision Process

The rationale of investigating pre-purchase behaviour is important because pre-purchase satisfaction can be treated a predictor of the purchase behaviour (Simintiras *et al.*, 1997). Pre-purchase satisfaction represents an affective state of affairs reflecting the feelings of satisfaction that occur in the present time before an anticipated purchase. Mood, for example, can greatly influence the pre-purchase satisfaction responses, since if the customer felt particularly happy at the time, this might override actual feelings towards the planned purchase. Mattila and Wirtz (2002) showed that subjective knowledge had a stronger influence on the utilisation of personal sources of information (e.g. memory and word-of-mouth), while objective knowledge played a more determinant role in affecting the use of impersonal sources (e.g. books, newspapers, and advertisements). The issue under examination is how the consumer behaves during this pre-purchase stage, as a logical antecedent of reaching a decision of purchasing an airline e-ticket.

3   Methods/Sampling & Measurement

With permission from the Taiwanese CKS airport authorities, an on-site survey was administered at the waiting lounge of the Taipei's CKS international airport, a large international hub that accommodates most long haul and regional flights throughout Taiwan. The participants were quota-sampled using criteria based along the guidelines of the age distribution (between 20 and 59) of outbound travellers, according to the annual report of the Taiwan National Tourism Bureau (Taiwanese Tourist Board, 1997). Primarily the rationale for this is that any results generated from this data can then be inferred to the Taiwanese travelling population. Secondly, the participants would all have at least one experience in purchasing a flight ticket, as they are about to leave the country. Thirdly, a high response rate can be easily achieved by interviewing travellers during what is normally a long and tedious wait to board their flight after a series of check-in and security procedures. A total of 206 responses were collected according to the proportions of the sample quotas, which breakdown as 44 returns (21%) from age 20-29; 67 returns (33%) from age 30-39; 60 returns (29%) from age 40-49; and 35 returns (17%) from age 50-59.

With Hoffman and Bateson's pre-purchase model of consumer decision process (1999) as a blueprint, the questionnaire was structured into four sections and the instrument applies mainly quantitative measures. Section A begins with the evaluation of alternatives stage, and scrutinises the purchase decisions in selecting an airline, including who, where, ticket price, service, seat comfort, food selections, flight safety, timing, flight routine and convenience of purchasing. The section closes with an open-ended option. Section B has been constructed to investigate problem awareness of the issues of "online shopping", word-of-mouth (WOM) and measures any past online purchasing experience on convenience, price and security issues using a five points-Likert scale. Section C addresses information research issues concerning the application of relationship marketing within the airline industry through Frequent Flyer Programme (FFP) and the feasibility of airline web sites in the Taiwanese Internet marketplace. Section D gathers demographic data of participants that may reveal characteristics of potential online ticket purchasers in the stimulus stage. The results will then be compared with marketing data gathered by the Taiwanese Market Intelligent Centre (Chen, 2002) to determine any similarities and differences with the general Taiwanese populace.

4 Results and Discussions

4.1 Stimulus

Buying e-tickets is still at an immature stage even though it appears that web site usage is more prevalent among fliers than the rest of the Taiwanese population, compared with the results of the Taiwanese Market Intelligent Centre (Chen, 2002). For example, in this survey 66% of respondents reported using the Internet as a tool and awareness of e-ticket sales also appears to be high, with 77% of respondents having heard about online air tickets. The results also show that having higher education, particularly at the postgraduate level may indicate a higher intention of using the Internet and more satisfaction with e-services than other groups (Table 1). These findings are also consistent with the results of the MIC.

Table 1. Channel Intentions by Education

Intention to Purchase from Airlines Own Web Site	No idea & do not intend to try	Still prefer to buy from agent	No idea	Maybe try	Make Decision after comparison	Total % of Response
% within educational attainment						
Below high school	59.1%	0	36.4%	4.5%	0	100%
High school	39.3%	3.6%	39.3%	12.5%	5.4%	100%
College or University	22.1%	1.8%	59.3%	14.2%	2.7%	100%
Postgraduate	13.3%	0	60%	26.7%	0	100%
% within sample	30.1%	1.9%	51.5%	13.6%	2.9%	100%

With regards to gender differences, female users indicated a greater likelihood to go online, although males comprised the majority of this research. However, this physical cue did not indicate any significantly different satisfaction toward airlines online.

4.2 Problem Awareness

Respondents were asked to rank eight factors associated with airline choice, including safety, ticket price, service, seating comfort, flight timing, direct flight route and convenience of ticket purchase. Safety is unquestionably the first concern for Taiwanese consumers, with 135 respondents (65.5%) ranking it in first position. The second most important factor was price, followed by the service provided. Food was ranked as the least important concern for 39% of respondents. Table 2 shows the full results as absolute frequencies (i.e. number of respondents), cross-tabulated with choice aspects and ranking.

Table 2. Overall Ranking of Airline Choice Aspects

Choice Aspect Rank	Price	Service	Seat Comfort	Food	Safety	Flight Timing	Direct Flight	Ticket Purchase Convenience	Total No. of Response
1st	24	20	3	0	**135**	11	10	3	206
2nd	47	41	15	5	32	22	33	11	206
3rd	38	**51**	29	5	13	28	25	17	206
4th	**40**	28	34	17	11	27	27	22	206
5th	25	31	**36**	29	6	27	29	23	206
6th	15	17	**50**	30	6	34	31	23	206
7th	8	11	36	39	1	42	26	**43**	206
8th	9	7	3	**81**	2	15	25	64	206

(Note: the highest count of respondents for each choice aspect ranking is printed in bold.)

Assigning a weighting factor of 1-8 (to reflect the ranking) to the counts within each choice aspect provides an overall ranking measure. The total weighted count for each airline choice aspect is shown in rank order in Table 3 below. This confirms safety as the most important airline choice aspect. Significantly, the research found that the one of the perceived benefits of going online to purchase tickets (i.e. convenience) does not appear of great concern to the consumer, as over half of the respondents ranked the convenience of ticket purchasing as either their last or second last concern. Among the 14.1% of respondents that have had an internet purchase experience, satisfaction appears to correlate with convenience of online shopping, price, and security of payment ($p<0.05$). This rather unsubstantial result casts some doubt on previous assumptions regarding the predicted growth of e-commerce development.

Table 3. Ranking of Airline Choice Aspects Using Weighted Measure

Ranking	Choice Aspect	Total Weighted Score
1st	Safety	1483
2nd	Price	1119
3rd	Service	1097
4th	Direct Flight	882
5th	Flight Timing	854
6th	Seat Comfort	842
7th	Ticket Purchase Convenience	624
8th	Food	515

4.3   Information search

The respondents have not yet shown a willingness to use the Internet for searching for the best ticket price as the results found that only 28% of respondents have visited an airline web site and only 16% of them has purchased goods online. Since e-ticket services in Taiwan are only offered on a limited basis with a poor selection, in practice the web sites have not appeared so user-friendly to the consumer as being on familiar terms with an agent. Once the barrier of accessing the web has been broken, the respondents expressed some intentions of buying flight tickets through the net. However, few actually gathered fare information on air tickets before travelling.

Even with Frequent Flyer Programmes (FFP), the respondents showed their loyalty to the airline via their travel agent. The result showed 98% of FFP members purchased their air ticket from a travel agent and no significant relationship was found between the willingness to purchase from an airline web site and being a member of a FFP. E-tickets through the airlines own web site may not have significant sale increases in the short term but may act to maintain competitiveness in the long term.

4.4   Evaluation of alternatives

So, who actually makes the ticket purchase decision? The results show a fairly even distribution among the users themselves, their company and travel agents, with a small proportion from family opinion, although 58.7% of trips were for a vacation and 28.2% of them for business. The majority of respondents (95.1%) purchased flight tickets from their travel agents. It appears that intermediaries continue to play an important role as the major distribution channel for air ticket sales in Taiwan. E-tickets have offered airlines an opportunity to interact directly with their customer with a paperless cost.

However, saving the commission costs of a travel agency may have additional impacts. This new communication channel could dilute the current market shares of travel agents and they may require new deals to be negotiated with new purchase behaviours in order to maintain their business at competitive margins. With service being ranked the second highest concern, the concept of Kuan-hsi appears to be rooted in the behaviour of Taiwanese travellers trading in a singular relationship with their travel agents. Giving mature respect (as *Mein–tsu*) and favour (as *Zen-chin*) to their travel agent may benefit both parties in understanding their needs and wants from the Chinese sense. The fact that many users rely on their agent to decide which ticket to buy merely reinforces the dependence of this relationship. That has been found to be very different from the views of western customers, who are perceived to be more sensitive to price and brand name (Gilbert and Tsao, 2000).

5 Conclusions

E-ticketing in Taiwan is still in its infancy. Air ticket purchasing behaviour appears to be concentrated in the first Stimulus stage and the second Problem Awareness stage of the four stage pre-purchase model, and then skips to the final Choice stage, with travel agents largely responsible for the Information Search. Hence, the primary distribution channel remains with the travel agent. This is a likely consequence of the Chinese concept of Kuan-hsi, which continues to influence buying relationships in the Taiwanese air travel market, with online behaviour remaining largely immature. Regardless of channel opportunities, safety remains the top priority in selecting an airline.

Whilst airlines make efforts to create seamless flight connections, provide more comfortable seats, improve meals, and connect directly through e-ticketing, these seem unlikely to meet Taiwanese customers' current needs. Searching out alternative purchase methods or even using the Internet to research airfares is not common within the Taiwanese travel market, while e-ticketing largely remains unused. However, as consumer web site usage becomes more widespread and network connections become faster and more convenient, the perceived benefits of this distribution channel may slowly overcome cultural factors such as Kuan-hsi and open this channel to the Taiwanese customer.

References

Butler, P. and Peppard, J. (1998). Consumer Purchasing on the Internet: process and prospects. *European Management Journal.* 16 (5): 600-610.

Chen, F. (2002) *The Online Shopping Habits of Internet Users in Taiwan.* Taipei: Market Intelligent Centre (MIC), (http://mic.iii.org.tw/english/).

Geissler, G.L. (2001) Building Customer Relationships Online: the Web Site Designers' Perspective. *Journal of Consumer Marketing.* 18(6): 488-502.

Gilbert, D. and Tsao, J. (2000). Exploring Chinese Cultural Influences and Hospitality Marketing Relationships. *International Journal of Contemporary Hospitality Management.* 12(1): 45-53.

Grönroos, C. (1999). Relationship Marketing: challenges for the organisation. *Journal of Business Research.* 46: 327-335.

Gummesson, E. (1997) Relationship Marketing as a Paradigm Shift: some conclusions from the 30R approach. *Management Decision,* 35(4): 267-272.

Gummesson, E. (1999). *Total Relationship Marketing: from the 4Ps - product, price, promotion, place - of traditional marketing management to the 30Rs - the thirty relationships - of the new marketing paradigm.* Oxford: Butterworth Heithmann.

Hoffman, K. D. and Bateson, J. E.G. (1999). *Essential of Service Marketing,* London: Dryden Press.

Kotler, P. (1997). *Marketing Management, Analysis, Planning, Implementation, and Control* (9th edn.). London: Prentice Hall.

Reardon, J. and McCorkle, D.E. (2002) A Consumer Model for Channel Switching Behaviour. *International Journal of Retailing & Distribution Management,* 30(4): 179-185.

344

Schoenbachler, D.D. and Gordon, G.L. (2002) Multi-channel Shopping: Understanding What Drives Channel Choice. *Journal of Consumer Marketing*, 19(1): 42-53.

Sharma, N. and Patterson, P.G. (1999). The Impact of Communication Effectiveness and the Service Quality on Relationship Commitment in Consumer, Professional Services. *The Journal of Services Marketing*, 13 (2): 151-170.

Simintiras, A., Diamantopoulos, A., and Ferriday, J. (1997) Pre-purchase Satisfaction and First-time Buyer Behaviour: Some Preliminary Evidence. *European Journal of Marketing*, 31(11/12): 857-872.

Simons, L.P.A., Steinfield, C., and Bouwman, H. (2002) Strategic Positioning of the Web in a Multi-channel Market Approach. *Internet Research: Electronic Networking Applications and Policy*, 12(4): 339-347.

Taiwanese Tourist Board. (1997) *The Annual Report of the Taiwan National Tourism Bureau*. Taipei: Taiwanese Tourist Board.

ICTs & Internet Adoption in China's Tourism Industry

Jennifer Xiaoqiu Ma
Dimitrios Buhalis
Haiyan Song

Centre for eTourism Research (CeTR),
School of Management , University of Surrey, Guildford, GU2 7XH, UK
{msm2xm; d.buhalis; h.song} @surrey.ac.uk

Abstract

The fast development of information communication technologies (ICT) and the expansion of Internet have affected all industries in many countries. Such technologies have been adopted in tourism industry in Europe and America for more than 30 years, and the trend is likely to continue into the future. China, as a fast-growing developing country in Asia, is gaining increasing importance in the international tourism market for its historical and cultural attractiveness as a destination and, more importantly, as a booming tourism source country with its huge population. This study examines how the ICT and Internet have changed the tourism development in China, how important such changes are, and to where such changes will lead China's tourism industry. This exploratory research is conducted based on information collected from airlines, hotels, tour operators, visitor attractions and the tourism authorities within China.

Keywords: ICT, Internet, tourism, China

1   Introduction

China has been one of the major international tourism destinations since early 1980s. The same period has also witnessed the revolutionary development of information communication technologies (ICTs), especially the recent development in Internet. However, little has been done in exploring the ICT and Internet adoption status in the Chinese tourism industry and the impact of ICT on the structure of the industry. This paper concentrates on describing the ICT applications and their impacts on the development and industry structure of tourism in China. Using the existing theoretical framework on ICT and eTourism development in other parts of the world, Europe and American in particular, this paper sets out to examine the current status of eTourism application in China.

Information communication technologies have been applied in tourism since the early adoption of Computer Reservation Systems (CRS) in airlines in 1950s. The evolution of ICT application in tourism is evident in the transformation of CRS to Global Distribution Systems (GDS). Hotel property management systems and CRS systems appeared shortly afterwards, which brought switch companies into the market as well. It is the development of Internet that brings the revolutionary changes to the structure of the industry by providing tourism principals, airlines and hoteliers, an opportunity to sell directly. Because of this development, the intermediaries between suppliers and consumers have faced a danger of being cut off and replaced. Travel agents felt the most pressure of such changes. The discussion of disintermediation and reintermediation has lasted for more than ten years and ever since the e-commerce model appeared along with the Internet booming. The tourism distribution system has been discussed thoroughly and extensively in many published studies including O'Connor (1999, 2000), Sheldon (1997), Inkpen (1998) and Cooper et al (2000), Palmer & McCole (1999), Lang (2000) and Standing & Vasudavan (2001).

The most recent development in eTourism application is at the destination level. By employing Internet, Intranet and Extranet, some DMOs (destination management organisations) have successfully integrated this function in promoting the destination, providing tourists with pre-trip and in-trip information, helping SMTE (small and medium tourist enterprises) to promote their products, and internal management within DMO into networking systems (Buhalis, 1993, 1998).

Dot-com entrepreneurs and other companies and investors have realised that the Internet should not cannibalise business models, but act as an enabler that empowers organisations to achieve their objectives. Technology itself cannot bring value to any company. However, the strategic and efficient implementation of technology within a business can reduce the operation costs and enable the delivery of better products or services to customers. This will help the company to gain competitive advantage by either maintaining its price leadership in the market or differentiating its product and services, which will eventually lead to the increase in value added to the company. It is from this point of view that strategic implementation of information technology and Internet become critical for all companies trying to survive in this new economy (Poon, 1993; Sheldon, 1997; Werthner & Klein, 1999; Alford, 2000, WTO, 2001).

Most published literature on eTourism has been based on the development and case studies in Europe and America with little effort being made to study the eTourism development in China. However, the huge market and great potential of the demand in China can be hardly excluded or ignored in any phenomenal studies conducted within a global scope. According to an online survey held by one of the official information websites (www.cnnic.com), the number of total Internet user in China doubled every year for the last three years. Therefore, eTourism applications in China are worth the attention of researchers considering the potential dynamic changes.

2 Research Methodology

This research aims to find out what happened, what is happening and what will happen in the eTourism application in China. As the research is exploring a new study area focusing on the supply-side of the tourism industry, qualitative approaches are used as the main research method with the quantitative research as a complementary method. The research starts by answering such questions as what, where, how and why China's eTourism has been developed, and then some analytical points and possible future trends are discussed. Most primary data were collected from interviews conducted with tourist companies of various sectors in Beijing, the capital city of China, as well as the heart of the country's tourism development. Twenty organisations were interviewed, ranging from airlines, hotels, tour operators to visitor attractions and tourist authority. A questionnaire survey was also conducted in the hotel sector exploring further supporting evidence.

The primary data were collected from 45 tourist organisations including airlines, hotels, tour operators, visitor attractions and the tourism authority. Given the unique government-oriented business structure of Chinese tourist industry, the research has been focused on interviewing the key representatives of these sectors. Quantitative data has only been collected within the hotel sector with a sample size of 30 hotels randomly chosen from 190 hotels in Beijing.

3 Research Findings

Although the study covers airlines, hotels, tour operators, visitor attractions and tourism authorities, only are the results related to Airlines and Hotels presented due to the constraint of space. As the research is explanatory in nature, key issues are highlighted for each industry sector and it also provides an initial understanding of the characteristics of the eTourism development in China. Evidently the tourism industry in China is heavily regulated with a high degree of government intervention. This is particularly the case for the air transportation.

The interviews were conducted with the a number of key decision makers in Chinese eTourism and included the marketing director of Travelsky and Beijing representative of Galileo; the director of computing centre of Beijing Capital International Airport; the yield managers of Air China; the PR manager of China Southern Airlines; the sales manager of United Airlines, Beijing office; and the ticketing manager of CITS (China International Travel Service) ticketing office. As the focus of the interview was to understand the current status of ICT integration with their businesses, and possible impacts of ICT adoption on the organisational structure, only those interviewees who understand both the enterprise's strategy and the ICT adoption status within the company were chosen in the study.

3.1 Computer Reservation System and Electronic Distribution

According to the information provided by the marketing director of the company, Ms. Jia Lin, the TravelSky system handles almost all business of the Chinese airline industry. This includes 97 % of domestic ticketing and 75% of overseas ticketing. It provides inventory management solutions for all 25 airlines in China, both local and national. The system also handles passenger transactions for all major domestic airports, and the total number of enabled airports will reach 100 by the end of 2002. According to the government regulation, Travelsky's reservation system is the only GDS within China allowed to issue tickets with its e-Terminals that connects all ticketing agents, of which 500,000 are domestic ticketing agents and 1 million are international. All major

GDS (four global ones – Amadeus, Sabre, Galileo and Worldspan, and three regional ones in Asia – Infini, Axess and Topas) are linked with Travelsky. Backed by the CAAC authority, airlines are not allowed to develop their own systems, and all GDS were banned for any commercial operation. In other words, any bookings and ticketing concerning flights within China or to and from China have to go through Travelsky, either via the system directly or via global or regional GDS and Travelsky as illustrated in Figure 1.

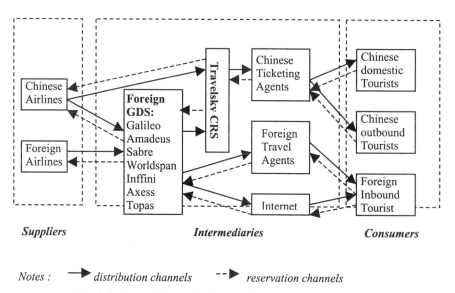

Notes : ➤ *distribution channels* --➤ *reservation channels*

Fig. 1: Chinese Airline distribution & reservation channels

Travelsky treats the opening up of the Chinese airline market to foreign airlines and foreign GDS as the major challenges after China entered the World Trade Organization (WTO). Foreign GDS rather than the Internet, will be the greatest threat to Travelsky. As indicated above, Travelsky believes that the real B2C (Business to Consumer) market based on the Internet is not yet mature and will not be the case for at least another five years. On the other hand, the opening up of the GDS market to foreign systems is totally up to a governmental decision, which may be implemented in the next a year or two. Ms Lin said, "it will take foreign GDS a certain period of time to get used to the market, but as soon as they are localised they are much more competitive than Travelsky in terms of information richness, processing speed and system stability."

3.2 Suppliers: Airlines

According to interview results with Beijing Capital International Airport, Air China and China Southern Airlines, the respondents are satisfied with the current ICT application situation monopolised by CACI, or Travelsky, and have little intention to change. On the one hand, they are not able to establish their own system because of both policy prohibition and restriction of cost and time in building the whole system and data network by themselves. On the other hand, as they cannot fully control their own operation, since most parts of their operation are handled by the government such as procurement, pricing, route design, and distribution, they have no incentive to explore the benefit of ICT or Internet applications because they may bring them more trouble.

Ms. Chen, the yield manager from Air China, the country's flag carrier airline, also the largest international airlines in China, gave a detailed description of the ICT application in airline industry from another perspective. The ICT applications in Air China currently focuses on distribution and inventory control system. This has little adoption within the enterprise except for an accounting system and an Intranet which is mainly used for internal e-mails. In terms of the ticketing system, it currently uses Travelsky for all domestic ticketing, and foreign GDS for overseas distribution of international routes. The yield manager stated that the airline sincerely wishes to realise direct online selling to cut out commission costs, however, there are quite a few problems that make this wish unrealistic.

First of all, the main problem as indicated already is the means in which the transactions are settled. In China, the majority (over 60%, according to the yield manager) of air passengers are business travellers whose travelling costs are covered by their employers or working units. These travellers prefer settlement with company cheques as no personal cheque is available in China. On the leisure market, consumers prefer cash transactions, and they normally chase for discount tickets offered by agents. As the enterprise is directly controlled by the government airline authority, CAAC, Air China cannot respond to the market as a fully functional business. This is demonstrated in the following:

- The company cannot determine its prices according to market conditions.
- The company cannot control its own procurement and costs. The fuel supplier, equipment supplier, airports, CRS and other air computing system suppliers are all appointed by CAAC.
- The company has no right to decide its own flight routes without permission of CAAC. The air routes within China as well as international routes are divided among various airlines by CAAC.

Due to the lack of autonomy, Air China does not have the incentive to develop its own ICT adoption strategy and no intention to change the current distribution status with online direct selling, tackling all the above difficulties. However, with the fast expanding Internet users in China and the e-Commerce booming, some airlines did try to implement Internet solutions in their business operation, mainly in distribution. Air China has jointly developed an auto-ticketing machine system in co-operation with the Bank of China. The ticketing machine is a web-based CRS terminal, which is linked to the bank's ATM cash point. The concept is that consumers can use their cash card to buy a ticket at the cash point. After the transaction is completed, the bank's ATM machine will issue a receipt for customer. The customer can then exchange the receipt for tickets at the airport before check-in. The system is still under development and may be fully functional in a year or two. China Southern Airlines and Hainan Airlines were the only two private airlines that considered to develop their own web-based ticketing and reservation system. However, the project failed due to the CAAC intervention.

Talking about the future, the yield manager of Air China expressed certainty that the Travelsky monopoly in China's airline ICT market will continue for at least another 10 years. First of all, the Chinese government will not open up this market for the next 5 to 10 years, according to her estimation. Secondly, even if foreign GDS is allowed to operate in China, it is not easy for Chinese airlines, airports, and air cargo companies to switch to another incompatible system. Thirdly, the image of a foreign GDS is not welcoming. Foreign GDS charge US$3.5-3.8 per transaction and is more expensive than Travelsky (US$0.85 per transaction). Furthermore, due to the strict control over foreign GDS operation and their eagerness to make profit from the Chinese market, the commission over-riding problem is serious among those GDS. Foreign GDS encourage ticket agents to make two reservations and confirm both, one with Travelsky, who issue the tickets, the other with a foreign GDS, which is later, cancelled. In that case, airlines have to pay two GDS transaction fees for a single ticket. Foreign GDS make money from a false booking, and refund a certain amount of commission to agents who make the reservation through them.

3.3 Traditional Intermediary: Ticketing Agents

In China all ticketing agents have to be authorised by CAAC with a certain amount of guarantee deposit. There are about 1.5 million air-ticketing agents categorised into two distinctive groups: A-type agents that can handle international flight ticket reservations, and B-type agents handling domestic ticketing. However, the boundary between the two has been blurred as many B-type agents issue international tickets from A-type agents and sell them to their customers. Some B-type agents develop their own agents, generally known as C-type agents.

The agent interviewed in this research is CITS Air Ticket Agent, an A-type agent. The agency is managed by the largest tour operator in China (China International Travel Service), and is the first IATA agent in China. In the year 2000, the agent adopted a web-based international flight ticketing system with its newly launched website www.citsair.com.cn. On this website, customers can view real-time flight information as well as prices (point to point only, not including transferring flights) updated everyday by a person in charge of data input. The cost is the agent discount ticket price plus the agencies' profit. So far, it is still the only website to provide the customer with both discounted prices and real-time flight information. When the customer makes a reservation on the website, after inputting certain information, the server automatically sends an e-mail to the agent's terminal. The agent will read the e-mail and call the customer to confirm the booking and wait for the money to be received before issuing the ticket. The system is manually maintained by the back-office staff, with flight information coming from Travelsky and Galileo. The price information is manually input into the database according to the contracts to different airlines and various documents issued by CAAC. After two years in operation, the online reservation system accounts already for one third of the agent's sales volume, with another one third of reservation by walk-in and call-in customers, and the rest by B-type agents coming to issue international tickets.

In comparing Travelsky with Galileo, the agent prefers Galileo as it provides more detailed information such as airport location, name of transferring city, transfer waiting time, flying time, correct prices etc. while Travelsky only displays how many en-route transferring stops and nothing else. Galileo is easier and more convenient to operate and is a much more stable system than Travelsky. However, due to the restriction of issuing BSP (bank settlement plan) ticket with only Travelsky's eTerminal, the agent has to use Travelsky for ticket issuing while browsing for information on Galileo.

When talking about the impact of the Internet, the sales manager of CITS ticket office, Ms. Tao Min, believes that there is no threat to the agent in China in the near future and for at least 10 years. The reasons she gave included:
1) The complexity of international ticketing with various flight and price arrangements will be too much for ordinary travellers Even agents have to take weeks to get trained;
2) Customers tend to change their reservation frequently. That will cost airlines a lot of time and money to deal with these changes; and
3) Online settlement will be a major barrier for any online transactions. Let alone the preference for cash transactions among Chinese consumers and the under-developed banking system, each credit card transaction is charged with 1% to 3% service fee by the local banks, making the online transaction unattractive. Neither customers nor suppliers would like to pay the bank charges. From the agents' perspective, for each international ticket reservation, agents can receive 7-9% commission from airlines. However due to severe competition, most agents give part of their commission out to customers as discounted points. Hence the margin left for agents is only 1-3%, barely sufficient to cover bank transaction charge.

3.4 Hotels

No central reservation system exists in the hospitality industry in China. However, property management system (PMS) has been widely spread in hotels ranked as three-star or above. The majority of domestic invested hotels use Huayi and Zhongran (China Software) HIS (Hotel Information System) or PMS, while those foreign investments or operated by overseas hotel chains tend to use Fidelio, Landmark or other imported application software systems in order to be compatible with other properties in their chain. IT applications in hotels has concentrated on front-office functions including reservation, reception, housekeeping, cashier, catering, marketing, customer relations management, etc. According to Peter, ICT manager of Holiday Inn Lidu Hotel in Beijing, the major benefits of using front office system (PMS) are: increasing the efficiency and accuracy of internal information transmission; saving the costs of paper work and information processing labour; and increasing the effectiveness of management, especially in terms of cash, account, stock and yield management. Few hotels use ICT in managing its back-office functions like procurement, security, engineering, etc. Though most hotels were aware of the importance of back-office system, few actually use such system. Interviewees explained that the primary reason is concerns over costs. It is much cheaper to purchase a front office system only, which is enough to keep the hotel running in good order. The ICT manager further mentioned about the training cost and possible position changes of certain personnel brought by the new system employment, as additional factors that the hotel has to considered when deciding whether to apply an information system in back office departments. That departments like engineering and procurement often benefit from under-the-table commissions by blocking information access. Such benefits would disappear if all information is transparent and can be shared among all accessible people. This caused resistance from within the department.

Hotel distribution in China has been heavily relying on sales personnel mainly, as no CRS or GDS exist as an alternative channel. The Internet has been used by most hotels as a marketing tool only. On-line reservation is not truly welcomed by hotels because of the lack of trust of consumers' credit. Without a deposit, many online reservations have been proved to be an invalid booking. Most hoteliers would rather trust telephone or fax booking. Hence, on-line bookings would normally require off-line confirmation. When talking about Internet application in hotel operations, Sam, the sales and marketing manager of Lidu Hotel pointed out that the hotel does not have its own website, and does not trust online reservation at all. Several reasons were quoted: 1) since most people can book a hotel room by using telephone or fax paid by company, why bother with Internet; 2) online bank settlement is a major barrier to realised online reservations; 3) hotels cannot offer any discount for online reservations while customers can get a discounted rate by booking via a tour operator, booking centre, etc.; and 4) lack of mutual trust between customers and hotels, as invalid booking does happen quite often.

In order to explore further the Internet sales conditions in Chinese hospitality industry, a telephone survey was held among 30 hotels in Beijing. Sales managers of these hotels were questioned using a simple questionnaire. The data analysis results showed that: Firstly, Internet sales percentage is very low in hotels, accounted for 4% of total sales in average. Secondly, there is no correlation between Internet sale volumes and factors such as hotel indicators suggested by the number of guest room, occupancy rate and whether the hotel has its own website. Thirdly, the Internet sales data distribution among different hotels is highly skewed, as percentages range from 0 to 40% and the mode is the lowest value 0%, meaning most hotels in Beijing do not have any Internet sales or reservations. Only one out of the 30 hotels reported 40% Internet sales, which come from hotel reservation centre development by travel websites such as Ctrip (www.ctrip.com) and elong (www.elong.com). Finally, 3-star hotels generate higher sales percentage from the Internet. Among the 30 hotels, 86% of the 3-star hotels (6 out of 7) reported certain percentage of Internet sales (5-10%) while only 55% of the 5-star hotels (6 out of 11) and 58% of the 4-star hotels (7 out of 12) reported Internet sales.

4 Discussion and Conclusion

The Chinese tourism industry has followed a similar pattern in the development of ICT applications as Europe and America. The development was first started in airlines, and then extended to hotels, tour operators, attractions and DMOs, with the first two sectors taking the lead. It is true that airlines and hotels are comparatively matured and developed in terms of their ICT adoption within the organisation and the awareness of ICT. Though some changes have taken places in tour operators, visitor attractions and DMOs, ICT applications in these sectors are limited. The research findings in this study show a general picture of Chinese eTourism development status. First, China's tourism development is very much inbound oriented. Companies involved in handling international tourists, including inbound tour operators, 4 and 5 star hotels that cater for foreign tourists, and airlines, are well established, and comparatively matured. However, the demand for eTourism or the driving force of ICT and Internet adoption in tourism lies in domestic and outbound tourist market. This is because that inbound tourism is heavily controlled by established intermediaries. The mismatching supply-demand market can partly explain the stagnation of eTourism developments in China. Secondly, the structure of Chinese tourism industry is different from that of the developed countries, as it is mainly inbound oriented. The whole industry in China is simply the supply section of the whole tourism system, while the consumers and intermediaries are all from source countries that are out of the scope of this research. Under such circumstances, upon examining China's tourism industry structure, only suppliers – airlines, hotels, inbound tour operators and attractions have adopted ICT and Internet, while intermediaries are lagged far behind. Only when looking at domestic and outbound tour market, a few newly set-up ticketing agents and hotel booking centres can be found. Thirdly, under strong government intervention and policy guidance, the market is not fully operational. Most tourist companies in the market are not qualified enterprises that can respond to market signals in pursuit of profit maximisation. Some or most parts of the business functions are controlled directly or indirectly by government agents, rather than by companies themselves. Individual companies therefore have no motive to search for ways to gain a competitive edge by reducing costs and increasing operational efficiency. In China's tourism industry, the research results show that most tour operators, visitor attractions as well as local tourism boards (DMOs) still simply use ICT as a data processing tool. Some have managed to use ICT as a management tool internally to control primary business activities defined in the value chain model (Porter, 2001). However, none of interviewed companies have ever strategically integrated ICT application with all activities in the value chain. Even less has been done in networked ICT facilitation among value chains with the industry value system. Hence, the e-business development in China' tourism today is equivalent to developed countries in the 1970s, i.e., 30 years ago if not earlier. The current ICT and Internet adoption level in China means that in order to catch up its counterparts in Europe and America, China has to pay much attention to interaction between the following parties: tourism suppliers, consumers and the government. Although the technology slipovers can be fast, its impacts on the ways in which business operates and on consumers' habit may take time. Furthermore, given the frequent government intervention in the Chinese economy, the role that the tourism industry and other related government authorities play will also determine the direction and speed of China's eTourism development. Although the e-boom from 1998 to 2000 did not bring revolutionary changes to the structure of China's tourism industry, it did shake the industry and attracted more attention to the use of ICT and Internet. The situation has changed already and at a growing speed. The growing ICT awareness and market demand will lead China's tourism into the e-Business era. Since there are huge potentials in the Chinese tourism market, we should be optimistic about the eTourism developments in China in the future.

References

Alford, P., 2000, E-Business in the Travel Industry, UK: Travel & Tourism Intelligence.

Buhalis, D., 1993, RICIRMS as a strategic tool for small and medium tourism enterprises, Tourism Management, Vol. 14, No. 5, pp.366-378.

Buhalis, D., 1998, Strategic Use of Information Technologies in the Tourism Industry, Tourism Management, Vol.19, No.3, pp409-423.

Buhalis, D., 2000, Marketing the Competitive Destination of the Future, Tourism Management, Vol.21, No.1, pp97-116.

Cooper, C., et al., 2000, Tourism: Principles and Practice, London: Longman.

CNTA, 1979-2001, The Yearbook of China Tourism Statistics, Beijing: China Tourism Press.

Inkpen, G., 1998, Information Technology for Travel and Tourism (2nd ed.), London: Longman.

Lang, T. C., 2000, The Effect of the Internet on Travel Consumer Purchasing Behaviour and Implications for Travel Agencies, Journal of Vacation Marketing, Vol.6, No.4, P.368-385.

O'Connor, P., 1999, Tourism & Hospitality Distribution and Information Technology, Oxford: CABI.

O'Connor, P., 2000, Using Computers in Hospitality (2nd ed.), London: Cassell.

Palmer, A. and McCole, P., 2000, The Virtual Re-intermediation of Travel Services: A Conceptual Framework and Empirical Investigation, Journal of Vacation Marketing, Vol.6, No.1, P.33-47.

Poon, A., 1993, Tourism, Technology and Competitive Strategies, Oxford: CABI.

Porter, M., 2001, Strategy and the Internet, Harvard Business Review, March, pp63-78.

Sheldon, P., 1997, Information Technologies for Tourism, Oxford: CABI.

Standing, C. and Vasudavan, T, 2000, The Impact of Internet on Travel Industry in Australia, Tourism Recreation Research, Vol. 25, No. 3, pp45-54.

Werthner, H., and Klein, S., 1999, Information Technology and Tourism – A Challenging Relationship, Austria: Springer-Verlag.

WTO, 2001, E-Business for Tourism: Practical Guidelines for Tourism Destinations and Business, Madrid: WTO.

ICT Adoption and Use in New Zealand's Small and Medium Tourism Enterprises: A Cross Sectoral Perspective

Carolyn Nodder[a],
David Mason[b],
Jovo Ateljevic[c],
Simon Milne[d]

[a] School of Computing and Information Technology, UNITEC Institute of Technology, Auckland, NZ
cnodder@unitec.ac.nz

[b] School of Information Management, Victoria University, Wellington, NZ david.mason@vuw.ac.nz

[c] Tourism Group, Victoria University, Wellington, NZ
jovo.ateljevic@vuw.ac.nz

[d] New Zealand Tourism Research Institute, Auckland University of Technology, Auckland, NZ
simon.milne@aut.ac.nz

Abstract

This paper focuses on New Zealand (NZ) small and medium tourism enterprises (SMTE) and their use and adoption of information and communication technologies (ICT), particularly the Internet. An overview of the research that has been conducted in this area reveals a number of gaps in our understanding of issues facing small operators. We argue that these limitations are due, in part, to an over emphasis on survey based research. The paper then reviews key findings from a qualitative, multi-sectoral, program of research based around over 250 in-depth interviews conducted between 1998 and 2002. The core aim is to provide a deeper understanding of the major factors that hinder or enhance prospects for ICT uptake that have remained largely hidden from the survey based research. The ability of technology to foster improved competitive performance through network and alliance formation is highlighted.

Keywords: Small and Medium Tourism Enterprises, New Zealand, ICT adoption, networks

1   Introduction

The way in which tourism organizations conduct their business can be significantly enhanced by ICT, as can the way in which all key stakeholders in the tourism sector access products through distribution channels (Buhalis, 2000; 2001). The capacity of

ICT to affect relationships, establish and build networks and communities (both virtual and real), and drive visionary business strategy development, is the focus of much current debate and discussion (Mason and Milne, 2002; Surman & Wershler-Henry 2001).

A number of challenges that reduce the uptake of ICT by SMTE have been identified worldwide, including: a lack of training and capital, limited understanding of the potential of technology, and a lack of clear business strategies (Buhalis 1996; Hull and Milne 2001). The degree to which these threats can be mitigated depends, to some extent, on the effective implementation of government policies (Grant et al 2001; Atkinson and Wilhelm 2002). Part of the role of government is ensuring that the infrastructure is available to enable full advantage to be taken of the opportunities offered by ICT (UNDP 2001). Additionally, public sector policy addresses areas such as training and network development (Buhalis 2000). While there is a growing body of literature dealing with technology uptake by the tourism industry in Europe and North America, there have been relatively few attempts to analyse these themes in New Zealand. Our understanding of ICT diffusion is especially poor in relation to the small and medium sized enterprises (SME) that comprise the bulk of the nation's tourism businesses.

The research that has been conducted on ICT uptake in the New Zealand tourism industry has a number of limitations: a lack of cross-sectoral coverage; limited coordination; and an over-emphasis on survey based, quantitative research. While much of this work has efficiently described the rather parlous state of ICT uptake and use, it has largely failed to provide a detailed understanding of the factors that create this situation and the tools that may be adopted to remedy it. This paper attempts to gain a deeper insight into SMTE and ICT use by synthesizing the findings of an ongoing program of research being conducted by the New Zealand Tourism Research Institute. The research revolves around in-depth interviews and case studies. The core aims of the research program, which began in 1998, are to:

- Gain an understanding of the competitive environment facing New Zealand's SMTE.
- Study the problems small operators face in trying to access tourists and to analyse the use of ICT as marketing and information management tools.
- Gain insights into management's knowledge of information technology.
- Understand the factors that hinder ICT uptake.
- Analyse the role that ICT can play in facilitating network and alliance formation.

Unlike the European context, (Buhalis 1996) researchers and policy makers in New Zealand have been slow to adopt any strict definition of small tourism firms. In this paper our emphasis is on enterprises that have 5 employees or less. Various studies have estimated that such firms account for anywhere between 75 and 85 % of all tourism businesses in New Zealand, though no firm figures are available (Ateljevic et al 1999; TIANZ 2001).

2 The New Zealand Context

The New Zealand Tourism Strategy (2001) is based on the principals of sustainable development and economic yield. One of the key elements of the Strategy is the need to build the ICT capability of the 13,500 to 18,000 SMTE that are estimated to operate in the industry. It is first useful to evaluate broader levels of SME usage, uptake and perceptions of ICT in New Zealand. In a survey of the adoption and implementation of e-business in New Zealand, Clark et al. (2001) found that computer use in businesses was high (92%). However, only 21% reported having a website capable of taking orders and only 8% being able to handle payments online with the use of a secure server. The study found that among companies with websites the most important concerns are the need to improve telecommunications infrastructure and consumer access to the Internet. Those companies without websites were more interested in seeing improvements in on-line security and the provision of e-business training. The results confirm a need for both training and management education. Only 25% of the SME surveyed engaged in business planning that explicitly incorporated e-commerce.

Two key caveats need to be attached to this work – it focused only on enterprises with more than 10 employees, and the survey achieved only a 20.9% response rate. Thus the study's value and applicability to a tourism sector dominated by small enterprises must be questioned. Nevertheless the fact that the key inhibitors to e-commerce development were a lack of knowledge and perceived high costs is of significance and reflects recent tourism research (Applebee et al. 2000; Hull and Milne 2001)

A study undertaken by the Ministry of Economic Development into the Net Readiness of New Zealand industries focuses on measures related to connectivity, hardware, and software (MED, 2001). While 91% of tourism businesses who took part in the survey had a website, only 10% of these were capable of sending bills. Respondents reported their highest "readiness" in the area of leadership, indicating a confidence among senior management that they can cope with the emerging digital economy. The report states:

> "the readiness score for Tourism was 73.63, which is at the top of the Net Savvy range; this is well above the New Zealand Industries mean score of 63.03. Further, 10.1% were in the top Net Visionary category"
> (MED, 2001:7).

Nearly 50% of the tourism businesses surveyed had more than 20 full time equivalent employees, a very different balance from the industry average. At the same time the confidence expressed by management in their 'net readiness' seems to be contradicted by the very basic functionality evident in the websites reviewed.

The Ministry of Tourism is aware of the need to conduct further research into ICT and strategic business planning and is working with the Tourism Industry Association of New Zealand (TIANZ) on strategies to enhance SMTE performance. The government has recently funded a study on ICT usage by tourism businesses though it is impossible to access further information as it is embargoed for confidentiality reasons. Such "closed doors" do little to enable the broadening of knowledge in this area and also restrict access to findings by the people who need it most – the SMTE. Similarly a new CD-ROM based tool to enhance SMTE management skills (including ICT dimensions) is only available to paid members of TIANZ at a cost of $500 per annum.

3   The New Zealand Tourism Research Institute (NZTRI) Research Program

Research into ICT adoption and use by New Zealand SMTE is clearly at a very nascent stage. Survey based methods are leading to an under-representation of the very firms that comprise the bulk of the industry. Concomitantly the paucity of in-depth, qualitative studies undertaken in New Zealand makes it difficult to evaluate the reasons for the reticence of many SMTE to integrate ICT into their day-to-day operations. The research has also tended to be aggregate in nature, paying limited attention to the cross-sectoral variations that can be so distinct in the tourism industry (Applebee et al. 2000).

In response to these issues the NZTRI has been conducting a program of SMTE research that focuses on cross-sectoral, in-depth studies on ICT uptake and use. To-date 272 in-depth interviews have been conducted with representatives from the following sectors (number of interviews completed): museums (33); hot pools/spas (6); transport (32); cultural products (10); accommodation (65); restaurants (31); tour operators (46); organic farming (12); travel agents (20); arts and crafts (17).

The work has been carried out throughout the country and has emphasized the need to look at both rural and urban contexts and the ways in which small businesses are embedded in broader community structures and networks. We were able to conduct interviews in person (220) and over the phone (30) with people who often show little desire to participate in mail surveys (positive response rate > 90%). We now present a broad overview of the key themes and issues raised, more detail on specific sectors can be found at www.tri.org.nz.

3.1 Diffusion/adoption and Net Readiness

Overall levels of adoption and net readiness reflect the general trends found in the survey-based research outlined above. The research does, however, highlight a number of key issues that warrant further exploration. Levels of uptake are highest in sectors that are directly related to the tourism industry, indeed there is a strong link between the amount a business uses the Internet and broader ICT and the percentage of its revenue that is dependent on tourism. The leading sectors in terms of uptake are travel agents, tour operators, transport (car rental, shuttle services) and accommodation.

Travel agents have the highest level of ICT uptake - driven by the challenges of disintermediation and commission cuts. Bed and breakfast operations and boutique hotels have also embraced technology but have a tendency to rely on larger portals provided by RTOs or private companies. As with tour operators and transport companies, the bulk of the accommodation sites are relatively static in nature and offer limited opportunities for on-line booking

Sectors that overlap with tourism (spa, organics, crafts, cultural products, restaurants) have a lower level of uptake and have more basic web-presence. This is due to the fact that many of these businesses do not really see the need to reach visitors and local clients directly through the Internet, preferring instead for core tourism operators to direct visitors and to use local word of mouth. Increasingly though there is a realisation that an Internet presence, replete with links to core tourism product providers, is a vital tool in enhancing revenues. The arts and crafts sector is particularly active in Internet use.

Adoption and effective use of ICT is higher in urban areas where infrastructure provision is greater, as is access to technical expertise and training. Nevertheless some rural areas are doing well, largely on the back of a few ICT savvy individuals who have been able, occasionally, to transfer technological knowledge and also lobby regional government for support.

3.2 Individual backgrounds and embeddedness

Small firms are shaped by the individuals that operate and manage them. A key focus for the research program has been to gain insights into what types of individual characteristics allow ICT uptake to occur and enable local 'technology champions' to emerge. The research also focuses on the degree to which small business owners are 'embedded' in the surrounding economy and culture, and how this impacts upon business performance. A number of people interviewed, particularly those originally from urban areas that moved to rural environs, see tourism ventures as an opportunity to "escape the rat race" rather than as major income generators. Retirement also draws people into more attractive locations that provide opportunities to occupy free time while earning a little money. This group is often less interested in the potential benefits associated with ICT. As one bed and breakfast operator noted:

> "Yes we have a got our 'bits and pieces' on a few websites (Regional Tourism Office, Jasons, AA) but we have decided not to do anymore than this – we don't want to be inundated with visitors, we came here for a simple stress-free lifestyle and we don't want to see this change, the Internet is instant and people are waiting for an instant reply – we don't want that type of pressure"

On the other hand SMTE operators who have moved into the business for economic reasons are more likely to embrace ICT, as one respondent stated:

"I am not going to get to my sixties and expect to be provided for by the government – I see the Internet as my saviour – it brings so many high paying visitors to my doorstep and it costs so little".

The flow of tourism operators into the countryside from urban areas provides another potential benefit. These new entrants can bring technological skills with them. Sometimes these people are transformed into 'champions' that facilitate the development of local websites and raise levels of ICT acumen. There are many more cases, however, of newcomers who have failed in their attempts to enhance local technology uptake. The reasons for this vary but a major factor is the lack of 'embeddedness' that these newcomers have in their surroundings. Until relationships of trust and reciprocity have been built the good ideas of newcomers are often seen as a threat rather than a way to generate more business.

3.3 Networks, Clusters and Community

A recent review of existing New Zealand SME studies (Wilson 2002) found the nation's businesses to be "stubbornly self-sufficient" (Wilson, 2002:11). The NZTRI research tends to support these findings. There is a lack of effective networking involving SMTE. Many operators claim that they simply do not have the time and money to become involved in such networks even though they realise the benefits that can accompany them. Networks are often formed due to an immediate need, for example a particular regulatory issue that demands collective lobbying. Once the initial catalyst dissipates, or a key member leaves, it proves to be very difficult to maintain an effective network on an informal basis. Unfortunately the move to more formalised structures often dissuades SMTE as they see time commitments and costs becoming an issue.

3.4 Relationships with Experts

In developing a website or purchasing software many operators seek and accept professional advice from ICT experts who sometimes have very little knowledge about the tourism industry and its needs. The issues that stem from this state of affairs were, perhaps, raised most clearly during our research with small museums. Small museums in New Zealand have not made very effective use of the Internet and broader ICT. They have either a very limited adoption of technology or they have embarked upon an over-enthusiastic adoption of Internet applications. The reason for this is basically the mismatch between the capabilities of the technology and the information acumen of the average curator/manager.

An example of the latter comes from a small rural museum that had received support from a university to set up a website. The developers at the university had an immense array of Internet tools and decided to use them all. These were all very exciting when viewed at the university but it turned out that many of the prospective visitors from the surrounding area had basic PCs, early browser software, and 14.4kb modems. None of the exciting add-ons were accessible, and people were irritated.

This case raises a series of lessons for the tourism industry as a whole. SMTE must:

- Recognise the value of building a relationship with web developers. Our research shows that this relationship holds the keys to success.
- Set a realistic development timeframe. Internet sites take a lot of fine-tuning and often information or layouts change as development occurs.
- Document the scope of the work accurately (avoiding specification creep)
- Agree on cost/terms of work

In only a few cases were instances where all of these factors had been covered.

3.5 Training

The research had no difficulty in finding managers who can use and navigate the web, however there is a worrying lack of basic information management skills and conceptual foundations. Part of this is tied up with a rather contrived phobia about computers generally, but beyond this we found an absence of vision and of any abstract understanding of the role information technology plays in the presentation and management of tourism products.

These findings appear to contradict the studies mentioned earlier that rank tourism as a highly Net Savvy/Visionary sector. They reinforce our concerns that the bias toward larger firms in these surveys is giving us a rather skewed portrayal of the industry as a whole. In simple terms our experience has been that until this information management culture is embedded into the organisation, time and resources spent on a great deal of IT training is likely to be wasted.

In terms of ICT support and information the more proactive operators have established informal networking with other businesses, business associations and local, regional and national tourism organizations. These networks are used to obtain information and advice on technology related issues and generally appeared to be quite successful in enhancing ICT acumen.

4   Conclusions

The lack of uptake and ineffective use of ICT represents a significant weakness in the development of New Zealand's SMTE. Small business owners are often intimidated by the 'hidden costs' of ICT adoption, such as training and upgrading software. They are also wary of time commitments and the problems of relying on external expertise. New Zealand does not currently possess the framework of policies and tools to assist SMTE to overcome these hurdles, however this research points to some approaches that can be taken to begin to turn this state of affairs around.

Any policy mechanisms developed to enhance uptake and use of ICT must include more than pure technological infrastructure (connectivity). The area of human capital development is particularly important. Any focus on operational skills acquisition

362

must be partnered with programs enabling owners to update their knowledge of strategic business planning.

Most important though is the facilitation of networks that can play a role in fostering technology uptake. The Canadian Tourism Exchange (CTX), for example, has been credited as one of the major reasons for the successful growth and expansion of that nation's tourism industry. This Internet based network of tourism operators (many of whom are SME) has provided a forum for tourism providers to establish connections and relationships, exchange ideas and access the latest innovations and market intelligence reports from a number of industry leaders (Gretzel and Fesenmaier 2000).

Of course such a network cannot achieve ICT uptake and improved use on its own. Whilst the Internet may improve the capacity and quantity of information available, "it is not yet clear that it will also change motivation and interest, let alone cognitive capacity" (Bimber 1998:138). Nardi and O'Day (1999) call for responsible, informed engagement with technology in local settings through what they call information ecology: an inter-related system of people, practices, technologies and values in a local environment. Social understanding, values and practices become integral aspects of the ICT tool itself. The research presented here makes it clear that if we are to understand the complex mix of factors that enable effective engagement with ICT by SMTE we will have to step beyond a simple reliance on survey based research, and begin to engage directly with the businesses we are trying to assist.

References

Applebee, A., Ritchie, B.W., Demoor, S. & Cressy, A. (2000). *The ACT Tourism Industry Internet Study*, Centre for Tourism Research, University of Canberra, Australia.

Ateljevic, J., Milne, S., Doorne, S & Ateljevic, I. (1999). Tourism Micro-firms in NZ, NZTRI, *Centre Stage Report No. 7*, Victoria University, Wellington.

Atkinson, R. & Wilhelm, T. (2002) *The Best States for E-Commerce* www.ppionline.org/ppi_ci.cfm?knlgAreaID=140&subsecID=292&contentID=250162[September 15, 2002]

Bimber, B. (1998). The Internet and political transformation: Populism, community and accelerated pluralism. *Polity, XXXI*(1), 133-160.

Buhalis, D. (1996), Enhancing the competitiveness of small and medium sized tourism enterprises at the destination level by using information technology, *Electronic Markets, Vol.6*(1), pp.1-6.

Buhalis, D. (2000) Tourism and information technologies: past, present and future, *Tourism Recreation Research* 25(1) 41-58.

Buhalis, D. (2001) Tourism distribution channels: practices and processes, in D. Buhalis and E. Laws (Eds) Tourism Distribution Channels: Practices, Issues and Transformations, Continuum, London, 7-32.

Clark, D., Bowden, S., Corner, P., Gibb, J., Kearins, K. & Pavlovich, K. (2001). *Adoption and Implementation of E-Business in NZ: Empirical Results, 2001.* University of Waikato Management School Research Report Series, ISSN1175-5571.

Grant, G., Louis, C., Maheshwari, M., Murty, D., and Tao Y. (2001). Regional Initiative for Informatics Strategies: Sector ICT Strategies Planning Templates (www.comnet.mt/CRIIS 2001, Documents/ICT2001-w.pdf) [August 3, 2002]

Gretzel, U. and Fesenmaier, D. R. (2000). *Assessing the Net Readiness of Canadian Tourism Organizations*, in Exploring New Territories in the New Millennium, Joppe, M. (Editor), TTRA-Canada Conference Proceedings, Whitehorse, Yukon, pp. 15-20.

Hull, J. & Milne, S. (2001). From Nets to the "Net": Marketing Tourism on Quebec's Lower North Shore, in Baerenholdt. N. E. and Aarsaether J.O (Eds.), *Coping Strategies in the North* (Vol. 2). Copenhagen: Nordic Council of Ministers.

Mason, D., & Milne, S. (2002). *E-Commerce and Community Tourism*. In P. C. Palvia & S. C. Palvia & E. M. Roche (Eds.), Global Information Technology and Electronic Commerce: issues for the new millennium (pp. 294-310). Marietta, Georgia: Ivy League Publishing Limited.

Ministry of Economic Development, NZ (2001) Net Readiness in NZ Industries: Empirical Results.(www.ecommerce.govt.nz/statistics/readiness/netreadiness-11.html) [Sept 20, 2002]

Nardi, B.A. and O'Day, V.L. (1999) *Information Ecologies: Using Technology with Heart*, MIT Press, Cambridge Mass.

Surman, M. and Wershler-Henry, D. (2001) *Commonspace: beyond Virtual Community*, FT.com (Financial Times/Pearson), London

TIANZ Tourism Industry Association of NZ (2001) *NZ Tourism Strategy 2010* www.tianz.org.nz/Current-Projects/New-Zealand-Tourism-Strategy-2010.asp [Sept 03, 2002]

UNDP, (2001). *Human Development Report 2001 "Making new technologies work for human development"*. New York: Oxford University Press.

Wilson, H. (2002). NZ Small and Medium-Sized Enterprises (SME), Research & Policy Paper prepared for Auckland Regional Economic Development Strategies (www.areds.co.nz/resources/New%20Zealand%20Small%20&%20Medium-sized%20Enterprises.doc) [September 18, 2002]

Service Quality of Application Service Providers: Perspectives from the Greek Tourism & Hospitality Sector

Marianna Sigala

The Scottish Hotel School
University of Strathclyde, UK
School of Economics
Free University of Bolzano, Italy
M.Sigala@strath.ac.uk

Abstract

Service quality is argued to be a crucial success factor for ASP's, but yet an empirically validated instrument for measuring the service quality of ASP vendors needs to be developed. This paper aimed to fill in this gap. After synthesising previous literature on the service quality construct within the context of IS, e-commerce, ICT outsourcing and ASP effectiveness, the paper proposes a set of dimensions and model for measuring ASP service quality. The model is tested within a set of Greek tourism and hospitality businesses. Directions for future research as well as suggestions for improving the practices of ASP suppliers and users are also provided.

Keywords: Application Service Providers (ASP); service quality; outsourcing; partnership.

1   Introduction

The ASP model has proven to be a promising solution to IT outsourcing and despite its potential benefits, firms are still cautiously considering it mainly due to lack of trust, reliability and availability of service (Sharma & Gupta, 2002). Thus, the identification of the service quality factors that could enable ASPs to enhance their services and customer satisfaction is crucially important. Research has so far been concerned with ASP's costs, benefits, vendor selection and contracting, while factors determining the quality of ASP have been neglected. This seems paradoxical as the service quality of ASP can vitally determine ASP's success (Lauchlan, 2000; Sharma & Gupta, 2002). Chen & Soliman (2002: 189) stressed that *"an empirically validated instrument for measuring the service quality of ASP vendors is urgently needed"*.

Although the service quality concept has been researched and adopted into the context of IS services, B2C Websites on PCs and mobile phones, there is no current research dealing with service quality of ASP. After synthesising previous studies, the paper proposes a set of dimensions and model for measuring ASP service quality. To that end, four major areas of literature are critically evaluated, namely service quality in IS, e-commerce, ICT outsourcing and ASP effectiveness. The model is validated by conducting research in the Greek tourism and hospitality industry. By focusing in a particular industry and country, limitations regarding the reliability of the service quality model that can rise due to contextual factors (Dyke et al., 1997) are eliminated. Suggestions for future research, ASP suppliers and users are provided.

2   Theoretical background on service quality

ASPs provide access to and management of an application that is commercially available. ASPs "sell" application access to customers and they claim to remove the day-to-day IT management by assuming total responsibility for application delivery, updates, ongoing maintenance/support. ASP manage applications in a central location and customers access the applications remotely through the Internet or via leased lines. So, ASP providers-customers relationships are characterised as a B2B business model, whereby businesses outsource applications via the Internet. Following Zeithaml's definition (2002) of e-service quality, ASPs' service quality is defined as the extent to which ASPs facilitate efficient and effective delivery, maintenance, support and use of an application. No previous research was found investigating the determinants of ASP service quality. However, as ASP involves ICT outsourcing over networks, ASP service quality should include issues of service quality in the context of Internet, other networks and ICT outsourcing. After reviewing the theory on service quality, the following section analyses how traditional theories and constructs of service quality have been adapted and applied into these contexts.

Parasuraman et al. (1985) identified ten determinants of service quality: access (approachability, easy of contact); communication (informing, listening customers); competence (skills/knowledge possession to perform services); courtesy (staff attitude demeanour); credibility (trustworthiness, honesty); reliability (performance consistency, dependability); responsiveness (service timeliness, staff willingness); security (risk/doubt); tangibles (physical evidence of service); understanding/knowing the customer. Later, Parasuraman et al. (1988) reduced these to five determinants: tangibles, reliability, responsiveness, assurance (staff knowledge, courtesy), empathy (caring, individualised attention) to produce the SERVQUAL model. Johnston & Silvestro's (1990) study added five determinants, helpfulness, care, commitment, functionality (serviceability/fitness to purpose), integrity, while Johnston's (1995) study complemented the list with flexibility (willingness, ability to amend the nature of service/product to meet customers' needs). Gronroos (1990) found six criteria of service quality: professionalism & skills; attitudes & behaviour; accessibility & flexibility; reliability & trustworthiness; reputation & credibility; recovery.

Pitt et al. (1995) illustrated the application of SERVQUAL for measuring the *service quality of the IS function*, but this raise several debates. Based on SERVQUAL studies in marketing, Van Dyke et al. (1997) and Smith (1995) raised two concerns about the SERVQUAL application to IS service quality: difficulties of using a single instrument of IS service quality across industries and contexts: the use of difference scores (perceptions-minus-expectations) to measure the expectation gap. Van Dyke et al. (1997) concluded that it is better to use perceptions-only, while Pitt et al. (1997)

and Kettinger & Lee (1997) argued that the theoretical superiority of an alternative IS service quality construct should be backed by empirical evidence in the IS context.

Although fundamental theories/findings in traditional outsourcing can be borrowed to guide research *in ASPs*, ASP differ in two major issues from traditional outsourcing whose implications must be considered. First, ASPs offer strategic applications, e.g. e-commerce, CRM, ERP (Dilger, 2000), while traditional outsourcing primarily focused on software development and IT operational activities. ASP services are not installed on the client's in-house servers, but delivered over a network. So, to ensure application delivery, five components need to work seamlessly in the ASP value chain namely software, hardware, implementation, data center/hosting and connectivity (Lauchlan, 2000). Consequently, ASP service quality must refer all five components.

The network delivering ASPs' products/services also raises the client-vendor relation to a new level. Being an inseparable part of their clients' business, ASPs become an extension of the client organisation rather than merely an outside service provider. Due to these differences, ASP service quality include bandwidth, security, trust, credibility, scalability, integration, flexibility issues (Chen & Soliman, 2002; Ring, 2000; Ekanayaka et al, 2002), that are linked to constructs (e.g. understanding clients' needs, openness, commitment, information sharing) defining partnership relations in clients-vendors *ICT outsourcing context* (Grover et al., 1996; Saunders et al., 1997; Kavan et al., 2002). Although research has shown that quality of partnership is a key predictor of ICT outsourcing effectiveness (Lee & Kim, 1999), existing literature fails to examine issues pertaining to partnering (Kern & Willcocks, 2002). Sharma & Gupta (2002) argued that research on ASP service quality needs to consider the latter.

Literature on *service quality on the Internet* has also been based on SERVQUAL, but it has empirically focused in the B2C context only. Although ASP service quality refers to B2B, the operationalisation of its construct can significantly benefit from insights that the application and adaptation of service quality revealed in B2C models. In building a construct for service quality in e-commerce, Cox & Dale (2001) claimed that the lack of online human interaction means that determinants such as competence, courtesy, cleanliness, comfort and friendliness, helpfulness, care, commitment, flexibility are not particular relevant in e-commerce, but determinants such as accessibility, communication, credibility, understanding, appearance, availability, integrity, trustfulness are equally applicable to e-commerce as in physical services. In adapting SERVQUAL to measure websites' quality, Barnes et al. (2001) developed and applied WebQual (including information quality, website navigation/ appearance, user empathy/mobility) for measuring service quality in PCs and mobile phones. Madu & Madu (2002) proposed a model for e-quality including: performance (easy of navigation & information quality); website features (e.g. search engine); structure (e.g. hyperlinks); aesthetics (website appearance); reliability (consistency of website functionality); storage capability (easy of data retrieval); serviceability (complaints handling/solution); security, system integrity; trust for data sharing; responsiveness (courtesy, flexibility to respond to customer needs); product/service quality differentiation and customisation; webstore policies; reputation; assurance; empathy in elements of human contact e.g. e-mail, call centres. Research in the e-tailing context (Zeithaml, 2002) showed that e-service quality (e-SQ) has seven dimensions that form two scales: *a core e-SQ scale* including efficiency (ability, easiness to get to and navigate a website), fulfilment (having products in stock and delivering them on time), reliability (technical functioning of a website) and privacy (assurance regarding data sharing and security); and *a recovery e-SQ scale* including responsiveness (provide appropriate data when problems occur, online guarantees and mechanisms for handling returns), compensation and contact (speak to service agent). The recovery scale was proved to become salient only when clients run into problems.

3 Research aims and methodology

The paper aimed to investigate the ASP service quality dimensions and determinant factors. The literature review showed that a credible and widely agreed measurement of service quality in the IS was absent, but several studies have proved that the e-SQ (Zeithaml, 2002) model can reliably and consistently measure service quality in the context of B2C e-commerce. Thus, this paper used a qualitative approach to determine how the e-SQ model can be used and adapted within the ASP context. The literature revealed the vital issues that the new instrument should address, while empirical evidence, support and enhancement were also sought from Greek operators.

Critical realism was selected as the appropriate philosophical orientation to guide data collection and analysis. This paradigm holds that replication and interpretation with reference to established bodies of literature are the means of advancing knowledge of the "real" world (Guba & Lincoln, 1984). A two stage qualitative research design, an exploratory and confirmatory research stage, was determined the best means to develop a conceptual model depicting the ASP service quality construct. Exploratory, in-depth, semi-structured interviews (based on the e-SQ) generated a tentative conceptual model, followed by a focus group with a similar objective. The focus group was also used to critique the model that was inductively derived from the earlier series of personal interviews. The focus group constituted the confirmatory stage of our data collection and was chosen due to its effectiveness in generating and critiquing data. Responses were collated and analysed into subject categories by use of broad data matrices (Miles & Huberman, 1994), which represented those features of data that identified the main areas of commonality in what interviewees perceived as determining ASP service quality. As Turner (1983: 334) elaborates *"these merging categories/concepts, derived from the data are then used as basic building blocks of the theoretical understanding of the topic"*. Replication logic, as results were not got from isolated cases but corroborated by other informants, ensured external validity.

Potential businesses using ASPs were identified by using the database of Ethnoplan (a consulting company for ICT outsourcing), access to which was granted due to a personal contact of the author. Due to time and resource limitations, a convenience sample of tourism and hospitality firms based in Athens, Thessaloniki and Santorini was compiled and targeted, but this did not entail serious limitations, since the two cities represent the two major Greek business centres, while Santorini represents SMEs, family run and leisure-oriented tourism and hospitality operations that are very typical in the Greek tourism industry. Seven IS managers and 11 general managers representing 18 companies were interviewed. A focus group with seven members of staff of a hotel in Thessaloniki (3 users of the outsourced applications, the general, IS, front office and marketing managers) was conducted. This hotel was chosen, because of the great number and type of applications it outsourced (Table 1).

4   Analysis and discussion of the findings

Findings represent different tourism and hospitality businesses (4 travel agents, 1 tour operator, 4 cyberintermediaries, 1 yacht tourism company, 9 hotels) as well as IS and general managers, but ICT outsourcing is focused on website hosting, design and online sales, while only three firms outsource their customer databases and one its ERP (Table 1). Outsourcing time of ICT applications varied from 1 to 6 years.

Research findings revealed that the e-SQ efficiency construct fails to truly capture the essence and multiplicity of critical issues of ASP service quality. Efficiency was directly linked with application functionality, but aspects highlighted by interviewees differed depending on the type of the outsourced application and the size of their firm. All interviewees noted the value of interactivity and user friendliness of ASPs' software, but issues related to the scalability, continuous upgrades and consistency of application performance appeared to be more important efficiency issues. Application scalability was considered vital by small firms that could not afford to outsource all necessary applications and/or functionality at once.

Firms outsourcing their website issues were concerned with the ability of the outsourced application to deal with increasing number of visitors and/or online reservations, to be continually updated with current and enhanced Internet tools (e.g. flash technologies), while firms outsourcing customer warehouse/mining tools and ERP related efficiency with future possibilities to expand applications for storing/analysing more data and integrating them with other systems. The latter firms also highlighted application customisation and flexibility and related efficiency with ASPs' ability to understand their needs and ICT architecture and with ASPs' willingness, knowledge and ability to customise solutions to their cases.

One interviewee highlighted *"an ASP might have an excellent software but this might not work for us. We spend more than four months for customising the ASP software to make systems compatible and avoid reentering data from current systems into the ASP application"*. Yet, a small cyberintermediary argued did not customise its outsourced database for cost efficiency reasons.

Table 1. Interviewees' profile

Interviewee	Company	Type & outsourcing time of ICT appl.
1 Gen. Mang	Travel agency	Website hosting & design (6 yrs),
1 IS mang	Tour Operator	Website hosting (4 yrs), custom. database (1 yr)
1 IS mang	Travel agency	Website hosting & online sales (3 yrs),
1 Gen. Mang	Travel agency	Website hosting, design & online sales (3 yrs)
1 IS mang	Cyberintermediary	Web host.+ sls (5 yrs), cust. datab, min. (4 yrs)
1 Gen. Mang	Cyberintermediary	Website hosting, design & online sales (2 yrs)
1 Gen. Mang	Cyberintermediary	Website hosting &design (3 yrs)
1 Gen. Mang	Cyberintermediary	Website hosting &design (3 yrs)
1 IS mang	yacht tourism operator	Web host + sales (3 yrs), cust. datab. Min. (2 yrs)
1 Gen Mang	Hotel	Website hosting &design (1 yrs)
1 Gen Mang	Hotel	Website hosting &design (3 yrs)
1 IS Mang	Hotel	Website hosting & online sales (5 yrs), ERP (1)
1 Gen Mang	Hotel	Website hosting &design (6 yrs)
1 Gen Mang	Hotel	Website hosting, design & online sales (4 yrs)
1 IS Mang	Hotel	Website hosting, design & online sales (5 yrs)
1 IS Mang	Hotel	Website hosting &design (2 yrs)
1 Gen Mang	Hotel	Website hosting &design (3 yrs)
1 Gen Mang	Hotel	Website hosting &design (5 yrs)
Focus group: Hotel/ Thessaloniki, Website hosting, design & online sales (5 yrs), guest database and mining (2 yrs), ERP (1 yr)		

To customise solutions, ASPs' qualities and abilities for mutual understanding, data sharing, communication quality, co-ordination, commitment to understanding and addressing business needs and operations become essential ASP service qualities. Although the contract was generally perceived as the formal indication of the ASPs' commitment, interviewees emphasised that more commitment is required than just the contractual terms, which was true specifically in ERP cases. According to an IS manager *"ASPs need to be more proactive than just reactive as the contract suggests"*. The former dimensions were also confirmed as highly important also in the focus group, whereby one stressed that *"ASP's commitment to best serve our needs enhances my confidence in developing and keeping a relationship alive, as I cannot supply company information unless I can trust him"*. Thus, ASPs' trustfulness and credibility were considered as consequences of ASPs' ability to understand and address firms' needs. Overall, interviewees' incline to link service quality efficiency with tangible ICT issues (e.g. functionality) is not surprising, since their impact on service quality is widely argued and reflected on the distinction of service quality between technical/output quality and functional/process quality (Cox & Dale, 2001). The functional/process quality dimensions are raised in the concept of fulfilment.

Interviewees highlighted as an important issue of ASPs' fulfilment their ability to successfully deliver and develop their solution on time and on budget. An interviewee claimed that ASP's service quality is illustrated when there are no *"hidden costs"* behind application implementation, which are important because *"as soon as you invest a huge amount of money it is difficult to go back"*. Four firms reported that to although services were delivered according to agreement, they did not satisfy their requirements and the additional adjustments increased costs. Cost control and clarity over financial duties are a main concern, since reduced ICT costs is a major drive for using ASPs. Two managers perceived that regular cost reporting enhanced their ASPs' reliability and assurance on transparency of service costs/operations. Type of ASP application and client size were not found to affect the important of the former.

All interviewees also stressed the importance of the technical functioning of the application arguing that the ASP service quality is determined by the speed and reliability of the network and the security/privacy provided for data transfers and storage. Three managers noted that their perception of the security qualities of their ASPs were affected by the brand name/credibility of the other firms that ASPs used.

It was suggested that the e-SQ recovery dimensions become only salient when clients run into problems. However, all interviewees highlighted that conflicts or problems frequently arose and that ASPs' staff ability, motivation and skills to resolve them are crucial determinants of ASP service quality. Two main problem types were identified that required different ways and people (operational/senior managers) to be handled: day-to-day problems; operational, cultural and contractual problems. However, all interviewees highlighted the importance of the following ASP service quality dimensions: provision of different contact points with service staff (e.g. e-mail, call centre) having seamless multiple channel capabilities and information; empathy, responsiveness, knowledge and skills of customer service staff; the professionalism, attitude and behaviour of ASPs' staff. Compensation was not considered an issue of ASP service recovery quality, because as interviewees claimed when a problem arises the aim is not to get a refund and change ASP, but to co-operate and find a common solution. However, two managers raised the issue regarding ASP responsibility for service recovery when ASPs go bankrupt and/or are bought by another firm. When their previous ASPs (for virtual hosting and website design) were acquired, the managers valued the service quality that the former provided by informing and providing them with the procedures to amend or find alternative solutions. That was crucial as technically their websites and booking facilities never went down.

5   Conclusions and suggestions

Research findings offer a rich set of insights about the criteria and processes firms use in evaluating ASPs service quality, which, in addition to serving as a starting point for developing a formal scale to measure ASP service quality, they also constitute a blueprint that ASPs can use to qualitative assess their strengths and weaknesses. Findings also identify the issues that ASP users should check when selecting and evaluating ASPs. It should also be noted that perceived ASP service quality goes beyond contractual agreements and commitments, as ASPs' ability to be proactive rather than proactive was stressed. Findings highlight ways on how ASP users and vendors can together develop and keep a better relationship. Overall, findings revealed not only the complexity, but also the diversity and determinant factors of ASP service quality dimensions. Data validated previous theory, yet they provided useful insight regarding the weight and meaning of ASP service quality dimensions, as the latter differed depending on the outsourced application and the size of the ASP-client. Figure 1summarises these findings in the proposed ASP service quality model that emerged from the extension and adaptation of the initially applied e-SQ to reflect issues noted by respondents. As ASP quality is contextually determined and findings reflect ASP clients in one country using a limited number of applications, future research must aim to construct and validate a SQ instrument for specific ASP solutions. This should also be subjected to testing in cross-cultural and cross-industry samples. The significance and association of ASP SQ with partnership quality issues also indicate the need to review the literature on partnership quality for enhancing and contributing to the dimensional structure and drivers of the ASP service quality.

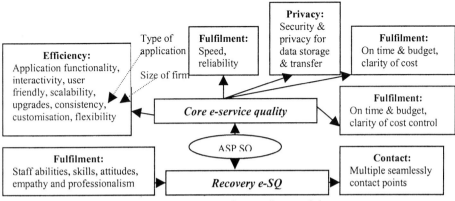

Figure 1. ASP service quality model

References

Barnes, S., Liu, K. & Vidgen, R. (2001) Evaluating WAP sites: the WebQual/m. *Proceedings of the 9th European Conference on Information Systems*. Bled Slovenia, 27–29 June

Chen, L. & Soliman, K. (2002) Managing IT outsourcing: a value-driven approach to outsourcing using ASPs. *Logistics Information Management*, 15 (3), p. 180 – 191

Cox, J. & Dale, B.G. (2001) Service quality and e-commerce: an exploratory analysis. *Managing Service Quality*. 11 (2), p. 121 – 131

Dilger, K.A. (2000) Application service providers: healthy growth foreseen for an already solution model. *Manufacturing Systems*, December, p. 76 – 78

Dyke, T., Kappelman, L.A. & Prybutok, V.R. (1997) Measuring information systems service quality: concerns on the use of the SERVQUAL. *MIS Quarterly*, June, p. 195 – 207

Ekanayaka, Y., Currie, W., & Seltsikas, P. (2002) Delivering ERP systems through application service providers. *Logistics Information Management*, 15 (3), p. 192 – 203

Gronroos, C. (1990) *Service management and marketing*. Lexington Books, Lexington, MA

Grover, V., Cheon, M.J. & Teng, J.T.C. (1996). The effects of service quality and partnership on the outsourcing of IS. *Journal of Management Information Systems*, 12 (4), p. 89 – 116

Guba, E.G. & Lincoln, Y.S. (1994) Competing paradigms in qualitative research. In Denzin, N.K. and Lincoln, Y.S. (eds) *Handbook of qualitative research*, pp. 105- 117

Johnston, R. & Silvestro, R. (1990) The determinants of service quality - customer based approach. Proceedings of the Decision Science Institute Conference, November, San Diego

Kavan, C.B., Miranda, S.M. & O' Hara, M. T. (2002) Managing the ASP process: a resource-oriented taxonomy. *Logistics Information Management*, 15 (3), p. 170 – 179

Kern, T. & Willcocks, L. (2002) Exploring relationships in information technology: the interaction approach. *European Journal of Information Systems*, 11, p. 3 – 19

Kettinger, W. J. & Lee, C.C. (1997) Pragmatic perspectives on the measurement of information systems Service Quality. *MIS Quarterly*, p. 223 – 239

Lauchlan, S. (2000) ASPs: are you ready to play?, *Computing*, 3 February, p. 29

Lee, J. & Kim, Y. (1999) Effect of partnership quality on IS outsourcing success: conceptual framework & empirical testing. *Journal of Management Information Systems*, 15 (4), p. 29 – 61

Madu, C.N. & Madu, A. A. (2002) Dimensions of e-quality. *International Journal of Quality & Reliability Management*, 19 (3), p. 246 – 258

Miles, M. & Huberman, A. (1994). *Qualitative data analysis*, California: Sage Publications

Parasuraman, A, Zeithaml, V. A. & Berry, L.L. (1985) A conceptual model of service quality and its implications for further research. *Journal of Marketing*, Fall, p. 41 – 50

Parasuraman, A., Zeithaml, V.A. & Berry, L. (1988) SERVQUAL: a multiple-item scale for measuring consumer perceptions of service quality. *Journal of Retailing*, 64, p. 12 – 40

Pitt, L.R., Watson, R. & Kavan, C. B. (1997) Measuring information systems service quality: concerns for a complete canvas. *MIS Quarterly*, June, p. 209 – 221

Pitt. L. F., Watson, R.T. & Kavan, C.B. (1995) Service quality: a measure of information systems effectiveness. *MIS Quarterly*, 9 (2), p. 173 – 187

Ring, K. (2000) *European Market Research: report to ASP Industry Consortium*, Ovum, March

Saunders, C., Gebelt, M. & Hu, Q. (1997) Achieving success in information systems outsourcing. *California Management Review*, 39 (2), p. 63 – 80

Sharma, S.K. & Gupta, J.N.D. (2002) Application Service Providers: issues and challenges. *Logistics Information Management*, 15 (3), p. 160 – 169

Turner, B. (1983). The use of grounded theory for the qualitative analysis of organisational behaviour. *Journal of Management studies*, 20, p. 333 – 348

Zeithaml, V. (2002) Service quality in e-channels. *Managing Service Quality*. 12 (3): 135—138

Formal Workflow Specification Applied to Hotel Management Information Systems.

José L. Caro[a]
Antonio Guevara[a]
Andrés Aguayo[a]
Sergio Galvez[a]
Antonio Carrillo[a]

[a] Lenguajes y Ciencias de la Computación
University of Málaga, Spain
{jlcaro, guevara, aguayo, galvez, carrillo}@lcc.uma.es

Abstract

Information Systems (IS) have evolved and adapted to the advances of hardware and software. However, rather than in technology itself, the innovations have taken place in two fundamental areas: IS analysis, design and development techniques; and changes in the philosophy of IS. Workflow management is a new IS vision than can be classified into the systems that coordinate and control the work of a real time system. Nowadays the information systems are fully implemented in several degrees in the hotel management. In this paper we develope a new workflow modelling technique that can be applied to hotel information systems for process improvement. Hotel information system complexity is related to multiple agents that are involved in its functionality and can develop a great number of coordinated processes. Thus, the applications in this area are fundamental for information system improvement.

Keywords: workflow, tourism information system, re-engineering, business processes, BPR.

1 Introduction

Today, the close relationship between Information Technologies (IT) and the tourism industry is fully consolidated. This industry implements the most recent advances in information technology to improve productivity, and the services offered to clients, as well as to increase customer satisfaction (Werthner & Klein, 1999). Currently, it is inconceivable to deal with the competitiveness of the market successfully without the proper technological tools and their correct implementation within the structure of any company. The tourism industry also follows this trend and it is well-known for using most the very latest Information Technology and other areas of engineering. Indeed, companies within the tourist industry seek ways to reduce costs and get closer to their consumers (Poon, 1993).

Most hotels have implemented their information systems (90% of hotels either have an IS or use computers for their management) (O'Connor, 2000). The use of an information system is indispensable to deal with the high competitiveness of the market efficiently, and offer the high quality services demanded by the clients of a hotel. Having it in mind, Hotel Management Information Systems (HMIS) evolved by implementing all the new technologies. This implementation ranges from the functional (e.g., computer networks) to the design perspective. It is fundamental for current HMIS to continue developing new systems, which are able to fully cover all the processes carried out in the hotel. Areas such as business processes reengineering (BPR), Internet, CSCW (Computer Supported Cooperative Work) or GroupWare have emerged with force, and can be applied to the HMIS.

2 Workflow

Computer supported cooperative work and, specifically, workflow technology aim at encouraging the work carried out in a company by the use of computer systems. The object of Workflow Management Systems (WFMS) is to provide a computer environment which efficiently supports the work carried out in any organisation. Furthermore, workflow technology not only includes the formal specification of processes, their monitoring and execution, but also reengineering tasks, evaluation and management of the processes themselves. The workflow management is of great utility for the companies since nowadays there exist a high competitiveness and the necessity of developing new products, offers, services, etc. These changes could not be produced if did not keep in mind the improvement of the processes of the information system and the applications. The workflow technology facilitates this constant evolution, providing methodologies and software in order to support the processes that take place in a company or organization (Guevara et al., 1997).

The workflow management includes the following aspects: process modelling and workflow specification, process re-engineering, and workflow automation. In order to put into practice these technologies we require models and methodologies for the capture and improvement of the processes, information system technologies and agents to develop the tasks and the activities related. Tourist organizations could benefit from this technology in order to get better efficiency and efficacy in their information systems. The activities developed in these companies involve a series of cooperative processes that are carried out by systems and people that work jointly or independently in a common goal: to offer a quality service (Caro et.al, 2000a; 2000b). These processes could be modelled in accordance with the workflow technology and they could be re-engineered subsequently in order to be improved.

The use of this technology has repercussions on the improvement of the company's information system, increasing the automation and control of the processes developed with a consequent improvement in the client satisfaction.

2.1 Workflow and WfMS definition

Workflow includes a set of technological solutions aimed at automating work processes that are described in an explicit process model called the workflow map. Workflow has a wide range of possibilities as demonstrated by group support and the automation of organizational processes. In general terms we can define workflow as workflow is comprised by a set of activities dealing with the coordinated execution of multiple tasks developed by different processing entities in order to reach a common objective (Rusinkiewitz & Sheth, 1994).

This definition of workflow does not indicate the nature of the processing entity, which, therefore, can be a person, a computer, a machine, etc. (Caro et al. 2000a). This technology is made tangible as information technology systems in the form of workflow management systems. The Workflow Management Coalition (WFMC) establishes a WfMS as a system that defines, creates, and manages automatically the execution of workflow models by the use of one or more workflow engines in charge of interpreting process definitions (workflow maps), interacting with agents and, when required, invoking the use of information systems involved in the work (Lawrence, 1997). Although any WfMS package is a complex system, often labelled as groupware and included in the field of CSCW, the main module is known as a workflow engine. The workflow engine is in charge of orchestrating the execution of the workflow model, by determining the agents involved (whether humans or not), the data, and the applications required to carry out the workflow.

2.2 Modeling techniques for workflow processes

The workflow system logic core is the specification of the process that successes on any organization. This is the modelling methodology or technique. Many authors agree on splitting workflow-modeling methodologies into two main categories:

- Activity-based methodology. Activity-based methodology focuses on modelling the activities that will take place during the development of the workflow (Lawrence, 1997; Aalst, 2002).
- Communication-based methodologies. Communication-based methodologies stem from Searle's theory, known as "speech-acts". This theory asserts that any action on the system successes via a communication between a client and server (Winograd, 1988; Action, 1996).

3 Multi-thread workflow modeling technique

The main workflow system components are: (i) organizational model (ii) resources, (iii) workflow map, (iv) Workflow Management System (WFMS). Figure 1 represents the workflow system general architecture.

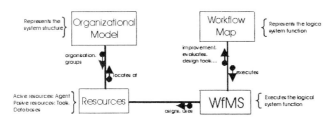

Fig.1. Workflow system architecture

3.1 Workflow primitives

In this section the basic components of our modeling methodology will be defined. This components will be used in the specification of any system and are classified in three groups: process, thread management, and connection constructors.

3.1.1 Process

These components correspond with the workflow specifications. There are three constructions:

- *Workflow specification panel*: This is the component that defines a workflow. The panel contains the specification of: workflow name, trigger, client, server, workflow description, objectives and resources to use.
- *Workflow*: Represents a task or workflow to be developed in the specification.
- *Atomic workflow*: A workflow simple.

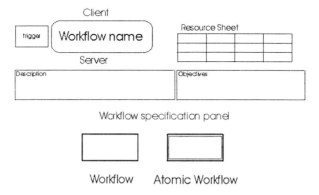

Fig.2. Process primitives

3.1.2 Thread management

This group evolves active elements entrusted to control the specification flow. The constructors are:

- *Thread start / end*: There are labels that indicate the initial and finish points of a thread.
- *Thread fork / join*: Using the thread fork constructor the main thread is subdivided into two threads that can be executed in parallel form. The join operator is used to determine the thread synchronization point.
- *Thread optionally*: The thread may take two ways depending on the expression value (true or false).
- *Wait box*: The wait box causes a pause in the thread progress until a trigger is produced.
- *Thread creation*: This block generates a new thread into the specification.
- *Trigger*: This element generates a trigger in the system.
- *Thread deletion*: This is a wait box extension that pauses the thread progress until a trigger is produced, unless a trigger or predicate expressed becomes true.
- *Trigger/Conditional repeat block*: The sub-thread that is repeated until the trigger successes or the predicate becomes true.

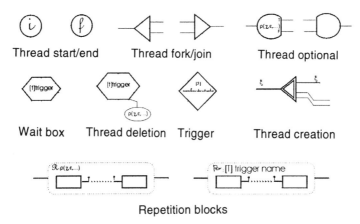

Fig.3. Thread management

3.1.3 Connection constructors:
These elements are used to simplify the specification notation.

- *Thread start*: It labels the start of a sub-thread in a specification.
- *Thread end*: It labels the end of a sub-thread in a specification
- *Label link*: Thread link that simplifies the specification.
- *Temporal break*: Indicates a skip.
- *Thread*: Temporal line.

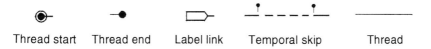

Thread start   Thread end   Label link   Temporal skip   Thread

Fig.4 Connection constructors

4   Hotel centered workflow process model

In this section we will define the workflow map corresponding to the processes that any HMIS may contemplate. For the modeling of the system, we will pay special attention to the activities a customer carries out in the hotel. That is, we will focus on either the interaction customer-hotel or the interaction of the staff with the information system. The initial workflow Wf_HMIS represents the entire hotel, and has been divided into other four workflows:

- Reservation: This is an optional workflow because clients may gain access to the hotel services without having to make a booking previously.
- Check-in: This process maps the stages involved in the client coming into the hotel for the first time.
- StayClient: Period in which the client is going to make use of the different services provided by the hotel.

- Check-out: The moment at which the client leaves the hotel, and the billing is carried out.

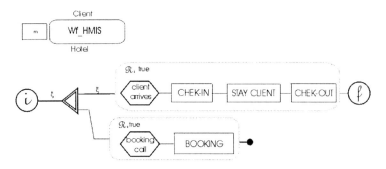

Fig. 5. Main workflow

In the following sections, we will develop the workflows for each one of these activities. We must bear in mind that not all the activities carried out are represented here. We have only implemented some basic client-oriented processes and the main processes carried out by the staff of the hotel. Also, triggers of the most specific tasks have not been specified.

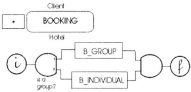

Fig. 6. Booking

4.1   Reservation

The reservation or booking process is optional. In this case (figure 6), we have made a distinction between individual booking (figure 7) and group booking (figure 8). If the hotel is dealing with a group booking, the room-listing will have to be broken down before the check-in process starts. An important point to remember at this stage is that all these actions end up with the updating of the hotel booking data.

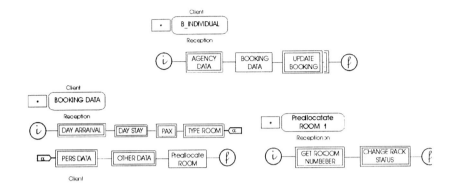

Fig. 7. Individual Booking

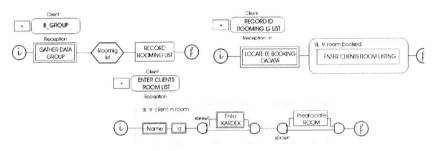

Fig. 8. Group booking

4.2 Check-in

The workflow for the customer check-in is conditioned by whether there was a previous booking or not. If there was no booking, a check-in for a transient client begins; if there was a booking, we check whether the client belongs to a group or it is an individual booking. All actions must end up with the updating of the RACK in order to keep this database consistent (figure 9).

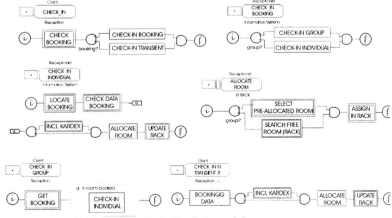

Fig. 9. Check-in workflow

4.3 Using the hotel's services (The running of the hotel)

In this section we model the actual running of the hotel, its day to day activities and tasks. In principle, the workflows will be instantiated separately because we are not dealing with a sequence of actions. For example, the room service, which shows the actions involved in cleaning and servicing rooms is activated by a temporary trigger; the restaurant service is triggered by the arrival of a client.

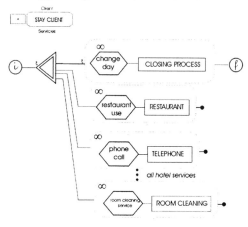

Fig. 10. Stay client

4.4 Check-out

The check-out process is made up by two distinct workflows. The first one involves billing the agency and the client. The second one, the RACK database, has to be updated in order to keep it consistent.

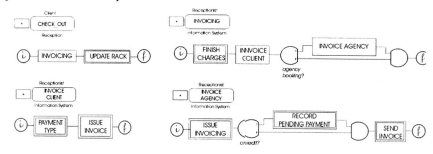

Fig. 11. Check-out

5 Conclusions and future work

In this paper we have carried out a hotel process modeling using a new methodology called multithread. This methodology improves the representation of cooperative concurrent processes that take in a real-time multi-agent cooperative complex environment like a hotel. The process modeling is very complex because it can take place at any moment, with the participation of any client in the system. Hotel information systems are very complex although they do not represent the total process that take place in the hotel. We agree with the fact of introducing new concepts into traditional information systems to get a global system representation. Workflow management systems offer an ideal framework for automating processes as well as supporting process execution in any organization. This technology can be applied in hotels because it presents highly competitive advantages, such as:

- Formal modeling of processes for better planning.
- The opportunity to apply re-engineering to the modeled processes (BPR).
- Process monitoring on the part of the agents.
- Process simulation prior to their implementation.
- Increase in service quality from the HMIS user's point of view and, consequently, better running of hotels.
- Better fulfillment of the client's requirements.
- Integration of older information systems.

References

Action Technologies (1996). Action Workflow Enterprise Server 3.0. User guide. Alameda.

Aalst, W. (2002) Workfow Management: Models, methods and Systems. MIT Press.

Caro, J.L., Guevara A., Aguayo A., Gálvez S. (2000a) *Increasing the Quality of Hotel Management Information Systems by Applying Workflow Technology*, Journal of Information Technology in Tourism. pp. 87-98. Cognizant Communication Corporation

Caro J.L., Guevara A., Aguayo A., Gálvez S. (2000b) *Workflow management applied to the information system in tourism: traveller - air line company, an example of application*, Journal of Travel Research (Information Technology).pp. 220-226. SGA Publications.

Guevara, A., Aguayo, A., Caro, J.L., and González, L. (1997). *Workflow technology: An application for tourism management.* In Tjoa, A., editor, Information and Communication Technologies in Tourism 1997, Springer Computer Science, pp. 307-317. Springer-Verlag.

Lawrence, P. (1997) Workfow Handbook. Ed. Wiley.

Medina-Mora, R., Wong HKT & Flores, P.(1993) Action workflow as the enterprise integration technology. IEEE Data Eng. Bull. 16(2) 49-54. IEEE.

O'Connor, P. (2000) Using Computers in Hospitality. Cassell Publications.

Poon, A. (1993). Tourism, technology and competitive strategies. Wallingford, Oxford: CAB International.

Rusinkiewicz, M. y Sheth, A. (1994). Specification and execution of transactional workflows. In Kim, W., editor, Modern Database Systems: The Object Model, Interoperability, and Beyond. ACM Press, Cambridge, Massachusetts.

Werthner, H. & S. Klein (1999). *Information Technology and Tourism - A Challenging Relationship*. New York, Springer-Verlag Wien.

Winograd, T. (1988). The language/action perspective. ACM Transactions on Office Information Systems, 6(2): pp. 86-91. ACM

Privacy and the Online Hotel Customer:
An Analysis of the Use of Fair Information Practices by International Hotel Companies

Peter O'Connor

Institut de Management Hotelier International
ESSEC Business School, Paris
oconnor@essec.fr

Abstract

Consumer privacy concerns are threatening to stall the growth of e-commerce. Although regulatory philosophies differ between Europe and the US, there is agreement on certain global principles, namely *notice, choice, onward transfer, access, security, integrity* and *enforcement.* A content analysis of the privacy policies of the 30 largest international hotel brands revealed that 25% fully complied with these guidelines, 69% partially complied and only 7% failed to include a policy of any kind. Transgressions most often occurred in terms of *Choice, Security* and *Integrity,* indicating an unsophisticated approach to the management of consumer information on the part of the hotel companies.

Keywords: Privacy, Fair Information Practices, Hotel E-Commerce

1   Introduction

A major benefit of the Web is its ability to individually customise sales messages. Many websites encourage users to register, define preferences, and subsequently add value by offering content specifically tailored to these needs (Metz 2001). Some e-commerce sites go further by subsequently tracking user actions – what pages they view, what products they buy – and using this "click-stream" data to refine profiles based on actual behaviour rather than stated preferences (Weber 2000). According to Internet & American Life (2000), nearly 75% of users find it useful when websites remember basic information about them and use this to provide better service.

However such personalised service comes at a price – a threat to privacy (Weber 2000). As Andy Grove (1998) pointed out "At the heart of the Internet culture is a force that wants to find out everything about you. And once it has found out everything about you and two hundred million others, that's a very valuable asset, and people are tempted to trade with that asset". Completing a retail transaction on the Web necessitates the disclosure of certain personal data (for example, name, address and billing information). However, as commercial use grows, consumers have become increasingly wary of who will have access to their information once the transaction is completed and what they will do with this data. Many websites try to reassure potential customers by publishing privacy policies – statements outlining what the site owners propose to do (or more importantly, not do) with personal data. Some websites have had their policies "certified" by third parties in an effort to add credibility and build trust (Gilbert, 2001). Such "trustmarks" do not prohibit the sharing of data, merely validate that the company is conforming to its stated privacy policy (Crowell 2000). The European Union (EU) has also introduced legislation regulating the collection, storage and use of personal data. Online consumer privacy has become a major political and commercial issue. A recent Forrester Research study found that privacy concerns continue to inhibit nearly 100 million people from shopping online (Gilbert 2001). Given the potential of hotel websites to collect personal data, either by encouraging site visitors to register, during registration for loyalty programs or while making reservations, such concerns must also be limiting the growth of hotel e-commerce. Yet little research has been carried out on the issue. Do hotel websites collect personal information, and if so, what reassurances do they give the customer about subsequent uses of their data.

2 Background

A key benefit of the Web is its ability to personalise the shopping experience. Technology exists to monitor the actions of website visitors, combine this with demographic and past purchase behaviour and customise the message presented in subsequent visits. For the consumer, this results in content more closely matched to their interests (Grover, Hall et al. 1998), while for sellers it essentially facilitates a one-to-one marketing approach, allowing them to target their most valuable prospects, tailor their offerings to individual needs and improve customer satisfaction and retention, all at a relatively low cost. Although such personalisation brings advantages to both parties, its use comes at a price – "the death of privacy". The Web, by its nature offers unprecedented opportunities to gather and disseminate detailed personal, demographic and behavioural consumer data (Opplinger 2000). In the paper-and-ink world, the sheer cumbersomeness of collecting, archiving and analysing data helped protect privacy (Blanchette and Johnson 2002). Technology based systems change not only the quantity and granularity of what can be collected, but also allow it to be analysed in increasingly sophisticated ways. Every site visit generates click-stream data, which can identify where the user came from, what was looked at and for how long, even where user's goes afterwards – all collected automatically, invisibly and without the user's knowledge. Consolidating such data with the personal and demographic data provided voluntarily results in a valuable marketing asset (Carroll 2002). Proponents argue that marketers have been gathering such information manually for years, that the Internet is simply an expansion of these efforts and that it allows companies to provide consumers with incentives that they are likely to use – an approach liked by many customers (Grover, Hall et al. 1998). Indeed consumers often willingly provide websites with highly detailed personal data – for example when completing an online sales transaction or when supplying information to facilitate the aforementioned customisation. Problems arise, however, when this is used for purposes other than the transaction for which it was collected – a process know as the "secondary use of data" (Hoffman, Novak et al. 1999). Secondary uses may be either internal, such as for marketing purposes or external if data is sold to / shared with third parties. Since information privacy is defined as "people's ability to control the terms under which their personal information is acquired and used" (Westin 1967). However when the website subsequently uses data for other purposes without either the knowledge or consent of the consumer, privacy clearly is compromised.

Studies have shown that consumers are concerned about the lack of privacy on the Web. Ryker et al (2002) quote a PriceWaterHouseCoopers study indicating that 92% of consumers are worried about privacy, with 61% concerned enough to refuse to shop online. Similarly, Forrester research has found that privacy fears inhibit nearly 50% of consumers from shopping online (Gilbert 2001). Other research shows that consumers often restrict the information they make available about themselves by declining to provide data requested by websites, or by providing false information (Georgia Tech Research Corporation 1997). Nearly one in five maintain a secondary email address to avoid giving a website real information (Phelps, D'Souza et al. 2001). In a recent survey, nearly 90% of respondents felt that privacy was the most pressing concern when shopping online, rating it more important than prices and return policies (EPIC Alert 2000). However consumers do realise that exchanging personal data for personalisation can be beneficial, and are open to providing information in certain circumstances (Hoffman, Novak et al. 1999). For example, Jupiter Research found that respondents would be more inclined to provide information online if they had a guarantee that it would not be misused, while other studies have shown that consumers would more readily cooperate if they subsequently had the right to force companies to delete such information (Gilbert 2001). In short, the issue comes down to one of trust. Trust is achieved when companies inform customers about how their personal data will be treated, and subsequently behave in a manner consistent with these disclosures (Culnan and Armstrong 1999). Many see this as one of the prime barriers to the growth of e-commerce, and forecast that its impact is likely to increase as less sophisticated consumers come online and are less able to distinguish valid threats from media hype and misinformation (Hoffman, Novak et al. 1999).

2.1 Fair Information Practices

Fair information practices are global principles that balance the privacy interests of individuals with the legitimate need of business to derive value from customer information (Culnan 2000). Strong arguments can be made for letting the market establish such principles. This assumed that consumers prefer to do business with firms that have implemented strong privacy protection, and will avoid firms that ignore / understate the issue. As a result, researchers argue that strong privacy norms will ultimately emerge as companies are forced to provide greater privacy protection, or at least the kind that consumers want in order to stay in business.

Currently, the most basic limitation on companies' behaviour is their privacy policy – a statement that describe the personal information collected and how that information is used (Metz 2001). While companies have no obligation to post online privacy policies, most now feel compelled to do so. In addition, the incorporation of privacy features (W3C Platform for Privacy Preferences or P3P) into Microsoft Internet Explorer 6 that automatically check for privacy policies mean that sites without privacy policies may be automatically blocked from being displayed. A key issue with privacy policies is quality. To aid companies develop policies that address user concerns, various industry bodies have proposed voluntary guidelines. The US government's principal consumer-protection agency (the Federal Trade Commission – FTC) guidelines are typical. These focuses on five core principles; *Notice /Awareness* implies that companies must disclose information practices before collecting data from consumers, must advise as to what information will be collected and how it will be used; *Choice/Consent* means that consumers must be given options as to whether and how the information is used for purposes beyond those for which it is originally provided; *Access/Participation* implies that consumers should be able to view and contest the accuracy and completeness of data, or delete that data if they so choose; *Security/Integrity* implies that companies must take reasonable steps to assure that personal data is secure during transition and storage, and is protected from unauthorised use; while *Enforcement/Redress* implies that facilities must be provided to resolve complaints about policy transgressions (Culnan 2000).

While having (and complying with) a privacy policy is the first step in building trust, many companies have sought third-party certification (for example, TRUSTe's "Privacy Seal" and Better Business Bureau's "BBBOnline Privacy Seal") to help reassure consumers as to their ethical behaviour with personal data (Opplinger 2000). Trustmarks such as these add credibility as their organisers scrutinize privacy policies to insure that certain minimal standards are met, audit companies periodically to insure continued compliance, operating dispute resolution schemes and only award their certification to those that meet its criteria. Such trustmarks also generate an additional layer of legal protection as sites displaying such symbols are bound to comply with their stated privacy policies under contract law (Endeshaw 2001)

2.2 The Legal Framework

While Europe and the US are in general agreement on the importance of privacy, philosophies differ substantially between the two continents. In the US, legal enforcement is based around a constitutional right to privacy, rather than specific data protection legislation (Camp 1999). While a patchwork of laws regulate the use of personal data in certain circumstances (such as credit, driver's licence, telephone and video rental records), the overriding approach has been to resist the introduction of comprehensive legislation in anticipation that the market will self regulate through adherence to voluntary codes. Companies are encouraged, but not required, to comply with standards such as the FTC's fair information practices discussed above. However this strategy is acknowledged to have been to a large extent unsuccessful. In a study of major consumer websites, over 80% failed to comply with the suggested practices, indicating that stronger action may be necessary (FTC Report 2000).

Europe considers privacy to be a fundamental right, and has introduced comprehensive legislation covering the electronic processing of personal data (Mayer-Schonberger 1998). Most recently, the European Union (EU) adopted *the European Union Directive on the Protection of Personal Data* (1995) – a mandatory and binding directive that requires member countries to incorporate its requirements. These include that personal data must be "only collected for a specified, explicit and legitimate purpose", that further processing without user consent is not permitted, that data must be kept "accurate and up to date" and users must be given access to their data as well as the name of the processor, the purpose for which the data is being collected and details of all recipients (European Community 1995). Such requirements place severe restrictions on how personal data can be used, and go beyond the fair information principles discussed above. While it could be argued that they do not apply to information processed outside the EU, such arguments are pre-empted by stipulating that data can only be transferred outside the EU if the recipient country has similar levels of data protection (Hinde 1998). While some jurisdictions, notably Hong Kong, New Zealand, Australia and Canada, have implemented legislation compliant with the Directive (Carroll 2002), the United States philosophy on self regulation makes transferring data to that region problematic. To overcome this, the US Department of Commerce and the European Commission have formulated the *Safe Harbour Agreement*, which regulates European personal information transferred to the US. This requires consumers to be notified as to the purposes for which data is being collected and used; that they are given the opportunity to choose whether and how data is disclosed to third parties; that the company must guarantee the currency of data; must protect it from loss and misuse and must give individuals the right to correct or delete personal data. Companies must also provide mechanisms for dispute resolution, and must publicly declare their compliance by registering with the US Department of Commerce (Zwick and Dholakia 2001).

Table 1, Comparison of Alternative Fair Information Practices

Principle	EU Directive	Safe Harbour	FTC	TRUSTe Guidelines	BBBOnline Guidelines
Notice	✓	✓	✓	✓	✓
Choice	✓	✓	✓	✓	✓
Onward Transfer	✓	✓			
Security	✓	✓	✓	✓	✓
Data Integrity	✓	✓	✓	✓	
Access	✓	✓	✓	✓	✓
Enforcement	✓	✓	✓	✓	

It is interesting to note a common thread flowing through each of the approaches discussed above. From the FTC's fair information practices to the trustmark company guidelines to the EU Directive and the Safe Harbour Agreement, common traits can be distinguished as to how personal information should be safeguarded. These are summarised in Table One. As the EU Directive / the Safe Harbour Agreement provides the most comprehensive (and an international) standard for data protection, this will be used as the yardstick for measurement in the remainder or the study.

3 Research Methodology

As has been outlined above, online privacy has become a critical issue. Hotel companies have traditionally collected large quantities of customer data for operational purposes. In addition, many are currently deploying Customer Relationship Management systems whose success depends upon the collection of personal data for subsequent interaction with customers. Given this reliance on personal information, the question must be asked as to whether they offer adequate levels of data protection to their customers? The primary research questions were therefore to establish the extent to which hotel companies collect personal information about visitors to their consumer websites, whether these sites contain privacy policies, and whether such policies reflect fair information practices. The population for the study was defined as the top 30 international hotel brands (by number of rooms) as defined in Hotels magazine (July 2001). This group was chosen based on the premise that larger companies have both the capital and technical expertise to develop their e-commerce capabilities in a professional manner, and thus an analysis of their behaviour should provide some insight into leading edge practices within the hotel sector. Whilst it means that the survey results are not generally applicable and indicative of the industry as a whole, it does provide a useful snapshot of the behaviour of key industry players.

A structured in-depth analysis of each website was carried out during July 2002. Sites were accessed and facilities for the entry of personal data explored. Where such facilities were available, seed data (i.e. false but functioning postal and email addresses) was entered to create a user profile. Privacy policies, where posted, were analysed using content analysis techniques to establish if they conformed to the guidelines discussed above. Because the study focused on disclosures, and because of the time delay needed to establish if personal information is used in inappropriate ways, no effort was made at this stage to establish if the site's practices differed from its stated policies. However, the use of seed data means that companies can subsequently be monitored and the issue explored at a later date.

4 Summary of Findings

4.1 Information Collected

The first question addressed was the range of information collected by hotel websites. The study adopted the definition of "personal information" used by the FTC which differentiates between personal information and demographics / preference information. The former is defined as "data that can be used to identify a consumer", while demographic information on its own cannot identify a specific person, but can be used in aggregate form for market research or in conjunction with personal identifying information to create consumer profiles (Culnan 2000). As can be seen from Table Two, the majority of sites collect at least one type of personal information with only two sites not collect personal information of any kind. As the majority of websites collect personal data, they therefore should also display appropriate privacy policies if they are to comply with fair information practices.

Table 2, Information Collected By Hotel Chain Websites (Base = 30)

Personal Information	% Collecting	Demographic Information	% Collecting
Surname	93%	Email Promotions	70%
First Name	93%	Bed Preference	70%
Email	93%	Smoking	60%
Street	90%	Wheelchair Access	43%
City	90%	Annual hotel nights	37%
State / Provence	90%	Language Preference	37%
Country	90%	Floor Preference	30%
Zip / Postal Code	90%	Business Title	30%
Phone	90%	Interests	27%
Title	77%	Favourite Destinations	27%
Middle Initial	77%	Traveller Type	23%
Company Name	57%	Email reservation options	17%
Fax	43%	Elevator Preference	17%
Credit Card Number	43%	Electronic statements	13%
Frequent Flyer Number	40%	Nationality	13%
Date of Birth	40%	Residence details	13%
Passport Number	13%	Email Format	7%
		Pillow Preference	7%
		Newspaper Preference	7%
		Number of Children	7%
		Currency Preferences	7%

4.2 Frequency of Privacy Disclosures

The second part of the study focused on establishing the extent to which websites incorporated privacy disclosures. As with the prior question, the FTC definition, which differentiates between privacy statements and policies, was used. The former are defined as "discrete statements that describe a particular information practice from which at least one potential use could be inferred". Policies, on the other hand, are defined as "comprehensive descriptions of a site's practices located in one place on the site and reached by clicking on an icon or hyperlink" (FTC Report 2000). No assessment of quality was carried out at this stage – sites were simply searched for the existence of a disclosure. Of the twenty-eight companies collecting personal information, twenty-five displayed a privacy policy meeting the definition; two others have more limited privacy statements while only a single failed to address the issue. Subsequent emails to the webmaster of this company revealed that the issue was being addressed "soon"! Of those displayed a disclosure, all had a link to it from pages where personal information is being collected, demonstrating an awareness of the importance of the issue in the minds of consumers. That being said, despite the role of trustmarks discussed above, only a single site included third party certification to show that their consumer privacy practices had been externally certified.

4.3 Nature of Disclosures

The third issue was to establish the extent to which privacy disclosures conform to fair information guidelines established earlier. Of the 27 websites that have posted a privacy disclosure, only one in five (21%) comply fully with the guidelines of notice, choice, onward transfer, access, security, integrity and enforcement. As can be seen from Table Four, it is with the principle of *Onward Transfer* that hotel websites had the most difficulty. Only a small number (25%) included statements informing consumers that that they would benefit from similar levels of protection if their data was passed onto a third party. Hotel companies to a large extent (>60%) also failed to meet the criteria for both *Choice* and *Security*. Considered as a whole, the disclosures present on hotel websites are clearly lacking.

Table 3, Summary of Compliance with Privacy Practices (Base = 28)

Category	Fully Comply	Partially Comply	Non-Compliance
Notice	61%	39%	0%
Choice	54%	14%	32%
Onward Transfer	25%	0%	75%
Access	36%	54%	11%
Security	57%	4%	39%
Integrity	50%	29%	21%
Enforcement	79%	0%	21%

Note: To achieve full compliance for a practice, a policy must receive a "Yes" for all associated questions. To achieve partial compliance, a policy must receive a "Yes" in at least one of the criteria. Policies in non-compliance with a practice must receive a "No" for all questions with that practice.

A more detailed analysis of each category highlights the limitations of current policies. While in general policies are strong in terms of *Notice* (particularly telling consumers what data is being collected, how it is collected and whether or not it is shared with third parties), *Access* (allowing consumers in the majority of cases to view and modify personal data on line) and *Enforcement* (providing contact information for queries and including a process to follow in the event of a dispute), it is in terms of Choice, Security and Integrity that policies are weakest. Only approximately 60% of policies mention if data is safeguarded in any way during either transportation or storage. An even small number (50%) mention procedures to insure data accuracy, comprehensiveness or reliability. However it is as regards *Choice* that the findings are most disturbing. While nearly 75% give users the option of opting-in or opting-out of having data disclosed to third parties, only 54% give users an option with regard to internal use of their data. The approach seems to be that once a consumer has surrendered data, it can be used for whatever purposes the company sees fit. This view is supported by the findings on the *Access* principle. Although most policies offer consumers the ability to view or modify personal data, only 36% mention the ability to permanently remove it, confirming the theory that companies feel that they own data once it has been given to them.

Table 4, Compliance with Privacy Practices (Base = 28)

Category / Principle	%
Notice – does the policy say anything about:	
What personal information is collected?	100%
How information is collected	93%
Whether communications are sent to consumers?	75%
Whether information is disclosed to third parties	82%
Use / non-use of cookies	79%
Choice	
Internal Use – Can the consumer opt-in or opt-out?	54%
3rd Party Use – Can the consumer opt-in or opt-out?	71%
Onward Transfer	
Mentions that data transferred to a third party benefits from similar protection	25%
Access – Can consumers:	
Review at least some personal information?	89%
Correct at least some personal information	89%
Delete at least some personal information?	36%
Security	
Is information secured in transit?	61%
Is information secured in-house?	57%
Integrity	
Mentions that data is relevant to the purposes stated	79%
Steps taken to ensure reliability and completeness	64%
Enforcement	
Contact Information	82%
Independent recourse for problems	79%

5   Conclusions

While the hotel sector is traditionally regarded as being conservative in terms of its adoption of new technology, an analysis of the websites of the major hotel chains reveals enlightened practices in relation to the collection and use of consumers' personal data. As might be expected, the majority collect extensive personal, demographic and preference data online. To reassure users and encourage them to enter data, most display a privacy policy clearly setting forth subsequent rights. While undeniably bias, the pervasiveness of policies indicates awareness on the part of hotels as to the importance of privacy in encouraging online consumer interaction. However, although the majority of sites have posted privacy notices, in most cases, these disclosures do not reflect fair information practices and do not adequate address consumer concerns. Despite their proliferation, the majority are weak, with only 25% meeting the requirements of the EU / US Department of Commerce *Safe Harbour Agreement* In particular, policies stumble on their treatment of *Choice*, *Security* and *Integrity* – a worrying combination given that their objective is to make consumers comfortable disclosing the personal data needed for relationship marketing!. As discussed earlier, when presented with uncertainty as to the use of data, consumers often react by limiting the data entered, or by entering false data. Combined with the hotel company's apparent lack of commitment to verify data accuracy, this has drastic consequences for data quality and its ability to be used for database marketing purposes, thus negating many of the benefits of collecting it in the first place.

References

Blanchette, J.-F. and D. G. Johnson (2002). "Data Retention and the Panoptic Society: The Social Benefits of Forgetfulness." The Information Society 18: 33-45.

Camp, L. J. (1999). "Web Security and Privacy" The Information Society 15: 249-256.

Carroll, B. (2002). "Price of Privacy: Selling Consumer Databases in Bankruptcy." Journal of Interactive Marketing 16(3): 47-58.

Crowell, G. (2000). "Let 'Em In: Profits and privacy: How the two can coexist on the Web." Business 2.0(14/11/2000): 290-292.

Culnan, M. (2000). "Protecting Privacy Online: Is Self-Regulation Working?" Journal of Public Policy and Marketing 19(1): 20-26.

Culnan, M. and P. K. Armstrong (1999). "Information Privacy Concerns, Procedural Fairness and Impersonal Trust: An Empirical Investigation." Organization Science 10: 104-115.

Endeshaw, A. (2001). "The Legal Significance of Trademarks." Information & Communications Technology Law 10(2): 203-230.

FTC Report (2000). Privacy Online: Fair Information Practices in the Electronic Marketplace, A Report to Congress, Federal Trade Commission. 2002.

Georgia Tech Research Corporation (1997). Seventh WWW User Survey, http://www.cc.gatech.edu/gvu/user_surveys.

Gilbert, J. (2001). "Privacy? Who needs Privacy?" Business 2.0 (August): 20.

Grove, A. (1998). Only the Paranoid Survive. New York, HarperCollins Business.

Grover, V., L. Hall, et al. (1998). "The Web of Privacy: Business in the Information Age." Business Horizons (July-August): 5-11.

Hinde, S. (1998). "Privacy and Security - The Drivers for Growth of E-Commerce." Computers and Security 17: 475-478.

Hoffman, D. L., T. P. Novak, et al. (1999). "Information Privacy in the Marketspace: Implications for the Use of Anonymity on the Web." The Information Society 15: 129-139.

Mayer-Schonberger, V. (1998). "The Internet and Privacy Legislation: Cookies for a treat?" Computer Law and Security Report 14(3): 166-174.

Metz, C. (2001). "What They Know." PC Magazine (November 13): 104-118.

Opplinger, R. (2000). "Privacy Protection and anonymity services for the World Wide Web (WWW)." Future Generation Computer Systems 16: 379-391.

Phelps, J. E., G. D'Souza, et al. (2001). "Antecedents and Consequences of Consumer Privacy Conecrns: An Empirical Investigation." Journal of Interactive Marketing 15(4): 2-17.

Ryker, R., E. LaFleur, et al. (2002). "Online Privacy Policies: An assessment of the Fortune E-50." Journal of Computer Information Systems (Summer): 15-20.

Weber, T. (2000). On the Internet, Everybody wants to be a Nobody. Wall Street Journal Europe. Paris: 26.

Westin, A. F. (1967). Privacy and Freedom. New York, Atheneum.

Zwick, D. and N. Dholakia (2001). "Contrasting European & American Approaches to Privacy in Electronic Markets: Property Right versus Civil Right." Electronic Markets 11(2): 116-120.

E-commerce Approach for Supporting Trading Alliances.

Pierre F Tiako

Department of Computer and Information Science
Langston University, USA
Tiako@ieee.org

Abstract

Traditional electronic commerce models just enable the online offering, ordering, payment, and delivery of goods. One common characteristic of the different models is their isolation. Though they are connected to the Web, they still are isolated islands in the online universe, as they cannot interact with each other without media-breaks. The isolation is enforced by the limited support of coordinating interactions among partners in today's marketplace. This work provides an infrastructure for modeling and coordinating ad-hoc relationships among partners involved in the market for buying and selling goods or services. Support for these relationships, called trading alliances here, is what this paper about. It provides opportunities to bundle complementary needs of partners according to unpredictable requirements to the market. This paper augments existing electronic commerce models by defining an infrastructure for collaborative electronic market. It considers an example of trading among a travel agency and partners. It also describes the underlying technologies for implementing the proposal.

Keywords: electronic commerce, alliances, travel agency, coordination, trading, chain supply

1 Introduction

As the Internet grew and evolved, it became more broadly used by everyone. What once was destined to military and academic purposes is now also destined to companies for e-business marketing strategies and kinds of alliances in the chain supply. Historically, companies have found many ways to work together, playing different roles with regard to manufacturing, supplying, selling, delivering, and buying, in the chain supply. Most of the time, according to role, members of each company get together in a shared space (i.e., marketplace) to work on a particular project (i.e., delivering quality goods or services to customers).

More recently several research work on software process technology found in (Tiako and Al., 2001) have proposed models for organizing distributed software development among autonomous and remote software enterprises. The suggested models of collaboration relied on Internet technologies, which offer a communication medium. They allow software development ventures in a much dynamic manner and help automate a variety of contract types among enterprise (Tiako, 2002). The ways e-commerce companies interact to build chain supply (Simchi-Levi and Al., 2000) over the Internet, dealing with constraints for delivering products to customers are about the same as those used by software enterprises, as described above. Related work in e-commerce, at some level, provide framework for negotiation (Benyoucef and Keller, 2000), or mediation (Moukas and Al., 1998; Guttman and al., 1998) to deal with e-business relationships, but they still limited in term of infrastructure for ad-hoc collaboration in the marketplace.

This paper is organized as follows: Section 2 presents an approach for trading alliance modeling and a typical scenario of usage where trading among a travel agency and its partners is considered. The underlying technologies used to develop this proposal are presented in Section 3. Section 4 discusses related work before conclude.

2   Methodology of Modeling Trading Alliances and Processes

Basically, there are two forms of e-commerce: business-to-business (B2B) and business-to-consumer (B2C). The first relates to electronic transactions that aim to automate business processes between two companies, without consideration to the final consumer. Here, companies interact using electronic data interchange. The second involves purchase of goods by the final consumer through the Internet, interacting with companies using electronic fund transfer. Fig. 1 presents B2B and B2C in the context of traveling agency.

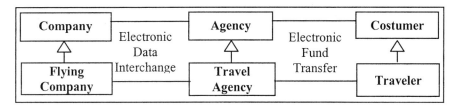

Fig. 1. Business-to-business and business-to-consumer transactions

An example that involves electronic data interchange transaction is the supply of flights by flying companies to travel agencies. An example of transaction that involves electronic fund transfer, is when a traveler, usually from home, sends electronic information on the checking account of a travel agency to purchase tickets. Let us present the different types of components, parts of chain supplies, that make up the trading community, before developing the model for supporting their e-business relationships.

2.1   Components of the Chain Supply

Types of components operating in chain supply (see Fig. 2) are described as follows:

Customer. It is the end user in the trading community that buys goods or services to suppliers from home, via the Web and using credit card for payment.

Supplier. It builds shopping malls to provide products to customers. For, it connects to banks for payment and to deliverers for delivering, besides linking with other suppliers or manufacturer, for supplying.

Bank. It manages the flow of money in the trading community using SET —secure electronic transaction (Panurach, 1996; Drew, 1999)— for secured payments.

Delivery. When needed —"*Tip:* Some itineraries may require paper tickets, and a $14.99 shipping and handling fee is charged for paper tickets." Expedia.com—, it allows the physical move of products between components.

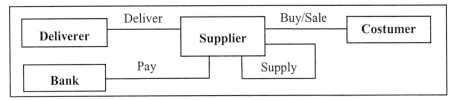

Fig. 2. Components of chain supply and their relations.

2.2 Modeling Trading Alliances and E-business Processes

Scenarios of chain supply (see Fig. 3) for travel packages in an agency are considered here. Components interacting are: (1) Customers —families and individuals— that make reservations for packages. A package has three elements: (i) Round trip flight, (ii) Accommodation, and (iii) Car rental. (2) Travel agencies *TAs* that interact with customers, flight companies, hotels and car rental companies for travel arrangements. (3) Flight companies *FCs* for traveling; (4) Car rental companies *CRCs* for transportation at arrival; (5) Hotels to provide accommodation. (6) Banks for managing payments, and (7) Shipper companies *SCs* for delivering tickets. This example is used to illustrate models of trading Alliances.

Trading alliance models are involved with several components while e-business (B2B or B2C) process models are to the responsibility of single component. Both models are interdependent. Alliances belong to family of organizational processes while e-business processes are in rapport with business processes. It is described below how the generic meta-model of federation introduced in (Tiako, 1999), also applied in (Tiako, 2002), can be specialized to support alliances and e-business processes.

The generic meta-model supports various types of process (software, business, and workflow) that an organization can plan to define. Each process has several activities —piece of work that must be done. The philosophy of defining both models is the same. The definition of a model starts by creating its entities and relations from the meta-model. An activity for e-business process is composed of a name and relationships with entities product, direction, tool, role, and its sub-activities. Precisely one role is assigned to an activity. An agent will perform this role. A product can form an input to an activity, or an output from an activity, or an intermediate result of an activity. An agent is a model entity that performs roles in the trading community; therefore carrying out activities. During the e-business process definition, some activities can be assigned to a role, and several agents can be identified to perform a role. A direction is a model entity that defines objectives of an activity, including constraints to be respected and may provide advisory guidance on carrying out activities. The only requirement imposed by the model is that direction provides instructions to complete an activity.

A trading alliance is composed of a name and relationships with entities events, direction, tool, competence, and eventually its sub-alliances. Competence is a model entity that defines the function (see section 2.1) a component must satisfy before involvment in an alliance. An alliance involves several competence. Several components can be identified to fulfill the same competence within an alliance. Event is a model entity that defines objects of any nature (goods, services, and information...) that can be exchanged in the marketplace. Sub-alliances function as alliances. Direction and Tool have the same semantics as defined above for e-business process. Each component can define models of alliance in its own ways, and according to purpose of the marketplace. A model is instanciated into alliances with appropriated partners. Enactment of an alliance allows establishing and maintaining trading collaboration. Its performance allows controlling the fulfillment of the commitment of each component during trading, including distribution of events among them.

2.3 Validation of Trading Alliance Model

In the example, TA makes a B2B inquiry for booking flights to FCs. FC responds by performing a process (defined in component Airline in Fig. 3) for supplying tickets as follows. Activity *Availability* checks flights availability before perform *Reject* for rejecting the inquiry if there is no flight. If available, *Booking* is performed to book flights and notify TA. Tickets can then be delivered to TA electronically or by mail. Then activity *Cash* is started to send invoice to TA with bank account to be credited. Before closing the process, FC starts activity *Flight* to transport Traveler; for, a check-in and check-out will be necessary. The trading alliance coordinating interactions between TA and FC is defined as follows in Fig. 3: (08) activity *Inquiry* for asking prices and flights availability; (09) *Notification* for informing TA to flights availability; (10) *Ordering* to order flights for TA; and (11) *Confirmation* to notify that flights are booked.

TA also makes B2B inquiries to reserve rooms for accommodation at destination. The basic process for supplying room and its activities are defined in component Hotel of Fig. 3. This process is about the same as the one used by FC to supply tickets to TA. The trading alliance coordinating interactions between TA and Hotels is defined as follows in Fig. 3: (04) activity *Inquiry* for asking prices and rooms availability; (05) *Notification* to inform TA to rooms' availability; (06) *Ordering* to order reservation by TA; and (07) *Confirmation* to confirm rooms' reservation.

TA continues by B2B inquiries for car reservation at destination. The basic process for reservation is defined in component Car Rental of Fig. 3. The alliance for coordinating interactions between TA and CRCs is defined as follows in Fig. 3: (12) activity *Inquiry* for asking prices and cars' availability; (13) *Notification* to inform TA to availability; (14) *Ordering* by TA to make reservation; and (15) *Confirmation* to confirm that cars are reserved.

At this point, TA is ready to interact with Traveler for delivering packages. Traveler, from home's Web system, defines or used an existing B2C procurement process for ordering tickets. The basic process *Travel* for buying packages and its activities (see component Traveler of Fig. 3) are defined as follows: activity *Request* is performed to inquiry TAs for package deals. Then *Deal* is performed to receive TAs' proposals, before performing *Evaluate* to evaluate them. Traveler through the process can receive a maximum of deals before evaluation, by just performing *Request* several times. After evaluation, Traveler can reject them, requesting other deals, or closing the process. If good deals, the process continues by contacting the selected TA for ordering. *Order* will then be performed, follows by *Tickets*, for receiving tickets. If SC delivers tickets, an acknowledgment is required. Performance of *Voucher* will provide information and instructions for the package. Then *Stay* starts to manage the trip. Back home, Traveler will close the whole processes after receiving and verifying the checking balance account from Bank, regarding trip's expenses.

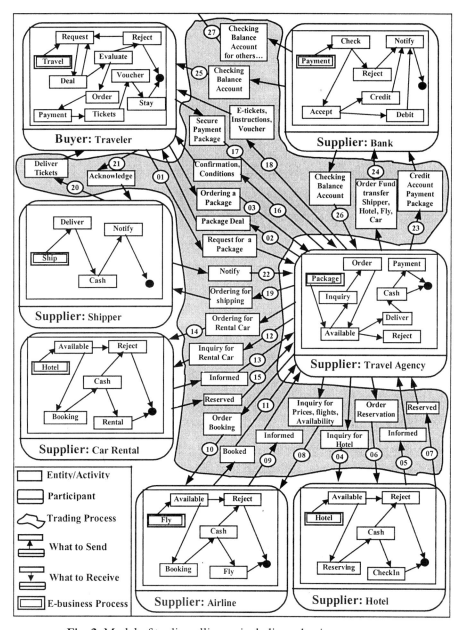

Fig. 3. Model of trading alliance, including e-business processes

The alliance coordinating interactions between Traveler, TA, Bank and SC is defined by the following activities in Fig. 3: (01) *Request* that inquiries prices and package availability; (02) *Deals* to inform Traveler to package deals availability; (03)

Ordering to order a package; (16) *Confirmation* to confirm package's reservation —here FCs, Hotels and CRCs supposed already supplied TA as presented above; (17) *Payment* uses SET to transfer account information to TA for payment; (18) *E-tickets* delivers electronic tickets to Traveler; as option, (20) *Deliver* delivers package via SC and (21) *Acknowledge* for delivery's acknowledgment; (25) *Checking* allows TA interacting with Bank to receive its checking balance account. Let us describe how TA acts with its own e-business process for selling packages.

After receiving an inquiry for packages, TA will start process *Package* for selling them, as presented in component TA of Fig. 3, and composed of activities that follows. *Available* that determines the availability of stock to honor the order. If not, *Inquiry* is performed to ask additional package's elements from suppliers, before beginning *Order* to make order. If the package still not available, Traveler's request is rejected by performing *Reject*. If available, *Deliver* is performed to deliver, electronically or by mail, the deal to Traveler. Delivery includes tickets, vouchers, and all the necessary instructions for traveling. Performance of *Cash* orders Bank to credit TA's account with Traveler's payment, the same time, and the activation of *Payment* orders Bank to pay suppliers, if any. Process *Package* will close when TA received correct balance accounts from Bank regarding package's transactions.

The Alliance coordinating interactions between TA, Traveler, Bank and SC is defined by the following trading activities of Fig. 3: (19) *Ordering* that orders SC to ship the package; (22) *Notification*, allows SC to notify TA after delivered; (23) *Credit*, orders Bank to credit TA's account; (24) *Debit*, debits TA's account to pay suppliers involved in the package deal; (26) *Balance*, returns TA's balance account from Bank. Other activities involve with TA has already been defined above.

3 Software infrastructure for Trading Alliances and Processes

The Internet and proliferation of distributing technologies provide basic infrastructure for e-commerce. Unfortunately, the corresponding infrastructure for properly modeling various e-commerce activities and their evolution is still missing. The software infrastructure for supporting the chain supply has to be open enough for integrating new components. Main underlying technologies for e-commerce and other distributed Web applications are CORBA, DCOM, and Java Virtual Machine (JVM) and its APIs (Jutla and al., 1999). CORBA specifies interfaces and protocols for a distributed infrastructure, working through ORBs (Object Request Brokers). DCOM allows writing its components in several languages, including C++ and Java. Java provides a mechanism for components discover each other's interfaces at runtime and may run on different platforms because of the JVM.

This infrastructure (see Fig. 4) for supporting trading alliances and e-business processes uses the protocol TCP/IP on which protocols HTTP and IIOP are grafted, basic element for distributing components of a trading community. Components functioning as customers must at least lay out a Web Browser including a JVM and the pluggins necessary for enacting and performing e-business processes, and for interacting with the community. It will interact with the different servers of the architecture through the HTTP protocol, which is the native protocol of communication between a browser and a Web server. Servers are designed using a firmware that allows querying infrastructure's repositories of e-business products (goods, services) and e-business activities (processes, alliances, artifacts...).

A generic server integrated into the firmware support e-business processes in trading community. It allows accessing local ORB and beyond CORBA architecture, ways to reach remote ORBs and then access other components of the community. To simplify

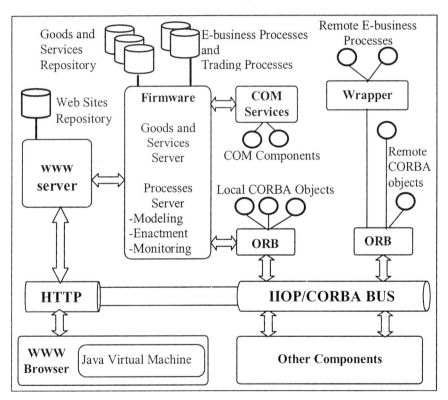

Fig. 4. Infrastructure for supporting chain supply in trading community

the prototyping of the proposed architecture and the portability of the resulting platform, we have adopted the Java implementation. On each remote computer connected to the trading community, resides a wrapper; piece of software that provides a generic interface for components that cannot communicate directly with the message protocol, that is the ORB. A concrete wrapper that is specific for each trading component is generated from the generic wrapper at runtime using the description of the function of the component in the marketplace.

Such infrastructure allows defining, establishing, enacting, and performing various types of alliances among the partners involved in the market. Alliance's performance allows controlling partners by requiring them to fulfillment their commitments in the marketplace.

4   Related Work and Conclusion

The approach by (Benyoucef and Keller, 2000) is close to this work because it coordinates activities in the marketplace. It is based on a combined negotiation support system that helps users conduct negotiations. This solution does not model negotiation and their evolution for unpredictable market. WISE (Alonso and al., 1999; Schuldt and al., 1998) and this paper in common, coordinate E-business processes. WISE provides a priori defined protocols for coordination, while it is explicitly modeled here by trading alliances. Supply chain process by QPR (see www.qprsoftware.com) includes phases from customer inquiries, purchase, and production, to product's delivery to the customer. Here, customer still not considered as an equal partner as supplier. For instance, customer just provides input information for processes. In this work, customer has an equal function as other components within the chain supply, and play roles in trading alliances. Strategic Alliances by (Hoffman, 1997) favor business relationship types for long term collaboration, but the approach is different because it does not provide a support for e-business processes modeling.

This work proposed an appropriated infrastructure for modeling e-business processes and coordinating them using trading alliances. The idea combines a support for modeling and coordinating relationships among trading components with an architecture for their distribution. Components are not only on a different location but also entirely autonomous. They can define, enact, and monitor all kind of relationships they would like to establish among them.

References

Alonso G., Fiedler U., Hagen C., Lazcano A., Schuldt H. & N. Weiler (1999). WISE: Business to Business E-Commerce. In Proc. of the 9th International Workshop *on Research Issues on Data Engineering: Information Technology for Virtual Enterprises* (RIDE-VE'99).

Benyoucef M. & R. Keller (2000). A conceptual Architecture for a Combined Negotiation Support System. In Proc. of 11th Intl. *Workshop on Database and Expert Systems Application.*

Drew G (1999). *Using SET for Secure Electronic Commerce.* Prentice Hall

Guttman R. & P. Maes (1998). Agent-mediated Integrative Negotiation for Retail Electronic Commerce. In Proc. . of the *Workshop on Agent Mediated Electronic Trading* (AMET'98).

Hoffman T. (1997). *Strategic Alliances For Competitive Advantage*. Alliance Ventures Group.

Jutla D., Bodorik P., Hajnal C.& C. Davis (1999). Making Business Sense of Electronic Commerce. *IEEE Computer*, March.

Moukas A., Guttman R. & P. Maes (1998). Agent-mediated Electronic Commerce: An MIT Media Laboratory Perspective. In Proc. of *1st Intl. Conf. on Electronic Commerce* (ICEC'98).

Panurach P. (1996). Money in Electronic Commerce: Digital Cash, Electronic Fund Transfer, and Ecash. *Communication of ACM* 39(6).

Schuldt H, Scheck H.J & M. Tresch (1998). Coordination in CIM: Bringing Database Functionality to Application Systems. In Proceeding of the *5th European Concurrent Engineering Conference* (ECEC'98), Erlangen, Germany.

Simchi-Levi D., Kaminsky P. & E. Simchi-Levi (2000). *Designing and Managing the Supply Chain: Concepts, Strategies, and Cases w/CD-ROM Package*. McGraw-Hill.

Tiako P. (2002). Maintenance in Joint Software Development. In Proc. of the 26th IEEE International *Computer Software and Applications Conference* (COMPSAC 2002) Oxford, UK.

Tiako P. (1999). Modeling the Federation of Process Sensitive Engineering Environments. *Ph.D. Dissertation*. National Polytechnic Institute of Lorraine (INPL) and LORIA Lab, France.

Tiako P., Lindquist T. & V. Gruhn (2001). Report on Process Support for Distributed Team-Based Software Development Workshop. *ACM Software Engineering Notes* 26(6).

Defining Internet Marketing Strategies for Alpine Tourist Destinations. Lessons from an Empirical Research Study of the Dolomites Area.

Mariangela Franch,
Umberto Martini,
Pier Luigi Novi Inverardi

Department of Computer and Management Sciences
University of Trento, Italy
{franch, martini, pnoviinv}@cs.unitn.it

Abstract

This paper presents a part of the results of a study undertaken in the Italian Dolomites region, focusing on the role and use of Internet in the different phases of the decision-making process of tourists choosing this destination. Taking place during the summer 2001 and winter 2001-2002 seasons, the study was conducted by means of 4910 interviews administered through local tourist organisations. The paper will put forth some recommendations for the development of effective activities of destination marketing, both online and offline, for Alpine tourist destinations.

Keywords: alpine tourism; tourist use of Websites; destinations' Websites' content; destination marketing strategy.

1 Introduction and Theoretical Background

It is safe to say that all tourist organizations now have Websites for promotion, information dissemination, or online booking services. Tourism is, in fact, one of the most highly represented economic sectors on the Internet, which is used not only as a means of providing information but also for e-commerce. In light of this, the need to develop and manage Websites has become an integral part of the overall marketing strategy of a tourist destination (WTO Business Council, 1999; Pan and Fesenmaier, 2000; Buhalis and Spada, 2000). Tourist destinations in the Alpine area have vastly increased their presence on the Web in recent years (http://www.cs.unitn.it/etourism/). The unique nature of the tourist market in which they operate, however, requires a more in-depth look at ways to improve the effectiveness of investments in Web technologies. Alpine tourism has some defining characteristics. Most accommodation facilities are small and family managed. This often implies limited resources to invest in new technologies and sometimes hesitation to change long-held traditions and practices. Alpine tourism also has a strong impact on the local community inasmuch as it provides jobs and other business opportunities for residents. Tourism has in fact become a pivotal economic basis for many communities, being integrated with other services or with agriculture. The local public entities, therefore, play an important role in organizing and promoting the tourist offering, especially in defining policies regarding tourism and management of their local destinations. In this context, the design and use of Internet for promoting Alpine tourist destinations must take into account the following factors (see previous researches of our group in Franch et al., 2001; Martini et al., 2000):

- the fragmentation of the tourist offering,
- the generally skeptical view local operators have of new technologies,
- the role of public tourist organizations in realizing Websites for the destination.

As for the demand side, most visitors engage in do-it-yourself tourism where they organize the visit autonomously by booking directly with the operator and traveling by car. This tourist segment was undervalued and relatively "unstudied" until recently, with the exception of a few studies carried out by local public tourist organizations. Changes in tourist demand have made it necessary now more than ever to develop models of analysis that can offer a solid conceptual base to actors engaged in incoming activities, so that they can then design appropriate marketing strategies. This need becomes more marked as innovations are continually developed in information and communication technologies. Little is known, in fact, about the degree to which these tourists use Internet for holiday planning, and the impact that the Web has on their ultimate choice of destination (see in general Fodness and Murray, 1999; Weber and Roehl, 1999; Blank and Sussmann, 2000; Govers and Go, 2001; Pan and Fesenmaier, 2002; Pechlaner, Rienzner, Matzler and Osti, 2002).

To study these issues in a more in-depth and systematic way, a research project was carried out during the summer 2001 season and the winter 2001/2002 season. The aim of the study was to build a behavioral profile and decision-making model of tourists who choose the Dolomites as their destination; particular attention was given to the tourists' use of Internet in gathering information in booking accommodations or services. The questionnaire focused on four key issues of research:

a) the way in which a specific location in the Dolomites was chosen;
b) how the vacation was planned and organized (bookings, journey arrangements, length of stay, lodging, expected cost, payment, availability of information);
c) what (which attractions) the tourists were looking for in the Dolomites;
d) the activities engaged in at the location and how tourists rated them.

Particular attention was given to how the Website of the location was used throughout the different phases of information gathering, choosing, and eventual bookings. As tourists were already present at the destination, the research also had to take into consideration the planning done before the journey phase.

2   The Research Method: Sampling and Data Gathering

When gathering data, we had to keep in mind some of the unique characteristics of tourism in the Dolomites: on one hand it was necessary to take into account the numerous tourists opting not to use hotel accommodations, and similarly it was necessary to ensure that the study covered what is a vast and differentiated territory. To this end, the sampling and data gathering was done in the following phases:

a)   N=5.000 was established as the total number of questionnaires to be administered within the sampling period, a number adequate to guarantee reliable results;

b)   we identified the rate of incidence to be 47% for summer and 53% for winter, accounting for the total number of presences in the region for the seasons studied (reference: the year 2000). Multiplying these rates by N, we arrived at the number of questionnaires to administer in the two seasons;

c)   the number of questionnaires to administer in the different areas of the Dolomites was determined by constructing a composition ratio of the presences in the different areas for the period of reference (summer or winter); for example, if N_E is the total summer presence and N_{Ei} is the summer presence in i-th area, the composition ratio for this area will be obtained by

$$\frac{N_{Ei}}{N_E}, \quad i = 1, 2, \cdots k.$$

d)   multiplying this result by N gives the number of questionnaires to administer in the i-th area of the Dolomites for the period considered;

e)   in each zone, for the period considered, we administered the questionnaires from within the individual Public Tourist Boards (PBTs) by means of a systematic sampling scheme. The questionnaires were administered at different times throughout the day so as to avoid periodicities and thus to ensure a sampling of the diverse subgroups of tourists present in the region. This sampling scheme can be considered an approximation of simple random sampling (in our case the order of administration and arrival was not pertinent to the study). Our sample excluded day visitors;

f)   the online administration of the questionnaires and the impossibility of completing the process if questions were left unanswered meant that some incomplete questionnaires could not be returned.

The data gathered were examined by means of simple tools such as analysis of frequency distribution as well as more sophisticated techniques such as correspondence analysis, thus allowing us to define descriptive behavioral profiles that are coherent with the results, as described in the following section.

3   The Main Results

The results of this study have made it possible to express some concrete considerations regarding the use of Internet among tourists when choosing the

Dolomites as their destination. The focus of the analysis was in fact on the way that Websites influence the decision-making process of these tourists when they are deciding where and how to spend their vacation. The data refer to the total number of questionnaires returned (4910), therefore to both the summer and winter seasons. The seasons are considered separately only when the statistical results showed a significant difference between the two periods. (The results of the research, with detailed tables and figures, can be found at http://www.cs.unitn.it/etourism-bin/etourism.cgi).

3.1   The origin of the idea to vacation in the Dolomites

A large part of visitors to the Dolomites are return guests, many for several years. It should therefore be no surprise that in fact the decision on a destination is informed 58% of the time by previous experiences, word-of-mouth testimonies play the key role in 28% of such decisions, while 8% of interviewees indicated advertising and 6% the Internet. The data indicate that Websites can indeed play an important role for potential tourists during their initial decision-making phase. Interesting findings emerge from a careful look at the percentage of interviewees who indicated that the idea to spend the vacation in the Dolomites was connected to their visit to a specific Web site (record of 310 interviews). The data showed that the importance of Internet in this phase is more prevalent among younger vacationers (76.2% from 21 to 45 years old) of a higher educational level (92% having high school or university degrees) working in fields commensurate with their educational background. A more equal use among all other sources of information was found among other age groups, leading to the conclusion that there is a strong association between age group and the role that Internet plays in the decision to vacation in the Dolomites.

The Internet appears to be particularly important in the initial selection of a destination for tourists who have never visited the Dolomites: 68.1% of those indicating the use of Internet in this phase were first-time visitors to the region; while fewer return visitors mentioned using Internet when deciding whether to return, (18.1% of those who with one previous visit, 10.3% with more than one). Those for whom Internet was a key influence in deciding to vacation in the Dolomites largely spent the vacation among family (40.3%) or friends (48.4%), they come in roughly equal numbers from Italy (41.9%) and the rest of Europe (51.3%) and report a willingness to have high financial outlays for the holiday (59% allowing for up to 1500 per person per week, 15.5% even more). As for length of stay, it is necessary to examine the two seasons separately because of the net prevalence of a two-week summer holiday (34.9% versus 26.6%) as opposed to the typical "winter week" (63.6% versus 50.8%). The percentages for short stays, on the other hand, are similar for both seasons. With respect to accommodation, hotels prove to be preferred option (hotels 49.4%, pension 21%). These data are in line with the other information obtained regarding the initial idea to choose this location. Worth noting is that those who were "inspired" to vacation here after visiting an Internet site stayed in camping areas and agritours almost twice as often as in other types of accommodation (5.8%).

3.2 Finding a vacation location in the Dolomites region

The second phase of the decision-making process regards the selection of the specific location for the holiday in the Dolomites. The data show that the use of Internet becomes more relevant at this point than it was in the initial phase, representing 7.7% of total "choices." The data are subdivided by season, given the nearly two-percent difference in use of Internet between summer and winter vacationers (Table 1).

Table 1. Means of finding a location

	Word of mouth	Advert	Internet	Previous Experience	Travel Agency	Other
Summer	20.9	15.3	6.7	46.3	1.8	9.1
Winter	26.8	12.2	8.6	42.7	2.8	6.8

Tourists using Internet to choose the specific location are predominantly between ages 21 and 45 (75.5%). They match the profile also with respect to educational level and profession, and as in the first phase they are mostly first-time visitors to the Dolomites (68.9%). Vacations among family or friends are most common (89.8%) and the willingness to spend remains high (59.2% willing to have a 1500 outlay per person per week, 13.9% even more); 50.8% prefer hotel accommodations and 20.5% pension. A significant 13.7% choose apartment lodgings. This group has more Europeans (52.1%) than Italians (43.2%) and the length of stay is predominantly two weeks in summer (36.8% against 24.9% in winter) and one in winter (68.4% against 51% for summer). A comparison of the data for the two phases leads to the conclusion that the decision-making phase can be realised by means of Internet even after the initial decision to vacation in the Dolomites has been taken. A look at the initial desires for the holiday for those using Internet to choose the destination revealed that 45.8% wanted to be in a beautiful natural environment and 24.5% were looking for a destination offering sport or relaxation (in equal percentages). Worth noting is that the interest in practicing sports is much higher than the other criteria for choosing a location. In winter, 25.3% were interested exclusively in a skiing holiday, 20% in relaxation in general and only 14.7% in the natural surroundings. In this case, the data are similar to those for other decision-making modalities.

3.3 Booking of services

Most visitors to the Dolomites fall into the category of "do-it-yourself" tourists, and this is no less true for bookings in both the winter and summer seasons: visitors contact the facility directly. A large number of tourists also lodge in a second house, either of their own or of friends or family, or simply search for accommodation upon arrival in the location. Bookings through agencies or consortia represent a small amount of the total. In this context the number of bookings via Internet becomes significant, particularly for the winter season (Table 2).

Table 2. Bookings

	Directly	Internet	Travel operator	No need to book
Summer	70.1	8.8	9.7	11.4
Winter	59.4	17.3	14.4	8.9

Internet bookings are more common among the 21-45 age group (69.8%) than for the 46-55 age group (18.7%), but nonetheless account for a notable 27.1% of all bookings for tourists over age 45. Educational level and professions practiced in this group are high. Once again the percentage of tourists who are first-time visitors to the Dolomites is quite high (55.2%), while even those who are familiar with the region indicated interest in the possibility to book online (20.4% of those who have visited the area at least one time, 8.7% several times). A measurable difference emerges between groups booking via Internet for summer vacations (63% are Italians) and winter vacations when non-Italian Europeans are the most frequent users (44%). Also in this phase the Internet users prefer vacations with family (60%) or friends (31.6%), even if those vacationing alone are a significant 7.4%. Willingness to spend remains high, even though 21.9% of those booking via Internet indicated a desire to spend less than 500 . Hotels and pensions are once again the preferred accommodation, representing 54.8% and 17.5% respectively, yet an increase was found in the number staying in apartments (14.4%) and agritours (15.4%). As for length of stay, tourists booking via Internet follow the general trend toward the one-week winter vacation (70.7% versus 44.1%) and two weeks in the summer (37.6% versus 21.3%).

4   Some lessons for the Web marketing strategies of the tourist destinations

The results of this research bring to light some useful considerations for Web marketing activities that can be implemented by local and regional tourist organisations as part of their overall strategy of augmenting their presence on the Internet. A profile emerged of tourists visiting the Dolomites who make use of Internet in deciding on the destination: young albeit sophisticated, well-educated professionals with a propensity to spend high sums, they vacation with friends or family, usually staying in hotels but also in other accommodations (pensions, apartments, agritours). They are mostly first-time vacationers in the Dolomites, even if a large portion of them has had some prior visit to the area. Sports, relaxation and enjoyment of the natural surroundings are the principal pursuits of these tourists in the Dolomites. A typical stay lasts on average one week in winter and two in summer. This niche group is taking on increasing importance for the tourism industry in the Dolomites, and will most likely continue to grow as Internet use becomes even more widespread, even representing the avant-garde for services linked to mobile phones.

This profile can be used to give shape to numerous activities aimed at destination marketing, such as:

- to reconsider the nature of the offering, verifying whether the services and attractions currently available are in line with those stipulated in the profile;
- to define appropriate communication strategies in terms of both message content (copy strategy) and the means used to reach the target audience (media strategy);
- to more precisely describe the contents of a Website, based on the socio-demographic and behavioral characteristics of tourists as described in the profile.

An interpretation of the results points towards some considerations for Web marketing policies. The fact that return visitors use Internet more for bookings than for research implies that the Internet is perceived mostly as a tool to simplify information gathering and bookings. Knowing this, it becomes possible to formulate some operative guidelines that can serve to improve the effectiveness of Internet for current users and to promote its use among other tourists. The findings take into account a parallel study that, starting from an analysis of the Web sites of Alpine destinations (Franch, Martini and Mich, 2002), went on to more analyse 45 Websites of specific locations in the Dolomites and in the three political regions that include the Dolomite territory (Trentino, South Tyrol, Veneto). Tourists unfamiliar with the Dolomites or going there for the first time use Internet to get the original idea, to identify the specific location or to make bookings. It seems logical, then, that Internet has a key role to play in attracting new customers to the region and can be used as an effective tool to expand the market. Some interesting findings emerged from the research on this point:

a) The need to integrate the information needed by tourists in a small number of sites that serve as the portal for the larger offering; in fact, the current situation with different types of Websites giving information at different levels (the Dolomites, regions, touristic valleys, single locations) tends to present the offering in a fragmented way, potentially disorienting the tourist unfamiliar with the territory and certainly not facilitating decision making. It becomes expedient, then, to make an effort (also in the political sphere) to favour the creation of portals for the Dolomites that would provide links to various single locations. The challenge is to coordinate the different actors present in the territory.

b) Considering the unfamiliarity of some tourists with the area, the site should have tools that make it easier to navigate, such as search engines and recommendation systems. The tourist can thereby identify the attractions and tourist facilities of interest from among all those available in the area covered by the site. The sites for the tourist locations then, would by nature tend towards a use of sophisticated navigation systems that incorporate the psychological rules embodied in case-based reasoning and human-computer interaction.

c) Because the aim is to reach new target markets, special attention should be given to the local "brands," that is, to the unique features of the single location. It therefore becomes necessary to avoid both the fragmentation and the overlapping of the territorial "brands," and at the same time to further integrate activities of online and offline marketing, in practice meaning that other marketing activities (fairs, advertising, brochures, etc.) would not only promote the brand but would also direct the potential tourist to the Website.

The numbers showed that only a small percentage of tourists are unfamiliar with the Dolomite region. While it is vitally important to reach new customers and markets, an effective marketing strategy would also include activities of customer relationship management whereby relationships with return customers are strengthened. To this end, further points should be made regarding online and offline marketing:

a) Give more attention to return tourists by setting up virtual communities at the Website where these experienced visitors can exchange information and ideas or chat with the operators or tourist organisations. Online promotional activities such as photo competitions can also arouse interest in the area. The aim of these Web marketing activities should not be limited to increasing sales—which could in fact be marginal—but to retaining customers. The information available in this section of the Website can be useful to potential first-time visitors, thereby creating a kind of virtual word-of-mouth advertising scheme.

b) Include in the Website information that reflects the tourist profiles, taking into account the specific needs and desires of a relatively young but sophisticated tourist segment (young families with children, for example). This requires a high degree of integration with offline marketing activities, starting with a clear definition of the tourist product.

c) Differentiate the contents of the Websites according to the variations in market demand. The tourist can thus be more quickly and efficiently directed to the attractions and services of immediate interest. Instead of overlapping, the different sites can link all of the information with a system of links or varying levels of search.

d) Facilitate online booking services: most Websites currently provide information on operating schedules or availability but far fewer allow for actual bookings. The high number of bookings taking place despite this limitation would indicate that results could be improved if this information were more complete and if the tourist could more easily consult the database and consistently get the results desired.

e) A large number of tourists surveyed indicated a general willingness to have sizeable financial outlays for the holiday, thereby affirming in effect that the Websites should not concentrate disproportionately on special offers and promotions: even where quality of services and facilities is guaranteed and it is secure and easy to make reservations, the tourist would in any case probably be interested in a product in the medium or high price range.

5 Conclusions

This study enabled us to draw up some guidelines for the development of effective policies for Web marketing. Firstly, we were able to build a profile of the characteristics and behaviours of Internet users whom we referred to as "do-it-yourself" tourists in the Alps, a tourist segment that has not until now been studied. Secondly, we identified some clear aims that tourist locations should have for their Websites, underlining their need to provide complete and coherent information that is also adaptable to the needs of the target market. Similarly effective would be a rationalisation of the variety of sites representing the different tourist locations; this could be done by creating portals which enable the tourist who is unfamiliar with the Dolomites to learn about the location and facilities more easily and without encountering a redundancy of information. These are the principal operative indications that emerged from the study. The acceptance and application of these suggestions by local operators and institutions depends on their understanding of the potential role to be played by new technologies in the creation of new demand and the cultivation of relationships with return visitors, and on their willingness to coordinate promotional activities in a larger scheme of destination management.

References

Blank D., Sussmann S. (2000). Destination Management Systems and Small Accommodation Establishments: The Irish Experience. In D.R. Fesenmaier, S. Klein, D. Buhalis (Eds.) *Information and Communication Technologies in Tourism 2000* (pp. 418-429), Wien, Springer-Verlag Wien.

Buhalis D., Spada A. (2000). Destination Management Systems: Criteria for Success. An Exploratory Research. In D.R. Fesenmaier, S. Klein, D. Buhalis (Eds.) *Information and Communication Technologies in Tourism 2000* (pp. 473-484), Wien, Springer-Verlag Wien.

Fodness D., Murray B. (1999). A Model of Tourist Information Search Behavior. *Journal of Travel Research*, February. 37(3): 220-230

Franch M., Mich L., Martini U. (2001). A Method for the Classification of Relationships and Information Needs of Tourist Destination Players. In P.J. Sheldon, K.W. Wöber, D.R. Fesenmaier (Eds.), *Information and Communication Technologies in Tourism 2001* (pp. 42-51) Wien, Springer-Verlag Wien.

Franch M., Martini U., Mich L. (2002). The quality of promotional strategies on the Web. The case of alpine regional destinations. *Proceedings of the 7th World Congress on Total Quality Management* (pp. 643-652) Verona (Italy), 25-27 June, Faculty of Economics.

Martini U., Jacucci G., Calzà D., Cattani C. (2000) Mentoring Small Destinations into Destination Management towards Electronic Marketing., In D.R. Fesenmaier, S. Klein, D. Buhalis (Eds.) *Information and Communication Technologies in Tourism 2000* (pp. 485-496) Wien, Springer-Verlag Wien.

Govers R., Go F.M. (2001). Virtual and Physical Tourism Destinations: How to Measure Consumers' Evaluations. In P.J. Sheldon, K.W. Wöber, D.R. Fesenmaier (Eds.), *Information and Communication Technologies in Tourism 2001* (pp. 251-261), Wien, Springer-Verlag Wien.

Pan B., Fesenmaier D. (2000). A Typology of Tourism-Related Web Sites: Its Theoretical Background and Implications. *Information Technology and Tourism* 3(3/4): 155-166.

Pan B., Fesenmaier D. (2002). Semantics of Online Tourism and Travel Information Search on the Internet: A Preliminary Study. In K.W. Wöber, A. J. Frew, M. Hitz (Eds.), *Information and Communication Technologies in Tourism 2002* (pp. 320-328), Wien, Springer-Verlag Wien.

Pechlaner H., Rienzner H., Matzler K., Osti L. (2002). Response Attitudes and Behavior of Hotel Industry to Electronic Info Request. In K.W. Wöber, A. J. Frew, M. Hitz (Eds.), *Information and Communication Technologies in Tourism 2002* (pp. 177-186), Wien, Springer-Verlag Wien.

Weber K., Roehl W.S. (1999). Profiling People Searching for and Purchasing Travel Products on the WWW. *Journal of Travel Research*, February. 37(3): 291-298

World Tourism Organization (1999). *Marketing Tourism Destinations Online. Strategies for the Information Age.* Madrid, WTO Business Council Publications.

Use of Interactive Television Promotional Tools as Information Sources in Long-Haul Travel.

Anika Schweda
Duane Varan.

Interactive Television Research Institute
Murdoch University, Australia
{aschweda, varan} @central.murdoch.edu.au

Abstract

Travel and tourism has for many years been on the forefront of communications technology. Entering the next technological era, this paper looks at what interactive television may hold for long-haul travellers as an information source. The tourism possibilities for interactive television are considered in the light of the traveller's experiences with travel, planning, and interactive media such as teletext, the Internet and current interactive television in the United Kingdom (UK). From this discussion, hypotheses are presented that form the focus of a current major study by the authors.

Keywords: interactive television; information sources; long-haul travel; destination choice.

1   Introduction

Travel and tourism has for many years been on the forefront of communications technology because of its intangible, perishable, global nature and subsequent reliance upon the movement of information. As we enter the next technological era it will be interesting to examine how new technologies such as interactive television will assist in communicating and distributing information. This paper will look first at the evolution of the interactive television paradigm. It will then review the literature on destination choice, travel experience and information search in the light of long-haul travel. Lastly, these areas will be drawn together into a conceptual framework currently under investigation by the authors.

2   Interactive Television (iTV)

Interactive television (iTV) is a major paradigm shift in television viewing. The challenge is that it is not just like television – it requires new narratives, content and revenue streams; it is more intrusive, and requires a more active audience. Neither is it just like the Internet – it is regulated, trusted and media rich, and is used for entertainment and relaxation rather than work and communication. In the past there have been attempts at developing the concept (eg. Warner Amex's QUBE and Time Warner's Full Service Network) however with limited success mainly attributed to high costs and an unripe audience (Freeman, 2000; Swedlow, 2000).

Nonetheless, over the years, small steps to condition viewers to interactive media have been made with technologies such as the remote control, VCR, home shopping, teletext and the Internet (Swedlow, 2000; Wildman, 2001). While the VCR and the remote control has accustomed viewers to be selective and interactive essentially on impulse (Swedlow, 2000), home shopping accustomed them to virtual shopping. Teletext services advanced the audience to goal-oriented behaviour, and this has proved very successful for travel in the UK with 1 in 10 holidays and 1 in 15 flights being booked through Teletext Holidays in the UK (Ryan, 2000). Moving away from the television platform, the Internet has also become a popular channel for information and distribution. The majority of travel users use it to gather information on potential or decided holiday destinations and products (Lang, 2000). However, those that do purchase on the web tend to book flights and last minute deals (Morgan, Pritchard & Abbot, 2001). Research has found that while the Internet offers the user increased control in an information rich environment, it lacks the personal and inspirational elements that a travel agent or a television production offers (Morgan et al., 2001). It is with these skills, experiences, and expectations provided by earlier interactive technologies, that the audience will use to evaluate and subsequently embrace interactive television.

2.1   iTV in the UK

The interactive television industry in the UK emerged in the late 1990s and is still developing.   Interactivity has manifested itself most readily in direct-response advertising with viewers choosing to receive a product sample or information via their remotes.   Other experiments have seen viewers able to choose the next part of the storyline in a commercial or go to a walled-garden site for more information. Programme applications have drawn from the TV-PC parallel use phenomenon, with game shows being successful in attracting home-players through remote control access.   Larger conceptual leaps in interactive programming have been fewer due to the increased costs of producing truly interactive-narrative based programming, as well as the technological and audience capabilities for dealing with such new media. The BBC's *Walking with Beasts* is a prime example of a successful interactive programme, allowing viewers to switch between the main programme and three parallel streams.

Travel related applications have been limited, with major travel operators and travel channels taking the lead with what are essentially databases of their televisual content on various destinations that the viewer can then call upon at their leisure.   The purchasing of holidays through the television is still dominated by phone numbers to call centres.   This has been attributed to backend technology costs and even more importantly the readiness of the audience to purchase such personal and large transaction products through a machine mediated environment (Wildman, 2001).

Given the highly competitive nature of the industry, little systematic research is available for analysis. However, there has been some conceptual research into interactive television and numerous industry case studies with limited disclosure of results.

2.2   Conceptual Research into Interactive Television

Conceptual research into interactive television has had interesting results. Studies have shown choice alone is not sufficient in satisfying interactive television viewers (O'Dea, 2001; Tanjic, 2001). Rather the choice must be meaningful and important (Tanjic, 2001) and the expectations formed while making that choice must be met (Yeo, 2001). Furthermore, preliminary research has found interactive advertisements to elicit more central thoughts than non-interactive advertisements (Yeo, 2001), implying that interactive viewing may be more involving and engaging than regular television viewing.   Such findings highlight the importance of knowing one's audience and customers and to be aware of and understanding their needs. This study will take such conceptual findings further into the travel and tourism context.

2.3   Industry Case Studies of Marketing through Interactive Television

The interactive television industry in the UK has created some impressive case studies.   One of the first was by *Pantene* which asked viewers to answer a few questions, gave them hair care tips on these answers and then sent out a sample of the *Pantene* product suggested (Dornan, Brooks & Carter, 2000). *Domino's Pizza* was one of the first to enable product sales through the television, and this channel now accounts for 7% of their overall sales (Leach, 2001). *Rimmel* lipsticks also had a winning campaign where 50,000 free samples were given away and data was collected on consumer use of competitor products (Howells, 2002).   However, in terms of the tourism industry, the results are a little sparse.

BSkyB has run a number of campaigns for travel and tourism players. From these, the trend for interactive television applications is towards brochure requests and viewing mini-sites for more information on an offer. A campaign for Butlins Resorts with enticed 7,878 viewers (0.8% response rate) to request a brochure (Leach, 2002). The Welsh Tourism Board added interactivity to a linear advertisement producing a 0.2% response rate (5,298 requests) for brochures; below average for travel products but respectable considering product branding was not evident till the last 5 seconds of the advertisement (Leach, 2002). The Canadian and Ontario tourism commissions have also used interactive advertisements to distribute brochures and to gather consumer information (Leach, 2002).

The *Travel Channel* has used the video-on-demand network, *HomeChoice,* as a platform for their collection of video brochures on destinations which viewers can access at their own convenience. If the viewer is registers their interest in a brochure for the destination, an email is sent to the distribution centre automatically. As for purchasing travel products, viewers are prompted to go to a website or phone the call centre. It is has been suggested that such call centres may one day be accessible through video-link up through the television (Scott, 2001), though it will be up to the consumers to drive this initiative.

Clearly, interactive television has a lot to offer the travel industry as it extends the characteristics of traditional television (media rich, inspirational, emotive, trusted, entertaining and social) with the interactive possibilities reminiscent of teletext and Internet technologies (consumer choice in the what, when and where access of information). Consumers have been conditioned through previous non-personal distribution channels and have even expressed interest in using interactive television for booking flights (Wildman, 2001). However, how would interactive television and its applications (eg. interactive advertising and enhanced programming) fit into the decision and planning process for long-haul tourists?

3   Destination Choice, Travel Experience and Information Search

Previous literature has adapted consumer behaviour models of decision making to the tourist experience. There have been numerous models offered (Hudson, 1999), all with a different emphasis for investigation. However, all concur with a number of gross elements and the notion of the destination decision process as a goal-oriented problem solving exercise. This exercise uses internal and external resources to evaluate alternatives, and subsequently choose a holiday destination that will best serve the tourist's identified needs. Two major influences in this process are the tourist's personal characteristics (in particular, travel experience), and information sources and search strategies.

3.1   Personal Characteristics: Travel Experience

As with many other products or services, a tourist's destination decision process has been cross referenced with their *personal characteristics*, including demographics, psychographics, motivations, attitudes, constraints and past experience. While all these personal characteristics are important considerations of the destination choice process, this paper will focus on travel experience since it includes past information source use and is itself an information source.

Both the travel career ladder theory by Pearce and the travel horizon theory by Schmidhauser maintain that the travel experience of an individual is held to be indicative of future travel (Oppermann, 2000; Schul & Crompton, 1983; Soenmez & Graefe, 1998, Mazursky, 1989). The travel career ladder posits that a traveller will become more experienced as they look to fulfil higher level personal needs (thus travelling up the ladder) through travel. Meanwhile, the travel horizon theory identifies the furthest distance travelled as the 'horizon' or benchmark for future travel. Both theories have sparked debate on their initial lack of integration of travel experience with other personal and trip characteristics. Revisions have included such consideration (Oppermann, 2000; Ryan, 1998), and it is now accepted that an individual's past travel must be analysed in the context of personal and trip characteristics to make any significant contributions towards predictions of future travel behaviour. Travel experience not only provides skills to tourists, but also acts as an information source. It is often the first source consulted by tourists due to its accessibility, personal relevance, perceived reliability, and link to decision criteria formation (Mazursky, 1989; Soenmez & Graefe, 1998), and it has been shown to influence the importance and the use of external information sources (Zhou, 1997).

3.2   Information Search

The tourism literature views information search as a risk minimisation exercise to satisfy practical needs of a trip and to minimise the financial, emotional and social risks associated with such a significant intangible purchase (Gitelson & Crompton, 1983; Schul & Crompton, 1983; Swarbrooke & Horner, 1999). The literature also acknowledges peripheral benefits such as self-enhancement, vicarious experiences, and to build expectations and anticipation of the event (Hyde, 2000; Vogt, Fesenmaier & MacKay, 1993; Zalatan, 1996).

The search strategy a tourist uses has been found to differ depending upon, amongst other factors, whether the trip is a short haul/domestic or long-haul/international trip. Long-haul trips stimulate more information search for a number of reasons. For instance, the greater physical and cultural distances from the tourist's home often results in less familiarity and more uncertainty of the long-haul destination (Zalatan, 1996), which then causes more investment of time in researching the destination. This greater investment of effort implies greater involvement in the travel decision process. This notion has been further substantiated by Fesenmaier and Johnson (1989) who found that the more involved planners were long-haul tourists which tended towards particular types of information sources: destination specific literature and hard to find sources. Further studies support these findings, adding sources such as travel agents, tourist bureaux and even mass media to the list used by long-haul or foreign tourists (Duke & Persia, 1993; Gitelson & Crompton, 1983; Hseish & O'Leary, 1993).

Information search typologies have been formulated in a number of studies. Snepenger, Megad, Snelling, and Worrall (1990) found three search strategies amongst destination naïve tourists to a remote holiday destination (Travel Agent Only, Travel Agent & Other, and Other Only strategies distinguished by trip behaviour). Schul and Crompton (1983) also found differences in international tourists' search and planning behaviours (active versus passive planners and the benefits desired from the trip). Although not using long-haul tourists but self-drive tourists to Florida, the work by Fodness and Murray on information source use took search strategies one step further.

Fodness and Murray (1998) developed a 7-category typology, based on the location of information sources along the spatial, temporal and operational dimensions. The strategies included pre-purchase mix, tourist bureau, personal experience, ongoing, on-site, automobile club and travel agency strategies and were created by looking at where the mix of sources the tourist had used to plan their holiday lay on the three dimensions. Thus a source could be plotted on an internal to external, pre-purchase to ongoing, and decisive to contributory continuum. Figure 1 illustrates how the sources included in the Florida study locate themselves along the operational and temporal dimensions. For example, the travel agent is considered by tourists to be a highly influential source in aiding their decision-making and they tend to use it closer to actually taking the holiday. This is in contrast to the more contributory role newspapers play long before the holiday.

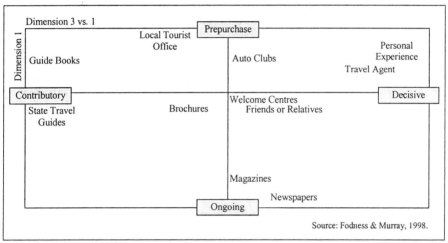

Fig. 1. Information source multi-dimensional scaling perceptual map: Dimension 3 (operational) vs. Dimension 1 (temporal). Source: Fodness and Murray, 1998.

Unfortunately, the Fodness and Murray work did not include broadcast or interactive media. However, given the use of mass media by long-haul travellers and the increasing importance of interactive media such as teletext, the Internet and interactive television in the planning process of travellers, it is important to discover how these may fit into such an environment. This grid offers an opportunity to compare the old with the new and how the new may work with existing sources to aid in tourist destination decision-making.

3.3 Travel Experience and Information Search

The link between travel experience and information search is largely untested. Some studies have looked at search strategies and destination experience (eg. Snepenger et al., 1990; Zhou, 1997), few have considered the effect of overall travel experience.

However, Vogt, Stewart and Fesenmaier (1998) give a fair overview on knowledge versus search theories and how these may be applied to tourism. Some theories hold experts to be greater information searchers than novices, while others hold the opposite view. Others still hold that the less and more extremes are more eager information gathers than the moderates. The authors of this paper posit that travel experience in general is an important factor in the information search strategy a tourist will undertake in deciding on a holiday destination, since many skills are transferable even if the destination is new.

4 Investigating iTV as an Information Source in Long-Haul Travel

From the exploration of the literature above, it is clear that there are a number of questions in the relationships between previous travel experience, interactive media experience, information search, destination choice and long-haul travel. The review has shown that as television and interactive technologies have evolved, so too have individuals' expectations of new technologies. Interactive television will need to be of the same quality of television but include interactivity on an increasing sophistication from teletext towards the Internet. Further research could look to how previous interactive media experience has influenced individual's attitudes towards and use of current interactive television technology. Within the travel industry, interactive television has already been used to address more traditional uses of television (travel show segments on destinations and brochure requests) with some success (Leach, 2002). However, whether this success has been due to the novelty of interactivity or due to actual added value is still to be established.

For long-haul travellers, interactive television theoretically stands to be a useful tool. It provides on demand, up-to-date, trusted, inspirational information often required by such travellers to reduce the risks of such a significant purchase. Due to the extensive information search that long-haul travellers tend to partake in, interactive television would be part of a strategy involving other sources. Within this strategy, interactive television would occupy its own position along the dimensions formulated by Fodness and Murray given its unique characteristics. It is also expected that the travel experience and related planning experience of the long-haul traveller would temper the attitudes and adoption of interactive television into the destination decision making process.

Interactive television stands to play a significant role in the communication and distribution networks of the travel and tourism industry. Results of research as suggested above promises to hold substantial implications for both the travel and tourism industry and the interactive television sector, as interactive television evolves to enhance the travel experience from the couch onwards.

References

Dornan, A., Brooks, G. & Carter, B. (2000). Advertising on iTV. *New Media Age.* (Oct. 19), 50.

Duke, C. R .& Persia, M. A. (1993). Effects of distribution channel level on tour purchasing attributes and information sources. In Uysal, M. & Fesenmaier, D. R. (Eds), *Communication and channel systems in tourism marketing.* New York: Hawthorn Press.

Fesenmaier, D. R. & Johnson, B. (1989). Involvement-based segmentation: Implications for travel marketing in Texas. *Tourism Management.* December, 293-300.

Fodness, D. & Murray, B. (1998). A typology of tourist information search strategies. *Journal of Travel Research.* 37(2), 109-119.

Freeman, L. (2000). Rites of passage: After decades of growing pains, identity crises and rejection, has interactive TV found itself? *Advertising Age.* 71, i84.

Gitleson, R. J. & Crompton, J. L. (1983). The planning horizons and sources of information used by pleasure vacationers. *Journal of Travel Research.* 21(3), 2-7.

Howells, A. (2002). How to get an extraordinary response from your iTV campaign - Case Study: Rimmel lipstick campaign. *5th Annual Interactive TV Advertising Conference,* London: Access Conferences.

Hsiesh, S. & O'Leary, J. T. (1993). Communication channels to segment pleasure travellers. In Uysal, M. & Fesenmaier, D. R. (Eds), *Communication and channel systems in tourism marketing.* New York: Hawthorn Press.

Hudson, S. (2000). Consumer behaviour related to tourism. . In Pizam, A. & Mansfeld, Y. (Eds), *Consumer behaviour in travel and tourism.* New York: Hawthorn Hospitality Press.

Hyde, K. F. (2000). A hedonic perspective on independent vacation planning, decision-making and behaviour. In Woodside, A., Crouch, G., Mazanec, J., Oppermann, M. & Sakai, M. (Eds), *Consumer psychology of tourism, hospitality and leisure.* (pp.177-191). Melbourne: CABI Pub.

Lang, T. C. (2000). The effect of the Internet on travel consumer purchasing behaviour implications for travel agents. *Journal of Vacation Marketing.* 6(4), 368-385.

Leach, R. (2001). The BSKYB experience Interactive TV advertising on the SKY digital platform. *Interactive TV Advertising Conference,* Sydney: dStrategy.

Leach, R. (2002). Interactive TV advertising on the SKY digital platform. *5th Annual Interactive TV Advertising Conference,* London: Access Conferences.

Mazursky, D. (1989). Past experience and future tourism decisions. *Annals of Tourism Research.* 16(3), 333-344.

Morgan, N. J., Pritchard, A.& Abbot, S. (2001). Consumers, travel and technology: A bright future for the web or television shopping. *Journal of Vacation Marketing.* 7(2), 110-124.

O'Dea, K. J. (2001). *Choice of advertising style: A digital television perspective.* (Honours Thesis), Perth: Murdoch University.

Oppermann, M. (2000). Where psychology and geography interface in tourism research and theory. In Woodside, A., Crouch, G., Mazanec, J., Oppermann, M. & Sakai, M. (Eds), *Consumer psychology of tourism, hospitality and leisure.* (pp.19-38). Melbourne: CABI Pub.

Ryan, C. (1998). The travel career ladder. *Annals of Tourism Research.* 25(4), 936-957.

Ryan, J. (2000). TV favourite adapts to the Internet age. *New Media Age.* (20th July), 32.

Schul, P. & Crompton, J. L. (1983). Search behaviour of international vacationers: Travel-specific lifestyle and socio-demographic variables. *Journal of Travel Research.* Fall, 25-30.

Scott, B. (2001). Understanding the technology behind iTV. *ITV for Travel Conference,* London: IQPC.

Snepenger, D. J., Megad, K., Snelling, M. & Worrall, K. (1990). Information search strategies by destination-naive tourists. *Journal of Travel Research.* 29(1), 13-16.

Soenmez, S. F. & Graefe, A. R. (1998). Determining future travel behaviour from past travel experience and perceptions of risk and safety. *Journal of Travel Research.* 7(Nov.), 171-177.

Swarbrooke, J. & Horner, S. (1999). *Consumer behaviour in tourism.* Oxford: Butterworth-Heinemann.

Swedlow, T. (2000). *Interactive enhanced television: A historical and critical perspective.* www.itvt.com Accessed: 22 May, 2001.

Tanjic, S. (2001). *Viewer choice and the mediation of the advertising experience.* (Honours Thesis), Murdoch University.

Vogt, C. A., Stewart, S. I. & Fesenmaier, D. R. (1998). Communication strategies to reach first-time visitors. *Journal of Travel and Tourism Marketing.* 7(2), 69-89.

Wildman, J. (2001). The interactive TV market in Europe: A new opportunity for the travel industry. *ITV for Travel Conference,* London: IQPC.

Yeo, C. K. F. (2001). *Persuasion through interaction: An exploration into the effects of interactive television advertising on consumer persuasion.* (Honours Thesis), Perth: Murdoch University.

Zalatan, A. (1996). The determinants of planning time in vacation travel. *Tourism Management.* 17(2), 123-131.

Zhou, Z. (1997). Destination marketing: Measuring the effectiveness of brochures. *Journal of Travel and Tourism Marketing.* 6(3/4), 143-158.

Multimedia Geoinformation in Rural Areas with Eco-Tourism: the ReGeo-System

Iris Frech
Barbara Koch

Department Remote Sensing and Landscape Information Systems
Albert-Ludwigs-University, Germany
{iris.frech, barbara.koch}@felis.uni-freiburg.de

Abstract

Rural regions, such as the national and nature parks involved in the ReGeo project, are often dependent on the tourism sector and thus show a high degree of interest in eco-tourism issues. How can such areas attract new visitors, how can they compete with the big players in the tourism industry and how can they make themselves known to a broad public? The following article presents the ReGeo project, which aims to help rural areas find solutions to these issues by working with modern technology. For most rural regions the solution is to set up an attractive Internet site presenting the region, establish a network of local information points and perhaps even provide information on CD-ROMs and for mobile and PDA devices. It has been proven that such tourist information systems are becoming more and more popular and attractive to the user (Pröll and Retschizegger, 2000). Such systems can be improved through the use of photographs, maps and animation. ReGeo also uses GIS-functions, a central interactive map and remote sensing data system, which not only improves the attractiveness but also the usability of an Internet-based tourist information system. It also emphasises the necessity of adapting this system to different devices. The ReGeo-project pays special attention to the specific problems of rural regions, including the lack of a central institution with the ability to administer such a system. Last but not least, ReGeo intends to act not only as a tourist information service, but rather as a multi-purpose digital geo-data service through which the test regions is attractively presented to a broad audience.

Keywords: GIS function; interactive map; virtual database; multimedia

1   Introduction

The majority of tourist information is spatially related. This means that nearly all information can be located somewhere on earth: A hotel is located in a certain place on a certain street. Hiking routes lead from one point to another. Spatial data are most likely to be analysed and presented in a Geographic Information System (GIS) format. The GIS system allows for the combined presentation of various types of tourist information, such as hotels, hiking trails and the location of events, and allows the user to determine the distances between various points of interest (Frech, 2001).

On the other hand, tourist information nowadays is generally presented in digital form - most likely in the Internet on semantic web pages. This means that the information is presented in the form of descriptive text and perhaps photographs. Hotels are described with their name and address, the number of rooms etc. It is mentioned where and when events are taking place. Hiking routes are described from start to end, and information about sight seeing spots is perhaps added. But how should one find out if a chosen hotel is near a preferred hiking route?

In order to solve this problem, some newer Internet sites try to combine semantic descriptions and map based representations of an area. One example of this is the web page of Berchtesgaden, Germany (http://www.info-bgl.de [October 3, 2002]). A central aim of the ReGeo project is not only to add completed 2D maps to every represented location, e.g. hotels, but to include a map where different services and attractions of the region can be combined interactively and shown together. Analytical functions enable users to answer questions such as 'what is near by?'. In this way the map becomes more interactive and is no longer purely a navigation tool. Objects like hotels, for example, can be shown in their spatial position, enabling tourists to make more informed choices. It can easily be checked, for example, if a preferred hotel is located near the train station or near the diving school where the tourist has booked a course. If the tourist is interested in sight seeing, he or she can add this to a search request and nearby events or hiking routes can be shown on the map. Of course, a main principle of the project is that such maps should be accessible and able to be used easily, without any special (i.e., GIS) knowledge.

The visual attractiveness of such a tourist information system can also be improved through the use of advanced visualisation technologies. ReGeo aims to open a new world of possibilities by integrating remote sensing data in two dimensional (2D) form and three dimensional (3D) presentations. Results of such integration range from a 2D map on the base of satellite data to interactive 3D maps for biking tours or a virtual flights over the region.

The ReGeo system is adaptable to different devices including the Internet, CD-Roms, local info-terminals and PDA's (personal digital assistants). The content of the system differs in relation to the advantages and disadvantages of each device.

Due to the existence of special problems in rural areas, the basis of the system is a *virtual decentralised database*. This means that the various data in the system remain mainly at their places of origin (i.e. the PC's of different administrations) and are maintained there. In rural areas, one of greatest difficulties in establishing systems such as ReGeo is to find someone who is willing to maintain a central database for all other parties. This problem of information system administration is one of the main differences between rural areas and bigger towns or cities, where central data administrators are easily found.

This initial introduction of the ReGeo system shortly comes to an end, but first it is important to consider one more thing. Once ReGeo has succeeded in establishing, or, in some cases, improving existing tourist information systems for the regions involved in the project, why should the infrastructure, and especially the geo-data, not be *used on other levels* as well? Let's have a look what is described above. Until now the ReGeo system consists of a virtual decentralised database maintained by different parties in the region. This database contains and connects different digital geo-data. These data can be used to access tourist information from various different devices. However, ReGeo goes further than the tourist market.

The ReGeo system is able to be used for various purposes by different administrations in the regions, including administrators of local places, small towns and national or nature parks. Different access rights to the different data is also managed by the database, and special interfaces are in place to enable the connection between administration and database.

The following paper is organised in three more sections. In section two the basic technical background of the project is discussed. Section three provides detailed information on how the user will see the ReGeo system, and emphasises, through the help of scenario examples, why GIS and tourist information systems should meet. The paper concludes in section four.

2 Basic Concept of the Technical Design

The introduction gave a short overview of the main ideas of the ReGeo system. From a technical point of view the single goals is listed in this section.

2.1 Main goals

The main technical objectives of the project are to:

- develop a decentralised virtual geo-mutimedia database and exchange infrastructure

- apply to existing tourist information systems new and improved ways of customising presentations for both offline and online devices, including the use of advanced visualisation techniques

- development of mobile applications to realise a location based service

- open the database for use by local or natural park administrations

- realise the whole system for four different regions in Germany, Austria, the Czech Republic and Poland

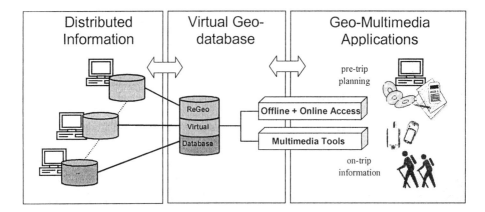

Fig. 1. The basic system idea

In the left part of Figure 1 it is outlined that the *data* for the system mainly comes *from separate and distributed sources*. Some data comes from the tourism sector of the region, some from local administrations or natural parks and some from private enterprises. Not all required data is available through these sources. Additionally advanced visualisation techniques are used to create new kinds of geo-multimedia objects, such as 2D- or 3D visualisations of the region (Nischelwitzer and Almer, 2000). All data, that is, the geo-multimedia content of the system, is collected and aggregated. They are made available through the *virtual geo-multimedia database*. Principally, the data in this database is not maintained by a central institution. The parties giving access to their data also continue to maintain their own data as they did in the past and are responsible in this way for the actuality of the database content. Single parties in the region, however, not purely act as data providers. Rather, some of these parties have the possibility to use the database for their own purposes. Interfaces are implemented to this effect.

The right part of Figure 1 outlines the workings of the tourist information system. Applications for different themes such as general tourist information, hiking and biking are designed. Innovative in this part of the project is that all are presented in the form of geo-referenced data. In this way all information is available not only in descriptive text form but can also be loaded into a map and be compared (geographically) to other (geo-) information. The different applications designed for the ReGeo system is described in section 3.

The following benefits can be expected from the ReGeo concept:

2. The combination of geo data, networking, and multimedia enables new applications for administrations, visitors, and enterprises by enabling:

- Improved use of existing data
- Optimal access to information
- Better planning and presentation capabilities

2. All provided information are geographically referenced and presented in an attractive way: e.g. interactive 3D visualisation of trails and routes (Figure 2).

3. Multimedia presentation by online and offline devices allow access from anywhere and at any time.

4. Mobile devices such as PDA's (personal digital assistants) and GPS devices provide a location-based access point to the information services. With such a visitor guide, tourists are able to be better aware of the fragile ecosystems in national park areas. (Figure 2).

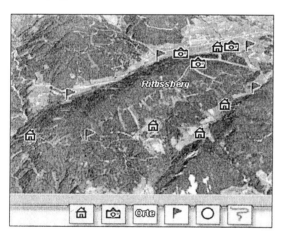 

Fig. 2. Screenshot from an interactive 3D representation of a biking route and tour-information on PDA produced by Joanneum Research, Graz (A) during the former BIKE-project (http://dib.joanneum.ac.at/Reportwebsite/index.html [October 3, 2002])

2.2   Main points in the technical design

Data: The data for the geo-multimedia database mainly come from existing remote databases. Missing data are collected or generated. To avoid high costs, or legal problems, remote sensing data are used as central data source instead of bought maps. Objects are extracted automatically. Therefore newest technology and remote sensing data types are applied.

Database and Interfaces: The virtual multimedia geo database is built from different geo-enabled data sources or services which are able to be accessed through so called Web Services. This means that the sources communicate with each other using XML over HTTP. The interfaces to the virtual geo-database conform to the standards of the OpenGIS Consortium (OGC), to be found on http://www.opengis.org [October 3, 2002]. Producers of the geo-multimedia content can plug into the ReGeo virtual database by using special multimedia metadata standards and/or OGC conform geo data exchange access, such as GML.

Application design: In the ReGeo system, there are client/server applications as well as offline systems. The client application processes the user input in a message format which complies to both the structure of the data base and the communication infrastructure of the networked servers holding the data base. The client/server application is also responsible for proper messaging and display of information sent from the servers in reply to these user requests. The server side application passes the aforementioned massage(s) to the appropriate database(s), processes the database output, render graphical 2D and 3D content, and sends the content back to the client. A WebGIS system is part of the application and provides geo-referenced information.
In contrast to the Internet based Client/server systems, which have limited bandwidth, the geo-multimedia-data also are stored on CD-ROM, storage devices of PDA's and PC's used as information terminals in the region. For CD-ROM and information terminals, there is a bigger emphasis on visualisation tasks. Applications for PDA's and GPS devices are implemented especially for the themes hiking and biking.

Visualisation tasks: The innovative presentation of geographically oriented information includes 2D and 3D visualisation techniques based on current standards like Web3D, VRML, Shockwave 3D, MPEG-4 and includes also interactive 2D and 3D-maps and Quick Time VR objects to provide panoramic views. 2D maps are generated anew every time a new request is given into the system. Remote sensing data play a big role, as is demonstrated in several 3D animation projects of the Joanneum Research Institute in Graz (http://dib.joanneum.ac.at/fe_proj_home.html [October 3, 2002]).

3   User Scenarios

This section describes some scenarios in which the ReGeo system could be used in order to shed some light on what potential system users can expect to see.

3.1   General Tourist Information on Web

A family is looking for accommodations in a rural area. They want to prepare a one-week holiday. In the Internet there is a lot of information about hotels, a descriptive list of events and suggestions of places to go with children. In a region using the ReGeo system the family could easily load all relevant information into one central 2D map and be able to compare the location of hotels and events. First of all, the family is looking for a hotel. A semantic web page gives the possibility to look for family hotels with horse riding. They then load the result into a map. The map allows for other themes to be loaded too. Using a sight seeing search, the family can find out about day trips and places to go with the children. On the map they can also see, for example, if there is a lake nearby to one of the hotels or the locations of restaurants and shops. As all of the information is presented on one interactive map, the decision of where to go is easy.

One month later, at the family hotel, they find a computer giving access to the tourist information system. Here they can spontaneously look for information, for example if any events or festivals are taking place near their planned destination for the day. They can also search for a handicraft shop along the way, since they want to take a present to grandma who is looking after the house during their trip.

3.2   Hiking Tourist

A hiking tourist wants to go on a weekend trip, hiking one day on one of her favourite routes and discovering a completely new route on the next. She uses the Internet as a search engine for hiking information in the region. Since she is often in this region, she has saved a individualised user profile and doesn't need to type in her special request. The list of her favourite routes appears and she can decide where to go this time. She is able to see the chosen route on a map and can look for a small hotel to stay in for the night at the end of the trail. She also requests for information about handicrafts to be shown in the map and decides to visit a cheese factory and a wine cellar the next day. There is no described hiking route from her hotel to these two points, so she is uses the route-query functions of the system to plot an interesting route for herself. At the end of the planning phase the hiking tourist downloads the appropriate information into files for her PDA. She then has a map showing the routes, a route description and 3D-representations of the hiking areas saved in her PDA.

3.3   Cross country skiing

Five young students like the location of their university, especially in winter, since it is not far to drive to the mountains where they can indulge in their favourite fitness training: cross country skiing. Before heading out, they first log into the Internet. On a map they can see which parts of the region have enough snow for skiing, since the system incorporates weather information and snow depths are represented by different colours.

3.4   Selling regional products with the aid of the ReGeo system

Two farmers from a particular region want to start to sell their products independently. The ReGeo system provides a possibility for them to become widely known and also acts as an access point to the market outside of their region. They are registered at the tourism administration centre of the region as interested in using the ReGeo system and have therefore received restricted access rights to the database.

The farmers can enter data into the database with an easy to use management tool. Here they have to type in the name of their farms, the address and their products. To geo-reference their location they have to mark the point on a digital map, where they are located. Every time their products or their prices change, they can update this information by themselves. One of the farmers already has his own homepage presenting the products, and thus chooses to establish a link to his homepage rather than re-entering all of his data.

4   Conclusions

This paper describes a work in progress of an international research consortium working together in an European IST project (ReGeo). The first prototype has already been designed and is being implemented for the tourism and the nature park administrations in the Thuringian Forest, Germany. This prototype will be intensely evaluated.

The second prototype will run in all four test sites of the ReGeo project – in Germany, as well as in the Thayatal national park in Austria, the Podyji national park in the Czech republic and the Konzienice landscape park in Poland. The implementation work in this stage of the project will concentrate on solving problems resulting from the different infrastructure in the different nations. The final evaluation will show the success of implementing the system in the different nations.

The project team must also be aware of problems other than differing national infrastructures. One problem with presenting information in map format on the Internet is the often slow transfer rate. With rates averaging only five kilobytes per second during times of high net-frequency, users have problems with interactive maps and 3D-Animations. A goal of the project is therefore to seek out the best possible solutions to this problem. It is also necessary to remember that the ReGeo system is also accessible through other devices. Distribution of information in the test areas by CD-ROM is especially important to reach customers with Internet problems.

The system is realised in many different modules. This enables enterprises not only to sell the system as a whole but to adapt it or parts of it to other thematic fields. The business concept of the project aims to access a wide market. Most applications refer to the tourism market. The intention to show all information in a spatial context and to implement easy to use GIS-functions will attract all people that like to think in spatial relationships.

Due to the fact that the tourism sector is always changing, just like the sector of digital media, the system is designed to be an open system. The database is open to new file formats and to additional interfaces. This will allow a wide range of future applications, such as new movie-file formats, thus keeping the system attractive in the future.

Acknowledgements

This work has been funded by the European Union's Fifth Framework Programme (under contract ReGeo IST-2001-32336). It is carried out by a consortium of eight research institutes and enterprises: Department of Remote Sensing and Landscape Information Systems, Albert-Ludwigs-University Freiburg, Germany; Joanneum Research, Graz, Austria; Institute of Geodesy and Cartography, Warzaw, Poland; Ustav pro hospodarskou upravu lesu, Brandis nad Labem, Czech Republic; Geo-konzept, Adelschlag, Germany; Agrolab, Oberhummel, Germany; Lesprojekt sluzby s.r.o., Martinov, Czech Republic; Taxus Information Systems LTD, Warsaw, Poland.

References

Frech, I. (2001). *Zur Nutzung raumbezogener Informationssysteme (GIS) und Methoden der Fernerkundung in der tourismusbezogenen Planung und Präsentation – Ein Beispiel aus den italienischen Alpen.* http://www.freidok.uni-freiburg.de/cgi-bin/w3-msql/freidok/ergebnis.html [Juli 10, 2001]; Dissertation at the Dept. Remote Sensing and Landscape Information Systems, Albert-Ludwigs-Universität Freiburg, 132-139.

Pröll, B., & Retschitzegger, W., 2000: Discovering Next-Generation Tourism Information Systems - A Tour on TIScover. *Journal of Travel Research*, pp. 182-191, Sage Publications, Inc., November 2000, Vol.39/2 [ISSN 0047-2875].

Nischelwitzer, A. K., & Almer, A., 2000: *Interaktives 3D-Informationssystem für Planung und Tourismus.* International Symposion CORP 2000 Archiv, http://212.17.83.251/corp/archiv/papers/2000/CORP2000_nischelwitzer_almer.pdf [October 3, 2002].

A Versatile Context Management Middleware for Mobile Web-based Information Systems

Martin Hitz
Stefan Plattner

University of Klagenfurt, Austria
{martin.hitz, stefan.plattner} @edu.uni-klu.ac.at

Abstract

A versatile middleware for the development of context sensitive Web applications is presented which is supposed to free developers from the great effort of managing several context dimensions such as location and time. The expected benefits are shorter time to market and reduced development costs. The middleware is currently employed in the course of the development of a prototypical campus information system at the University of Klagenfurt.

Keywords: mobile Web information technology; location based services; personalized services

1   Introduction

Due to the forthcoming introduction of the third generation mobile networks such as UMTS and the growing success of wireless LANs, the domain of mobile and personalized software applications is receiving more and more attention. Experts and business managers consider them to become the so badly needed "killer applications" which are expected to return the heavy investments taken so far (cf. Zipf et al. 2001).

To be able to build smart personalized applications or deliver personalized content in general, detailed and accurate data about the context in which the target user employs her application is needed. There are virtually infinite areas for meaningful application of personalized services, and the field of travel and tourism is among the most promising, where many practical uses can be envisaged within different application categories: mobile personalized services, location based services, geographic information systems, billing services, mobile e-commerce, or tracking services.

The gathering, management and especially the provision of context information presents an additional burden to the development of such applications. This clearly indicates that there is a strong need of a common, versatile software middleware which takes care of the details of the basic technical and administrative tasks inherent to the field of personalized and context based applications, thus significantly reducing the effort required for application development.

In this paper, the main concepts and building blocks of such a middleware (XdC – conteXt deduced Content) are presented which is currently being employed for a context-sensitive, Web-based campus information system for the University of Klagenfurt.

The most significant features of XdC can be summarized as follows:

- Tracking of a system user's *context* considering detailed information regarding four orthogonal context dimensions (cf. Section 2).

- Personalization and filtering of *content* by matching a user's context with *context pattern* augmented content (cf. Section 3).

- Selection of context dependent content: XdC provides an interface for the user to navigate and select from the content that is associated to the user's current context (whenever this association is ambiguous).

- Explicit navigation in the context space: By default, the user is presented a selection of the content that is available for the current context. But if the user wants to know what content is available at adjacent points in the context space (i.e., nearby locations, upcoming dates or other roles the user may take), the user is able to change his current context at will (cf. Section 4).

- Maintenance of user preferences with respect to context and content categories. This gives a user the possibility to restrict context dependent content to preferred categories, subjects or fields of interest only. This feature can be used as the basis to manage and select personalized tours through museums or similar sites.

- Explicit consideration of hardware independence regarding context tracking technologies and the client devices (cf. Section 5).

- Modular, distributed, Web-based design.

2 Context Information

To be able to deliver personalized content, the context information on which the customisation process of personalization is essentially based needs to be administered.

In particular, XdC distinguishes between four different context dimensions:

1. The spatial dimension   ("Where are you?")
2. The temporal dimension   ("When did you ask for information?")
3. The social dimension   ("Who are you?" i.e., the user's current role)
4. The technical dimension   ("Which device are you using?")

The combination of instances of all four dimension classes establishes a *fully qualified context* (cf. Figure 1) and characterizes an user's current situation. The following subsections deal with the individual context dimensions of XdC.

2.1 Spatial Dimension

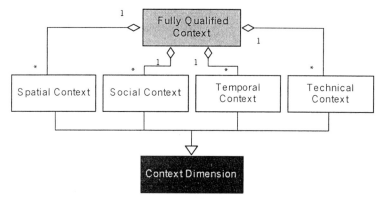

Fig 1: Context Dimensions in XdC

The perhaps most significant context dimension is the spatial dimension. The knowledge about a user's location opens manifold opportunities and represents the base for a whole class of applications already well known as "location based services" - a class of services to which very high expectations are linked.

Due to the importance of the spatial dimension, much effort and research has already been invested into the manifold problems of location tracking. However, to date no ultimate solution and no technical standard has emerged yet – thus, a context management middleware must be flexible enough to deal with several competing technologies.

For outdoor applications with moderate requirements of spatial resolution, the Global Positioning System (GPS) has become the de facto standard to gather positional information. If there is a clear view to the sky and signals from at least 3 satellites can be received, a GPS-receiver's position can be determined within 10 meters in all three dimensions and – besides the rather cheap client hardware – at no additional cost. Unfortunately, GPS does not function within buildings or other scenarios in which no free sight to the sky is possible. For such situations (e.g. indoors) several other technical approaches already exist, but vary widely in cost, scalability and accuracy. To make this clearer, an exemplary selection of already well known location tracking systems (cf. Hightower and Borriello 2001) and their major aspects is summarized in Table 1.

Technology		Accuracy	Scale	Cost	Limitations
GPS	Radio time of flight lateration	1-10m (95-99%)	24 satellites worldwide	100 / receiver	Not indoors
Active Badges	Diffuse infrared cellular proximity	Room size (100 %)	1 base per room	Administration costs, cheap tags and bases	Sunlight and fluorescent light interferences
Active Bat	Ultrasound time of flight	9 cm (95 %)	1 base per 10 m²		Ceiling sensor grid required
Cricket	Proximity lateration	2 x 2m (100 %)	1 beacon per 4m²	10 beacons and receivers	No central management of position information
MSR Radar	802.11 RF (WLAN) scene analysis and triangulation	3-5m (50 %)	3 bases (access points) / floor	802.11 WLAN installation, ~ 100 / NIC	802.11 infrastructure required

Table 1: Location tracking systems (exemplary selection)

Thus, a context management middleware which is to be employed in different application domains must support not only one but many different tracking technologies.

The data managed in the spatial dimension of XdC is independent of the location tracking approach used and represents information about *semantic locations* (e.g., "Office E.2.80", "in front of painting "Mona Lisa", "inside St. Steven's Cathedral"). The association of semantic locations to tracking system dependent *physical positions* is maintained by a so-called "spatial driver" module. It is also possible to use multiple such modules to integrate different location tracking technologies within one application. This could be used to overcome the often rigid limitations of specific technologies (e.g., GPS only outdoors).

2.2 Temporal Dimension

From a technical point of view, the temporal dimension is relatively easy to handle. The problem of tracking time has been solved since ages and there are multiple effective solutions with nearly arbitrary grade of accuracy.

However, similar to the spatial dimension, XdC distinguishes semantic times, as for example "work day" and "free time". It is the mapping between semantic times and absolute time information which represents the main problem in this case and may be defined in manifold ways. This mapping is handled by a so-called "temporal driver" module which in the most trivial case uses the system time of the application server for reference.

2.3 Social Dimension

Since information systems are used by people, XdC of course also manages information about the users of the system. The framework in the current state considers users with all their common properties (e.g. name, nationality, age, gender etc.) and also the different roles (e.g. tourist, guest, technician, employee etc.) a user may play.

XdC in its core however does not determine the way of user-authentication. Techniques like simple login with a password, smartcard or even biometric authentication can be used by integrating a corresponding "social driver" module.

2.4 Technical Dimension

The nowadays strong growing variety of heterogeneous mobile clients (e.g. organizers, mobile-phones, notebooks etc.) and their various networking capabilities cannot be ignored by serious personalized information services. It makes no sense at all to send a video-stream in CIF resolution and 5.1 surround sound to a mobile phone with a postage stamp like display and only very limited bandwidth. Personalized applications should only deliver content that is compatible to a user's technical context and in essence the technical context determines the way information content is to be rendered or if it – or parts of it - is to be rendered at all.

To give context-sensitive information services based on XdC the chance to suit their content to a user's mobile device, the middleware has to keep track about a user's technical capabilities. In particular, XdC manages information about a user's hardware (device type, screen size, audio capabilities etc.), the software (operating system, browser type and capabilities etc.) and since it is targeted at Web-based information systems, there is also information about a client's networking capabilities (bandwidth, latency etc.).

3 Context and Content

The central idea behind XdC is to associate a user's – in this case four dimensional – context with an information system's content. As XdC is Web based, content can be virtually everything which is presentable to the user by the client specific rendering software. This can range from a static text document to a full featured multimedia enhanced Web application which may contain complex logic and also exploits the context information gathered, managed and provided by XdC (cf. Figure 2).

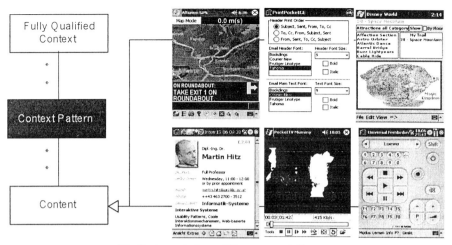

Fig 2: Association of Context and Content

3.1 Association of Context and Content

As the content is seen as external to XdC, content is only loosely coupled to its context information. In essence, XdC only keeps the information needed to access the content, i.e., the content's **Uniform Resource Identifier** (URI, aka URL). The major advantage of this approach is that beside the requirement that the content be Web-addressable, no further restrictions apply (cf. Pradhan et al. 2001).

However, having only information about the content's Web address, it would be impossible to decide whether the content is suitable to the user's current context or not. To this end, XdC features a flexible way to associate context and content. While it would be possible to directly associate a user's *fully qualified context* with content that should be available in this very specific situation, this approach in practice is very cumbersome. The reason is that a fully qualified context (cf. Section 2) describes the context state of a user in the very detail and this in most cases is really too restrictive. In the normal case, there is the need to control the availability of content only with respect to instances of some of the four context dimensions. For example, there could arise the demand to restrict the availability of a video to a specific location and only to client's that are technically capable to display it, without specifying further requirements regarding the social or the temporal dimension. To be able to do this in a versatile and flexible way, the concept of *context patterns* was introduced into XdC. Context patterns allow to describe such partial context states in a way that it is possible to decide whether a context pattern matches a user's current fully qualified context or not. So, when integrating new content into XdC's information base, it is crucial – and one important step of personalization – to augment the information about the content with reasonable context patterns.

3.2 Provision of Context Information

The main task of the XdC middleware is to determine which content out of a large content-pool is suitable to the current context of a particular user. This is done by trying to match the user's fully qualified context with the context patterns associated to the content. After this filtering process, the resulting selection – the so-called *conteXt deduced Content* (hence the name) – is presented to the user who then decides which content he likes to see or consume. So a user selected application is started by the user from within XdC. To give the content – in this case active Web applications – the opportunity to access and exploit context information managed by XdC, a handle identifying the user's current session is passed to it at start-up. With this session handle the content is capable of accessing the context information over a well defined API.

Beside information about a user's current context, it is also possible to access a user's context history which directly provides information about the way a user got into his actual state. As the middleware especially features the possibility to also change the current context manually – e.g., shift from room A to room B without actually moving – (cf. Section 4), the system also explicitly maintains information about states which are reached by this convenience function of the system. At least for security sensitive applications it may be important to know if a user really is in the context currently claimed or not.

The eventual degree of personalization depends on both, the amount of context patterns augmenting the XdC information base and the content's application logic which is external to the system but may exploit the context information of XdC.

4 Context Graphs

One of the main features of XdC is the user managed shift of context. This feature allows a user to choose from content available not only for his current context, but also for "adjacent" context. In the spatial context dimension this may be content associated with a nearby location or object. Similar effects can be achieved in other dimensions, although this feature of the XdC user interface has only limited usage with respect to the technical dimension, but at least it is very useful for testing different technical configurations in the phase of content development.

Whether two context instances are considered adjacent to each other is a non-trivial issue. Considering the spatial context dimension it might on the first glance be possible to determine nearby locations by just examining their physical coordinates, but unfortunately there are multiple problems with this approach. For instance, some location tracking systems do not provide world coordinates on which a neighbour search can be based. But even if full GPS-coordinates (latitude, longitude, altitude) are used for every location in the system, a topologically nearest neighbour can in practice be far away if there is such a nasty thing like a concrete wall separating the two. Such semantic problems are even more obvious if one is thinking about neighbourhood relationships in the remaining context dimensions where in general no such information like physical coordinates is available. To tackle these problems and to be able to deal with the notion of adjacent context, the conceptual model of context graphs has been introduced into XdC.

Context graphs are undirected, hierarchical graphs which not only model the neighbourhood but also a hierarchical parent-child or contained-in relationship. Thus, it is possible to express hierarchies of context instances which are used by the framework for organization and smart navigation in these graphs. Furthermore, the edges of the a context graph may be weighted which is useful to order the relation and a nice way to further influence the graph based navigation.

5   Dealing with Heterogeneity

Two major requirements which influenced the design of XdC are the following:

1. XdC should not be bound to a specific application domain.
2. XdC should not be bound to specific tracking technology or client hardware.

To converge to these difficult requirements, several particular techniques are used. A brief overview regarding these techniques is given in the following subsections.

5.1   Tracking Technology

As context tracking is an inherently hardware dependent task, XdC has to feature a way to access the technology dependent data without introducing static dependence to the tracking hardware (cf. Section 2). To be able to do this, the hardware dependent parts are integrated into XdC as interchangeable "driver modules". There is at least one driver module for each of the four context dimensions whereby in essence each driver fulfils two major functions. Obviously one function is to encapsulate the hardware dependent code which is needed to gather the technology specific data. After data acquisition, the second function of each driver is to abstract the raw (or physical) data by translating it to semantic context instances recognized by the XdC core. For example, a driver for the temporal dimension could gather precise data about the current time by querying a nearby ntp-server and then translating this raw data to corresponding semantic time instances which may be "Wednesday afternoon" or "holiday".

In addition, to support several complementary technologies, it is also possible to have several driver modules for one single context dimension.

5.2   Client Support

Adapting to different client types is based on XLS transformations. To be able to support multiple client types, there has to be at least one XSL style sheet for each device type. If multiple style sheets for one device type are available, the user may choose a specific style sheet according to her preferences.

As the market of mobile devices is constantly booming, XdC also features the possibility to serve heterogeneous clients within one single application domain. In

438

principle, the only hard requirement to a client device is the availability of a Web-browser. However, the main focus regarding client support in XdC lies in the domain of the "pocketable" devices. For devices of this class, typically some sort of Web browser is available, but these browsers vary widely regarding supported protocols and the corresponding markup languages. Beside software compatibility to existing Web-standards, of course also the physical characteristics like for example the size of the display have to be considered. To be able to offer smart support for these devices, XdC strictly separates data and the data processing logic from the representation specific to client device. In this connection, representation not only means the look of a document, it also includes the underlying markup language which may be some subset of HTML, WML or similar. The expected advantage of this approach is that whilst data and logic are retained unchanged, only the representation part has to be exchanged to support various device types. To achieve this functionality, the actual implementation of XdC heavily depends on XML based technology. Thereby, the eXtensible Markup Language serves the purpose of structuring the data independently to its eventual representation. The translation of the XML coded data to client presentable documents is done by using *XSL style sheets* (cf. Figure 3) which are based on XSLT, a transformation language for XML documents (http://www.w3.org/TR/xslt [September 20, 2002]).

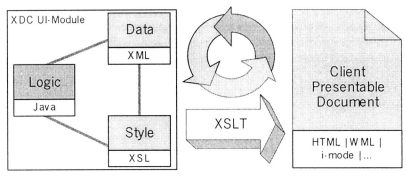

Fig 3: Separation of Concerns

6 Conclusion

Building personalized Web applications, i.e., applications that are responsive to the individual needs of each user, is a challenging and time consuming task. The goal of the versatile middleware presented is to free developers of context sensitive applications from the great effort of context management. The expected benefits are significantly reduced development times and thus also noticeably reduced development costs. At the time of writing, the design of the framework is fine-tuned on the basis of the experiences gained during the development of a prototypical campus information system. The fact that the conceptual model of the middleware is closely related to models already used in similar projects (cf. Hitz et al. 2002), together with its modular and versatile approach of context management qualifies XdC to be a stable base for rapid service development.

References

Hightower, J. & G. Borriello (2001). Location Systems for Ubiquitous Computing. *IEEE Computer*, 34 (8), 57-66.

Hitz, M., G. Kappel, W. Retschitzegger & W. Schwinger (2002). Ein UML basiertes Framework zur Modellierung ubiquitärer Web-Anwendungen. *Wirtschaftsinformatik 44 (3), 225-235.*

Huber, A.. & J. Huber (2002). UMTS and mobile computing. Artech House, Boston – London.

Poslad, S., H. Laamanen, R. Malaka, A. Nick, P. Buckle & A. Zipf (2001). CRUMPET: Creation of User-friendly Mobile Services Personalized for Tourism. *Proceedings of 3G 2001 - Second International Conference on 3G Mobile Communication Technologies. 26-29 March 2001. London, UK. http://conferences.iee.org.uk/3G2001/.*

Pradhan, S., C. Brignone, J. Cui, A. McReynolds & M. Smith (2001). Websigns – Linking Physical Locations to the Web. IEEE Computer, August 2001, 42-48.

Zipf, A. & R. Malaka (2001). Developing Location Based Services (LBS) for tourism - The service providers view. *ENTER 2001, 8th. International Congress on Tourism and Communications Technologies in Tourism. Montreal, Canada. 24-27 April, 2001.*

Individual Information Presentation based on Cognitive Styles for Tourism Information Systems

Hildegard Rumetshofer
Franz Pühretmair
Wolfram Wöß

Institute for Applied Knowledge Processing (FAW)
Johannes Kepler University Linz, Austria
e-mail: {hrumetshofer, fpuehretmair, wwoess}@faw.uni-linz.ac.at

Abstract

During the last years a broad spectrum of different Web-based tourism information systems (TIS) has been established. The acceptance and consequently the competitiveness of a TIS is mainly determined by the quantity and quality of data it provides. Most existing TIS already try to fulfil the tourists request for an extensive data collection. Moreover, tourism industry now realised also the potential to increase data quality by providing individual and specialised information about tourism objects. For this, appropriate search as well as presentation features are required. The advantage of individual information presentation is twofold. It increases the quality of information and on the same time it decreases the amount of data presented to the user. In order to cope with these demands, in this paper concepts of learning systems are applied to tourism information systems. Cognitive styles are used to describe the preferred way of information processing of each user (tourist). A rule system covers meta information about the mapping of user profiles in combination with cognitive styles and the corresponding query results of a specific TIS. Information presentation depends on the individual preferences and user skills, thus satisfying the specific expectations.

Keywords: tourism information system (TIS); cognitive styles (CS); extensible markup language (XML); extensible style sheet language transformations (XSLT); learning styles (LS).

1 Introduction

Current Web-based information systems are primarily content centred. This "one-size-fits-all" concept leads to users lost in cyberspace. A user has to deal with a huge amount of data that is presented without considering his individual information perception. The consequence is that in many cases users are not satisfied or bored with the kind of information representation. In case of a tourism information system (TIS) this lack could even detain a tourist (user) from booking touristic offers. For example, if a visually oriented user is supplied with less visual information or if a linguistic oriented user is overstrained with multimedia information. According to recent market studies online travel sites are among the most popular and frequently visited sites in the Internet (TIA, 2002). Tourists access online travel sites to get comprehensive tourism information and to make reservations of hotels, airline or train tickets and cars for both leisure and business purposes. Most TIS support different ways for information retrieval ranging from hierarchical navigation to flexible search capabilities extended with online booking and powerful access to tourism information and products.

In most cases state of the art TIS like TIScover (TIScover AG, 2002) are based on a centralised database in combination with a decentralised maintenance approach on the basis of an Extranet (Pröll et al. 1998) to offer high quality data with the demand to be up-to-date, accurate and comprehensive. TIScover offers a broad variety of flexibility to customise the appearance of the systems with its multi-layout concept to make the look and feel compatible with the corporate identity of different information providers like hotels, cities, regions or countries. Normally, TIS are flexible in presenting information according to their needs, but they are not flexible to present the information as a reaction on the skills of a user and the ways users process information.

The approach in this paper focuses on a methodology to present tourism information individually adapted according to the tourists kind of information processing known as cognitive styles. In contrast to existing Web-based information systems, tourism information systems which apply the introduced concept are optimised in order to cope with the broad spectrum of skills and individual information perception of their users. The presented approach aims to improve the user acceptance of a TIS and consequently, it increases its competitiveness.

2   Cognitive Styles

Cognitive styles characterise the preferred way an individual processes information. It represents a person's typical mode of realising, thinking, remembering, and problem solving. They are considered as bipolar dimensions whereby having a certain cognitive style determines a tendency to behave in a certain manner. It influences attitudes, values and degree of social interaction (Kearsley, 2001). "Cognitive style" is used for the more popular term "learning style". Generally, cognitive styles are more related to theoretical or academic research, while learning styles are more related to practical applications. The terms cognitive styles and learning styles have been widely used by educational scientists for the past decades. Terminology has varied from scientist to scientist. They tried to "catalogue" the ranges of learning styles. "Kolb's Theory of Learning Styles" and "Howard Gardner's Theory of Multiple Intelligences" are perhaps two of the best known theories that try to link personality characteristics to ways of acting and thinking. The discussed aspects are also worth considering for optimising information presentation and user interaction with information systems.

2.1   Kolb's Theory of Learning Styles

Kolb's theory is based on the learner's preferences for one or more of the four stages from learning by doing (Blackmoore, 2002):

- Feeling and sensing (concrete experience) – being involved in experiences,

- Watching (reflective observation) – watches others or observation about own experience,
- Thinking (abstract conceptualisation) – creates theories to explain observations, and
- Doing (active experimentation) – uses theories to solve problems and makes decisions.

The learner's preferences are always a continuum that one moves through over time. To cope with this Kolb defined the following learning styles:

- Activists (Accommodator) like to learn using concrete experience and active experimentation. They want to be personally involved, dealing with people (group work) or leading people
- Reflectors (Diverger) like to learn using reflective observation and concrete experience. They gather information by reading and brainstorming and listen with an open mind. In uncertain situations it is typical that they imagine the consequences.
- Pragmatists (Converger) like to learn using abstract conceptualisation and active experimentation. They stand out through experimenting with new ideas, making decisions, setting goals on a technical information basis.
- Theorists (Assimilator) like to learn using abstract conceptualisation and reflective observation. Characteristics for them are: Testing theories and ideas, designing experiments and analysing quantitative data.

2.2 Gardner's Theory of Multiple Intelligences

The theory suggests that there are a number of distinct forms of intelligence that each individual possesses in varying degrees. Gardner hypothesises that human beings are capable of seven independent means of information processing (Winters, 2002):

- linguistic (uses words),
- musical (uses sound and rhythmic),
- logical-mathematical (uses questions),
- spatial (uses pictures),
- body-kinesthetic (plays with moving),
- intrapersonal (plays alone) and
- interpersonal (socialising e.g., social skills).

According to Gardner, the implication of the theory is that learning should focus on the particular intelligences of each person. Different intelligences represent not only different content domains but also learning modalities (Kearsley, 2001). Each individual uses some of these styles when learning, but everybody tend to prefer a small number of methods over the rest. Both approaches give a categorisation of users. Most existing tourism information systems do not support content adaptation based on cognitive styles. In order to optimise the user support for realising, thinking, remembering and deciding, they have to attract their tourists with more individualised information.

3   Individualised Tourism Information Systems

According to Brusilovsky (2001) there are three kinds of adaptation in hypermedia systems: first, content, second, layout and third, navigation. Content adaptation in tourism information systems depends obviously on a user's preferences. The decision for layout and navigation has primarily been the task of computer scientists implementing information systems or in the Web era the success of Web designers producing creative applications. Unfortunately, this cognitive overload interrupts people having different intelligences, abilities and ways to process information. It seems obvious to provide users who are visually strong with images, graphs and tables additionally to common text. The visualisation and animation of content, combination of text, sound, video and images is heavily influenced by the kind of users' learning styles.

The approach presented in this paper primarily focuses on the methodology of adapting available TIS content with respect to cognitive styles belonging to individual users. Corresponding cognitive styles are represented as XML meta data information and are part of the user's profile. This profile can be determined by passing the user through an assessment centre responsible for identifying his cognitive style (Figure 1, 1.1 Assessment Centre). The complexity of this centre can range from simple questioning to more sophisticated cognitive tests. The presented concept proceeds on the assumption that the cognitive style related to a user is well known resulting from previous interactions between a specific user and the system and stored respectively in the user's profile (Figure 1, 1.1.1 User Profile).

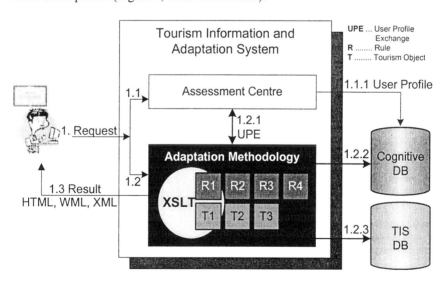

Fig. 1. Individualised tourism information system (TIS)

In the following, the workflow behind the individualisation process is illustrated with an example, whereby it is assumed that the profile of the current user (as illustrated in Figure 2) is already known in the system and requests can directly be passed to the adaptation system (Figure 1, 1.2 Request). According to Howard Gardner's Multiple Intelligences, the user focuses mainly on spatial and secondly on linguistic intelligence. Additionally, according to Kolb's Theory of Learning Styles the current user is an activist or accommodator. He likes concrete experiences and active experimentation. Suppose that the user requests some information about a specific destination such as Austria.

After an individual's cognitive style was identified, the tricky point is to find an intelligent mapping between the kind of cognitive style provided by the assessment centre (Figure 1, 1.2.1 User Profile Exchange) and the requested TIS content. Although, psychologists have already identified best practices of how to represent information with respect to an individual's cognitive style (Conner et al. 1996), the question of how to integrate and adapt it technically in a TIS still remains, but might be solved with this approach.

```xml
<userprofile>
   <username>tourist</username>
   <cognitivestyle>
      <theory>
         <expert>Howard Gardner</expert>
         <title>Multiple Intelligences</title>
      </theory>
      <characteristics>
         <name>spatial</name>
         <occurrence><percentage>80</percentage></occurrence>
      </characteristics>
      <characteristics>
         <name>linguistic</name>
         <occurrence><percentage>20</percentage></occurrence>
      </characteristics>
   </cognitivestyle>
   <cognitivestyle>
      <theory>
         <expert>Kolb</expert>
         <title>Theory of Learning Styles</title>
      </theory>
      <characteristics>
         <name>activists</name>
         <occurrence><fact>true</fact></occurrence>
      </characteristics>
   </cognitivestyle>
</userprofile>
```

Fig. 2. Individual user profile in XML

To meet this challenge, the system requires a set of rules, which perform the mapping of pure cognitive styles and content to individualised information. A single rule represents meta data by describing the combination of a cognitive style and a certain instruction. This instruction determines the kind of preferred content and layout. A tourism object will be appropriate as content, if it is able to fulfil the conditions defined in such an instruction. Conditions in this connection might be multimedia types such as text, image, sound or video, classifications of tourism objects such as pre-defined holiday packages or activity-rich destinations. Rules are executed by referencing meta information (cognitive styles) as well as plain information (tourism objects).

The awareness of the user's profile and the destination constraint enable to select on the one hand, an appropriate set of adaptation rules from the database (Figure 1, 1.2.2), in this case being relevant for spatial, linguistic and activist individuals and on the other hand, a matching set of tourism objects from the tourism information system (Figure 1, 1.2.3), namely fitting to the selected location Austria. Both results are provided in form of XML documents. Figure 3 introduces exemplarily different rules. The delivered adaptation rules, for instance, determine that people who strengthen spatial intelligence have to be supported mainly with images and persons who favour linguistic intelligence should get additional textual descriptions. There might be a rule which specifies that people with spatial intelligence should not be supported with spoken messages or any videos. A further rule may define, that activists should be provided with tourism objects suitable for active participation such as sport or activity offers.

```
<rule>                                    <rule>
   <cognitivestyle>                          <cognitivestyle>
      <theory>                                  <theory>
         <expert>Howard                            <expert>Kolb</expert>
         Gardner</expert>                          <title>Theory of
         <title>Multiple                           Learning Styles
         Intelligences</title>                     </title>
      </theory>                                  </theory>
      <characteristics>                         <characteristics>
         <name>spatial</name>                      <name>activists</name>
         <occurrence>                              <occurrence>
<percentage>80</percentage>                           <fact>true</fact>
         </occurrence>                             </occurrence>
      </characteristics>                        </characteristics>
   </cognitivestyle>                         </cognitivestyle>
   <instruction>                             <instruction>
      <layout>                                  <content>
<style>sample.css</style>                         <classification>
         <multimedia>                              skiing regions
            <image>1.0</image>                     </classification>
            <sound>0.0</sound>                     <classification>
            <video>0.0</video>                     mountain trails
            <text>0.2</text>                       </classification>
         </multimedia>                          </content>
      </layout>                                </instruction>
   </instruction>                            </rule>
</rule>
```

Fig. 3. Rules for individualisation in XML

This kind of meta data (rules) can be attached to any existing TIS since it is totally independent from the content or the structure of the internal database. The introduced concept is appropriate for most TIS, because their content is provided by tourism objects represented as a collection of multimedia data, and not only as static, unclassified or textual data. Additionally, rules can primarily be provided by people who are familiar with cognitive styles and know the impacts on individual's information processing.

The adaptation from common to individualised information is achieved by using XSLT (extensible style sheet language transformations). As input are acting XML documents such as an individual user profile including a cognitive style description and content requested by the corresponding user and returned by the TIS database as well as appropriate rules retrieved by the cognitive database (Figure 1). The rule system for the meta data information is realised as XSL transformation specifications. A single XSL represents therefore the appearance of the desired result, whereby the adapted result delivered to the user depends on the selected rules (R1 – R4), available content (T1 – T3) and requested Web page (Austria). There are two alternative XSL transformations:

1. The XSL transformation generates as output an XML document containing the individualised content. The resulting XML document acts as input for the application performing dynamic site generation.
2. The XSL transformation generates the desired output directly depending on the target device such as HTML (hypertext markup language), WML (wireless markup language) or PDF (portable document format).

To conclude the example, the response presented to the user depends on the XSL specification which is responsible for visualising information concerning destinations, in this case Austria (Figure 1, 1.3 Result). Suppose that the *common* appearance of the Web page for destinations consists of a textual description, an image and several links providing some details about the country. The *individualised* Web page in contrast considers the proposed rules and might therefore visualise several images supporting access to details about Austria, an interactive Austria map, a short textual description as well as a set of pre-selected links to present activity-specific locations in Austria such as skiing regions or mountain trails.

The key advantages of considering psychological aspects, especially an individual's cognitive style, in TIS are twofold. On the one hand, the amount of data a person is able to process and memorise is increased and consequently the acceptance of the system is improved. On the other hand, this information – that might be determined by the TIS provider (salesmen) as most important to be delivered to the user – can be presented most effectively and probably achieves the expected result.

4 Conclusion and further work

Existing tourism information systems are flexible in presenting information according to the needs of tourism industry, but they are not flexible enough to present the information corresponding to users skills and the ways users process information. To compensate this lack, in this paper an approach for individual information presentation based on cognitive styles for tourism information systems is introduced. Cognitive styles are used to describe the preferred way of information processing of each user (tourist). A rule system covers meta information about the mapping of a user profile in combination with cognitive styles and the corresponding query results of a specific TIS. Information presentation depends on the individual preferences and skills of a user, offering a twofold advantage: it increases the quality of information and on the same time it decreases the amount of data presented to the user. Consequently, tourism information systems are then able to satisfy the specific user expectations and therefore they increase their acceptance and competitiveness.

As further work it is planned to develop a prototype application in order to evaluate the concept in cooperation with the TIScover project. The determination process of a user's cognitive style will be a further challenge in this context. An alternative approach is to store the meta information about cognitive styles not within a central adaptation system but encapsulated in each information object. A comparison of both approaches considering different application scenarios will be the aim of the next research project.

References

Blackmoore, J. (2002), Telecommunications for Remote Work and Learning - Learning Styles, http://cyg.net/~jblackmo/diglib/styl-a.html.

Brusilovsky, P. (2001). Adaptive Hypermedia, User Modelling and User-Adapted Interaction 11: pp 87-110.

Conner, M., Wright, E., Devries, L., Curry, K, Zeider, C., Wuknsneyer, D., & D. Forman (1996). Learning: The Critical Technology, Wave Technologies International, Inc.

Kearsley, G. (2001), Explorations in Learning & Instruction: The Theory Into Practice Database – Multiple Intelligences, http://tip.psychology.org/gardner.html.

Pröll. B., Retschitzegger, W. & R.R. Wagner (1998), Extranet-Based Maintenance and Customization of Tourism Information, EMMSEC '98 (European Multimedia, Microprocessor Systems and Electronic Commerce), IOS Press, pp 331-338, Bordeaux, France.

TIA, (2002). Homepage of TIA – Travel Industry Association of America, http://www.tia.org.

TIScover AG, (2002). Homepage of TIScover – The Travel Network, http://www.tiscover.com.

Winters, E. (2002). Seven Styles of Learning - H. Gardner's ideas in a capsule, http://www.bena.com/ewinters/styles.html.

Tourism Information Systems based on Trail Network Information

Owen Eriksson,
Leif Åkerblom

Dept. of Computer and Information Science
Dalarna University, Sweden
{oer; lak} @du.se

Abstract

The purpose of this project is to develop methods for organizing and collecting information about trail networks for future use in a Tourism Information System. Some questions that are answered concern how to organize and structure the basic information as well as the location based data so that it can be processed, stored and presented in digital format. The practical work has resulted in an evaluation of equipment and methods for collecting location based data using a GPS-receiver. The project has resulted in: a topological data model for the collected information with focus on describing the trail system, a data base with a number of tables containing the collected data from the GPS measuring, a simple GIS application for presenting the available information in a number of different themes.

Keywords: Global Positioning System (GPS); topology; tourism information system; trail network.

1   Introduction

Särna, Idre and Grövelsjön are three minor villages in the northwestern mountain region of Dalarna that are to a great extent relying on tourism for their future economic outcome. Since 1997 local businesses in cooperation with local and regional authorities have run an extensive destination development project to improve conditions for the area to remain a living part of the region with a good service level, increased occupation and strengthened competitive power. Most of the tourism activities are oriented towards physical activity and wilderness experience in the surrounding mountains and forests. A large number of trails exist in the area and they are used for a number of organized tourism activities in both summer and winter. Hiking is a major summer activity while in the winter guided ski tours and dog-sledding are offered at some of the tourist facilities. Snowmobile activities are also offered where local organizers provide snowmobile adventures, guided tours and snowmobile rental.

These activities, using the trail network in the area, are highly dependent on the availability of maps and other location-based information. Today information about the trails is available in brochures with detailed descriptions and simple maps for over 60 suggested hiking trips and 30 cross country skiing trips. For snowmobile tourists the local snowmobile clubs have prepared a map showing all the marked snowmobile trails. Maps and brochures are available for sale at any of the three local Tourist Bureaus or at local shops. Some of the proposed trips are also presented on the official Internet home page of Idre Turism.

1.1 Problems

There are a number of problems that concern the availability of tourism information in the area. One problem is today's limited possibility to update the information at regular intervals. For example, information about trail conditions must be updated regularly to maintain an acceptable information quality.

Another problem is that much of the information is not available from one single source. Often the information has to be collected from a large number of unrelated sources - in written form or as knowledge from staff or local inhabitants.

A third problem is that a majority of visitors tend to stay only on the largest and most well-known trails. This has a negative influence on both the experiences of the visitors due to crowded trails and the impact on the sensitive wilderness environment. This problem is related to the lack of detailed descriptions of the trails.

This lack of detailed descriptions of the trails in combination with present maps getting out-of-date and trail signs and markings not always being consistent throughout the trail system makes snowmobile tourism one area that urgently needs further development. One reason for the planned development of snowmobile tourism is the fact that snowmobile tourists are estimated to be willing to spend much money with local tourism operators. Another reason for the development is the environmental aspect. By developing a well defined and maintained trail system tourism authorities hope that snowmobile traffic will be limited to these specific snowmobile trails. This would result in a lower impact on the sensitive wilderness environments.

In order to accomplish these goals trail maps and a consistent system for signs must be developed for the 1000 km trail network in the area. The comfort for the tourists using the network must also be increased by developing resting places with a guaranteed level of maintenance. Responsibilities must also be resolved; one organization must have the full responsibility for the maintenance of the trail network which also includes the responsibility to provide accurate information about the network. Today tasks and responsibilities are shared between regional authorities and several clubs.

1.2   The need for a tourism information system

In order to continue development of the destination Särna-Idre-Grövelsjön the local tourism authorities have several needs for the future. Among other things they need to offer consistent and updated information about the total existing trail network. They also want to increase their offers by including several minor trails that today are known only by a small number of local experts.

One way to accomplish this is to develop a Tourism Information System which has many possible uses e.g.:

- To facilitate the maintenance of the trail network.
- To produce detailed, updated and high quality maps of the trail network.
- To provide accurate information about service facilities related to the trail network.
- To provide weather information related to the trail network.

This type of information is of interest both to the people who are responsible for the maintenance of the network, to personnel at Tourist Bureaus and to tourists. In order to develop the tourism information system it is important to collect and organize information about:

- the structure of the trail network;
- different types of trails, e.g. snowmobile trails, ski trails;
- trail conditions, e.g. how the trail is prepared and how hard it is to travel on;
- locations and places related to the network;
- service facilities, e.g. shops and fuel stations related to the network;
- resting places.

The purpose of our project is to develop methods for organizing and collecting the information described in the list above and these methods include:

- modeling and structuring the information;
- collection of digital location-based information using GPS-receivers;
- storing the information in a data base management system;
- distribution to suitable software for presentation.

2   Business Analyses and Data Modeling

The project started with a business analysis which showed that today's information about the trail network is inadequate. There is a lack of: names on all trails, information about types of trails (width, trail conditions) and other objects of interest (huts etc.). In order to develop the Tourism Information System a mapping of the trail network in the area was performed as a first step. The reason for this is that the trail network is the basic business object for the tourism activities in the area and therefore high quality information about the network is a basic element of the system.

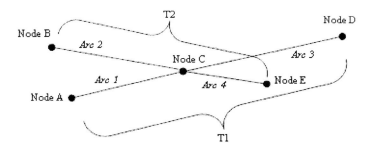

Fig. 1 Trail network

The trail network can be seen as a transport network (Bernhardsen, 1999) through which people travel, either by walking, skiing, dog-sledding or driving a snowmobile. The network can be described on a conceptual level as a number of arcs and nodes (see Figure 1); this is called the topology of the network. The nodes represent start- and end nodes and junctions in the network. This means that the basic arcs show how navigation can be performed through the network. Transport network models are used for describing roads, railways, paths and trails by combining the arcs in different sequences, e.g. Figure 1 shows that Trail T1 is constituted by Arcs 1 and 3, and Trail T2 by Arcs 2 and 4. This implies that the basic information structure of the tourism information system will be the trail network (see Figure 1) and most of the information that is gathered and communicated has to be related to the network.

The objects needed to model the trail system and to maintain the topological relations are described in the following table:

Table 1 Objects of the data model

Object	Description
Point	The exact coordinate of a geographic position given with a meter precision – the value is shown with x-, y- and z-coordinates.
Node	The start- or end of an arc. A node is something that is of importance for navigating through the network, e.g. start, end or junction in the trail-network system. A node can belong to many arcs.
Arc	The distance between two nodes. An arc always consists of a start- and an end node which means that it has a direction. An arc can belong to many trails.
Trail	Consists of a number of arcs. Has a name, e.g. Valley Trail. The trail may have attributes to describe something of interest to the users, e.g. day trip, easy, hard.
Tourism object	Something of interest to the target group that doesn't concern the navigation (examples are huts and rest areas). Has descriptive attributes and is related to a location-based point (coordinate). In this context the term "tourism object" is used as a collective name to exemplify the possibility to relate any kind of information to the trail system.

The objective of the data modeling process has been to find a suitable adaptation of the topological model to use as a base for the trail system. The basic idea behind using this topological network model is to separate the topology from the geometry (Lundgren, 2000). This separation is important because the topology describes the basic structure of the trail network and the geometry how it is located in the geography. This also means that information can be related to the network by using the nodes and arcs that constitute the topology. The resulting data model is shown in Fig. 2.

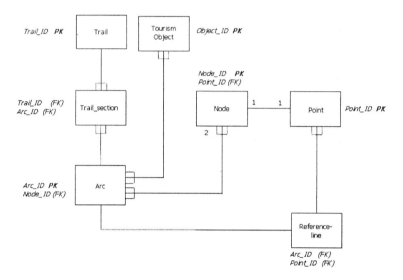

Fig. 2 Data model

3 Data Collection

3.1 Data collection using GPS-receivers

To collect data about the trail network and tourism objects related to the network GPS-receivers has been used. With the help of the GPS-receivers has information about geographical positions been stored as x- and y-coordinates and saved in the WGS84 format (latitude/longitude). It is also possible to use the z-coordinate to show elevation. There are two alternate ways of collecting geographical positions with the help of the GPS-receiver: manually or automatically.

- **Automatically:** The first way is to store coordinates automatically. While the GPS-receiver is turned on geographical coordinates are automatically saved in a track file. Tracking can be adjusted and set to a desired interval between 50 meters and 2 km's – or be used in the *Auto Mode*.

While using the *Auto Mode* coordinates are automatically saved at longer intervals while travelling along a straight line and at closer intervals when changing directions more often.

- **Manually:** The second way of saving coordinates is to store them manually. The points that are collected this way are called waypoints. A waypoint is stored by manually pushing a button on the GPS-receiver. The position will then be stored in a separate waypoint file.

3.2   The data collection in Grövelsjön

The initial mapping of snowmobile trails has been conducted in a limited area maintained by one of the local snowmobile clubs.

- **Mapping the nodes:** Coordinates for the nodes in the trail system have been measured and saved manually as waypoints.
- **Mapping the arcs:** On collection of track coordinates for the arcs in the trail system the GPS-receiver has been set to store coordinates at a 200 meter interval.
- **Mapping the tourism objects:** No clear directives have yet been worked out for the mapping of different types of tourism objects. At this early stage of the project these objects (mainly huts) have been mapped as if they where situated on the track. These objects have been measured and saved manually as waypoints.

4   Data Storage

4.1   Extracting data from the GPS-receiver

The coordinates stored in the GPS-receiver have been extracted using a program called Kartex. This program saves the information from the GPS-receivers into separate files: one waypoint and one track file. In both files the information about the coordinates is represented with an index number and the coordinate values (x-, y- and z). Also stored in the track file is the actual time for when the coordinates were saved. Upon transfer the positions are in this case converted from WGS84 to the local reference system (RT90). At this initial stage values transferred to the data base have been only the x- and y-coordinates. The basic information collected in the two files – the waypoint file and the track file – has to be processed in a number of ways to store the information in the database adhering to the data model (see Fig. 2 above).

4.2   Storing the information in the database

The coordinates from the track-file have been merged with the coordinates of the nodes from the waypoint file and all this has been stored together in the *point table*. This table is important for describing the geometry of the trails.

In order to generate the topology all the information from the track and waypoint files has to be matched. The coordinates of the track-file have been scanned sequentially and at the same time matched with coordinates of the waypoint file which contains information about the nodes. Every new node along the track has created a new entry in both the *node table* and the *arc table* because every new node implies the start of a new arc.

In order to resolve *many-to-many-relationships* in the database some additional tables have been added to the database. These tables consist of two foreign keys from adjoining tables that together make up a new unique combination. Additional information about the relationship between the topology and the geometry of the trail network has been stored in the *reference-line table*. This table includes information about every point along the arc and is important for the final drawing of each arc. For every arc an ordered list of points has to be created to describe the geometry of the individual arc. To be able to deal with arcs that contain several parallel trails the *trail-section table* has been added in the final data base.

The original waypoint file from each mapping session also includes geographical coordinates for other objects of interest to tourists. These coordinates are stored and treated separated from the point, arc and node tables. Additional information about these objects will be also supplied as attributes in the resulting *tourism object table.*

5   Presenting the information in GIS software

To present information from the database, the GIS software ArcView has been used. Here information from the database is presented in different layers called themes. In Fig. 3 we can see that there are five different themes presented.

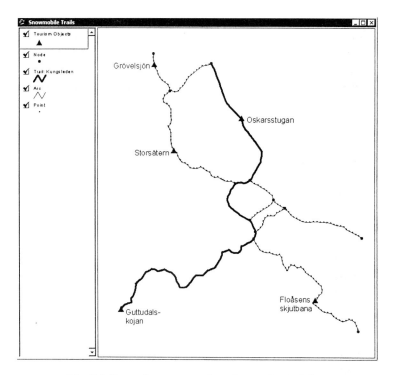

Fig. 3 Information presented as themes in ArcView

The themes used in this presentation are:

1) The point theme
2) The node theme
3) The arc theme
4) The trail theme
5) The tourism object theme

All themes have been constructed by selecting information and joining tables from the database. This has been accomplished by using the SQL-Connect function. This function makes it possible to use a SQL-query for selecting the desired information and the output of the SQL-statement is presented in each theme.

- In the point theme the points are presented as small dots and the information presented is collected from the point table.
- In the node theme the nodes are presented as large dots and the information presented is selected from the node table and the point table.
- In the arc theme the arcs are presented as thin lines. To produce the arcs a special program extension has been executed to convert the original point theme to a polyline theme by joining all coordinates to a number of line segments that make up the individual arcs. This theme is based on information from the reference-line table.
- The trail theme shows the selected trail as a thick line. The same procedure as for the arc theme is used to produce the resulting polyline theme. The information presented is selected from the trail table, the trail-section table, the arc table and, the reference-line table.
- The tourism object theme shows the objects as triangles and the information is selected from the tourism object table.

5.1   GIS and Information Systems

The big advantage with the GIS-software is that it can be used for presentation and analysis of location based information based on co-ordinates. This means that software like ArcView can present information in different themes and matched based on geometrical relationships. This implies that the software can be used to answer questions like "Show all huts that are situated within a buffer zone of 500 meters from the X trail".

This is a major advantage, but to be able to build a flexible information system based on location-based information, the information has to be structured from a business object perspective.

It is worth emphasising that location based information is something more than just maps and coordinates. This is important to recognise because we do not normally communicate locations and positions in terms of coordinates (Couclelis 1992), instead we use geographical identifiers (ISO/DIS 19112) like addresses and names of places. Another reason why this is important is that information about tourism objects has to be combined with other information (e.g. trail network information, see above) and in many cases geographical identifiers (e.g. node-ID, arc-ID) are easier to use than coordinates to combine different information sources (Eriksson, 2002).

One problem experienced when performing the data modeling in the project was that the analysis of location based information tend to be focused on the geographical aspect and not the business aspect.

- With the geographical aspect in focus geographical objects (points, lines, areas) and their coordinates are stressed, while things like huts, trails etc. are primarily seen as attributes to the geographical objects.
- With the business aspect in focus business objects (huts, trails etc.) are stressed and the geographical aspects are primarily seen as attributes to the business objects.

Our experience is that both the business and geographical aspect are important, but in order to build a flexible information system based on location-based information it is important that the analysis and data modeling are based on business objects and their relationships.

Another important experience from the project is that the information collected should be stored in a data base management system. The reason for this is that it must be possible to restructure the information in an easy way and it must be possible to access and present location based information in other types of systems than GIS-software. It is also important to be able to present the information in different clients, e.g. using mobile terminals.

6 Conclusions

The purpose of our project has been to develop methods for organizing and collecting trail-network and location-based information, not building a complete tourist information system. In the project we have reached a number of important conclusions and experiences that are presented in the list below. These conclusions and experiences imply that it is important:

1. To stress that the tourism system described in this paper is not intended to be a GIS-system – it is an information system built on location based information, which is another matter.
2. To build the mapping of the trail network on a good topological data model. The topological model is the foundation for good quality information about the trail network. The model should describe and identify the basic arcs and nodes in the trail network.
3. To use efficient and well proved methods to collect data about the trail network.

4. To store and make the information available in a relational database. This means that information about the trail system can be related to other information not only based on geometrical relationships but also based on object relationships.
5. To integrate knowledge from both GIS-experts and the system development area. This is necessary to create conditions to build information systems based on topological data structures and geographical information.

All this shows that the tourism information system discussed in the paper cannot be fully developed using only traditional GIS (Dueker & Ton, 2000). Instead we have to develop an information system where the information is integrated and accessed with a data base management system. This implies that there is a need for methods which integrate knowledge from different areas. Today there are specialized methods oriented towards GIS, transport modeling and information systems development, but there is a lack of methods which integrate knowledge from these different areas. This is a major problem because knowledge and systems integration will be a key factor in order to develop tourism information systems because transport (trail) networks constitute the basic business object of many information systems in the travel and transport business. Thus there is a need for extended knowledge about how information about these basic business objects should be structured, collected, stored and presented.

References

Bernhardsen, T. (1999) *Geographic Information Systems, An Introduction* Second Edition, John Wiley & Sons Inc., New York

Couclelis C. (1992) *People Manipulate Objects (Cultivate Fields): Beyond the Raster-Vector Debate in GIS, Theories and Methods of Spatio-Temporal Reasoning in Geographic Space*: proceedings / International Conference GIS - From Space to Territory: Theories and Methods of Spatio-Temporal Reasoning, Pisa, Italy, September 21 - 23, 1992

Dueker, K.J. & Ton, T. (2000) *Geographical Information Systems for Transport* In: Hensher, D.A. & Button, K.J. /eds/ *Handbook of Transport Modelling* Pergamon, Amsterdam: 253-269

Eriksson O. (2002) *Location Based Destination Information for the Mobile Tourist*, In: Information and Communication Technologies in Tourism, Wöber K.W., Frew A. J., Hitz M. (eds.), Springer-Verlag Wien, Innsbruck 22 - 25 Jan 2002

Lundgren M-L. (2000) *The Swedish National Road Database – Collaboration Enhances Quality*, In Proceedings of the Seventh World Congress on Intelligent Transport Systems, 6-9 November 2000, Turin, Italy

ISO/DIS 19112. *Geographic information - Spatial referencing by geographic identifiers.* [WWW document], URL http://www.statkart.no/isotc211/scope.htm#19112

Tourism Destination Analysis & Planning: An IT Application

Chris Vasiliadis[a]
George Siomkos[b]
Adam Vrechopoulos[b]

[a]University of Macedonia, Department of Business Administration
Thessaloniki, Greece
chris@uom.gr

[b]Athens University of Economics & Business Administration
Athens, Greece
{gsiomkos, avrehop}@aueb.gr

Abstract

The tourism industry is dynamically evolving, responding to significant environmental changes. Information Technology (IT) applications constitute a major contributor to the industry's evolution. This paper reviews several popular tourist destination analysis and planning instruments. Specifically, evaluation matrix techniques and simplified diagrams for landscape analysis are covered. It then attempts to develop such an instrument, namely the Tourism Destination Analysis Card. The card belongs to the section elevation diagrams type of simplified diagrams for landscape analysis and its philosophy is based on the presentation of related geographic areas. The data base management system, the architecture of the proposed IT application and the information flow diagram are finally presented analytically. The overall objective of this paper is to introduce a "coordination and communication" scenario among the involved players of the tourism industry (the public sector included) based on technology challenges upon which information flow and transactions will be accelerated and the value of the end consumer enhanced.

Keywords: Tourism Destination Analysis Card, Tourism Landscape Analysis.

1 Introduction

Advances in Information Technology (IT) are widely acknowledged as causing fundamental changes in organizational and market structures (Malone et al., 1987). The advent of Inter-Organizational Information Systems (Johnson and Vitale, 1998) and the Internet have resulted in previously unthinkable ways of conducting business, (Rockart and Scott-Morton, 1991). The Internet does not only provide a more efficient method of performing existing processes, but can also define entirely new methods of performing existing processes (Cronin, 1996; Kalakota and Whinston, 1996). This ability to redefine the essential activities of a firm has made the Internet and the World Wide Web valuable for businesses in the travel industry (Bloch et al., 1996). On the other hand, improving the quality of customer service is a key for achieving a competitive advantage (Coyne, 1989) and information technology can act as a powerful catalyst for improving and advancing customer service (Earl, 1989). In parallel, the pro-active tracing of customers' interests, needs and wishes contribute significantly to the identification and exploitation of business opportunities that have not yet been detected by competitors (Kotler et al., 1993).

The World Tourism Organization has predicted that the travel industry will grow in real terms by 3.7% annually until 2010 (The Economist, 1995). Technology and financial necessity are forcing a transformation in the industry (Helm, 1995). Tourism destinations' daily administrative activities are often oriented towards tracing the tastes and interests of potential customers. Following the wide success of the Internet, developers of major Computer Reservation Systems (CRS) realized that their added value would result from the information they process, rather than the technology they can implement across the world. However, the terrain is changing fast. Many applications of today will not survive tomorrow without substantial rethinking and restructuring (Poulymenakou et al., 1998).

The paper is structured as follows: Some important techniques for analyzing and planning the geographical areas' tourism offering are presented. Then, the tourism destination analysis card is introduced and recommended for the optimum product/service planning and design of the tourism destination attractions. An Information System for the effective operation of the destination analysis card concept is presented at the end.

2 Tourism Destination Analysis and Planning Instruments

Middleton (1998) developed a basis for the investigation of tourist destinations contributing to a better understanding of the analysis processes of destination management. The destination analyses were categorized into supply data analyses (the traditional philosophy) and market-oriented analyses, which is based on the recent marketing management philosophy adoption by tourism management. The emerging trend of the market-oriented analyses was further intensified by appropriate destination planning instruments, which were incorporated into the existing supply data analysis techniques.

Kotler et al., (1993) developed an analysis instrument for the development and planning of tourist destinations, called the "audit instrument for infrastructure, attractions and people." The audit matrix has the capability of simplifying the presentation of basic criteria for the assessment of destinations under review. The development of relevant criteria follows the strategic market planning process. The Destination Analysis instruments includes evaluations and measurements of the following: (1) *Infrastructure:* The infrastructure's aspects include elements such as the civil services (e.g., health centers, athletic establishments, security services) and the transportation networks and facilities, (2) *Attractions:* This element includes mainly the unique natural, cultural and technical resources of a geographical area (e.g., historical and archeological features of the destination). Infrastructure and attractions are evaluated based on their contribution and significance towards the development of tourism activities tailored to the wishes and interests of specific domestic or foreign visitor segments and (3) *Segments:* They are constituted by marked segments of domestic or foreign customers characterized by different behaviours, preferences and wishes.

2.1   Evaluation Matrix Techniques

The Evaluation Matrix for Tourist Attractions (Inskeep, 1991) constitutes an alternative technique for analysing tourist destinations attractiveness. It relies on experts' assessments as well as on primary and secondary data analyses. The limitations of this type of analysis are related to the lack of the appropriate information and software support.

An analysis and evaluation of tourist attractions through the use of matrices, could lead to a basis for the development of perceptual maps and the consequent determination of appropriate strategies for each specific development area. A practical application of such analysis techniques for perceptual mapping of tourist attractions was conducted in 1973, in Hawaii islands by Belt, Collins & Associates, Ltd. (Inskeep, 1991) presents this application in a detailed fashion. Such perceptual maps have usually two dimensions, and could include the following tourist attraction elements in their two-axis: Axis A - landscape, natural elements and recreation elements, and Axis B - cultural, historic and archaeological elements of attractions.

The evaluation matrices present information regarding the following dimensions: (a) accessibility potential, (b) potential for economic development, (c) environmental influences of development, (d) socio-cultural influences of development, (e) national or regional importance of the attractions and (f) international importance of the attractions. The attractions are assessed on 1-5 or 1-10 point scales.

2.2   Simplified Diagrams for Landscape Analysis

The simplified diagrams for landscape (architectural space) analysis can become even more detailed when applied on selected points and geographic areas. This is possible through the use of section elevation diagrams or functional diagrams. The two techniques are complementary of each other. They can analyse various functions in tourist areas (e.g., service stations, access to road networks, activity zones, etc.), presenting corresponding section elevations or ground plans of points in the geographic space (Reid, 1987). The applications of these techniques are thus far concentrated mainly in the architecture practice.

Section elevation diagrams can be transformed into useful strategic instruments for destinations management. Except topographical and geographic information, the following dimensions are also included in the analysis: (a) attractions (e.g., physical and cultural attractions), (b) existing infrastructure and technical elements (e.g., hotels, services, lodging), (c) possible vacationers' groups (e.g., possible segments of foreign tourists) and (d) tourists' preferences with respect to other competitive destinations. Section elevation diagrams constitute an improved instrument for enhancing the strategic development of tourist destinations. In general, however, their application is limited to smaller geographic sections. Therefore, there is need to use the Analysis Cards, which would include more than one section, as analytically presented below.

The diagrams for Landscape Analysis are characterized as wider geographic areas planning techniques and belong to the Spatial Analysis Techniques category. A specific area can be more effectively analyzed taking into account the attractions included in the wider geographic area framing it. After that, the specific area can be more successfully examined with respect to the special characteristics that it incorporates using Diagrams for Landscape Analysis.

This paper closely examines the Section Elevation Diagrams. These diagrams belong to the architecture applications for landscape analysis with the use of diagrams' utilities. Their use is recommended in the cases of presentation of activity occurring in the area (e.g. frequency of customer visits, access to visiting points and sights). In addition, their use provides the capability for a brief visual outlook of the area (side sight of the area) and the characteristics of the tourism product incorporated within the analysis areas (e.g., sights).

As discussed above, areas can be more easily evaluated regarding their suitability for tourism activities development by the use of some specific evaluation criteria. These criteria can be incorporated into the diagrams and convert them into useful elements for the planning and elevation of tourism destinations.

3   Towards the Development of the Tourism Destination Analysis Card

The underlying philosophy of the Tourism Destination Analysis Card is based on the presentation of related geographic areas of the "model of thermometrical users" type.

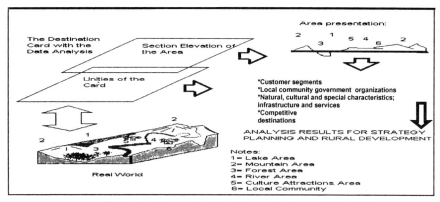

Fig 1. An Example – The Case of a Hypothetical Lake

The process begins with the distribution of the sections in the space. The sections of special interest to the analyst are identified, based on such criteria as view sites' number of attractions. They are named according to the view sites. For example, Figure 1 presents several views of a hypothetical lake area. The next step in the process is the area planning. In the example, the total view presents names of places and the geophysical condition - morphology of the area. In other words, Figure 1 presents the destination card as a GIS management tool. After geographic data are placed, information regarding strategic planning is recorded (e.g., physical, cultural and special site seeing and services, potential profitable customer segments, infrastructure, economic development potential, environmental consequences of development, sociocultural consequences of development, national and international importance of sites and competitors).

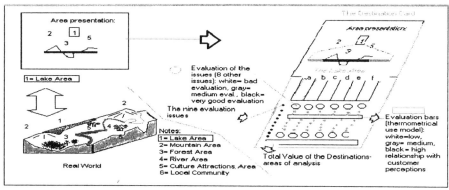

Fig. 2. The Analysis and Evaluation process for the creation of the Destination Card for the "Lake"

The analysis process for the creation of a destination card continues with the evaluation of the infrastructure and the matching with visitors' needs as well as the needs of the local society. Figure 2 presents an example of a hypothetical Tourism Destination Analysis Card. In the destination card, a series of six geographic view sites of the eastern view of the lake is presented, i.e., a,b,c,d,e,f. The aforementioned information areas or strategic analysis criteria are evaluated on the basis of a 3-level scale (black=high relationship; grey=medium relationship; white=low relationship), which denotes the level of relationship between the criteria and the current situation in the areas. As far as the customer segments are concerned, their evaluation follows the same method.

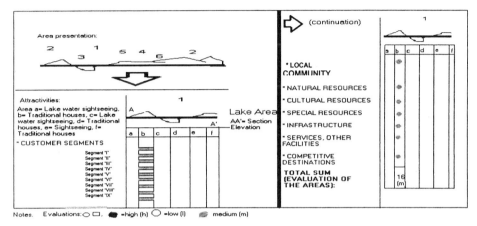

Fig. 3. A Tourism Analysis Card for the Lake Area

Figure 3 presents the final Tourism Analysis Card for the lake area. At the last step, the areas under investigation are evaluated, summing up the values of each across criteria (e.g., Area "b" receives 16 "medium" evaluations). After the final evaluations, the analysts can make their decisions, possibly under certain limitations.

4   The Proposed Data Base Management System and the Architecture of the Application

Figure 4 presents the proposed Data Base Management System for Tourist Destination Management applications.

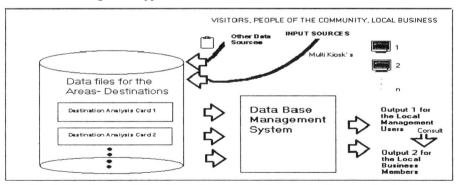

Fig. 4. The Proposed Data Base Management System for Tourism Destination Management Applications

464

The Data files for the Areas-Destinations are taking valuable input from Multi Kiosk's and other Data Sources. These data are being processed through a Data Base Management System providing output (i.e., consultation) for the Local Management Users and for the Local Business Members.

The application requires the adoption of a model architecture that will enable it to be analyzed and implemented. The architecture should consist of three fundamental layers and includes a vast amount of techniques and tools of contemporary technology, which should be considered as imperative by the designers of the application in order to fulfil their goals. Figure 5 presents the proposed model architecture, followed by a deeper analysis of each particular layer.

The use of IT would decisively contribute towards an easier classification of information through electronic area maps. The electronic maps could present diagrams of 3-D section analysis. In addition, the use of just-in-time information selected from visitors could help the local management of the destinations to determine the appropriateness of the area's characteristics and their "fit" with the preferences and interests of the visitors (customers).

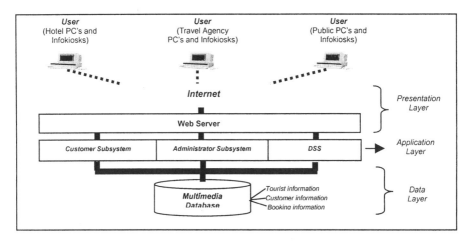

Fig. 5. Architecture of the Application

Hotels, Travel Agencies and Customers can utilize the Analysis Card and interact with the system. In Hotels and Travel Agencies there will be trained employees to support customers during the whole process while on the info-kiosks there will be a step-by-step guide to provide all the necessary information/support in a user friendly manner. There are also many other services offered to the users (e.g., booking information) based on the data coming from the database and implemented through the operation of the application layer. Users can combine the information/services offered by the Analysis Card with the other types of information offered (e.g., a customer can select a specific area from the analysis card and check out from the customer subsystem if there are available hotels/beds for booking in a specific area).

5 Conclusions and Managerial Implications

As digital technology and consumer behavior evolve, marketers need to continuously enhance the value of their digital marketing offering. In an interactive two-way, addressable world, it is the consumer, and not the marketer, who decides with whom to interact, what to interact about, and how to interact at all. To that end, marketers have to earn the right to the digital relationship, and they have to do so by continuously enhancing the value they offer to consumers. Thus, the importance placed on continuously studying and monitoring consumer behavior (preferences, complaints, etc.) by providing value added services, constitute the most essential characteristic of the proposed application.

Through this application many unexploited tourism destinations will be identified and supported, enriching on one hand the alternatives provided to customers and supporting business activity on the other. In other words, the proposed application is based on a "win-win" relationship where all the involved players have mutual benefits. Furthermore, it should be noted that the public sector could actively participate and even coordinate the proposed application. Besides, the public sector is the content provider of many products in the tourism industry (e.g., museums, archaeological sites, etc.).

Currently, the system is under development. The next step of the research is to apply the proposed application within a field experiment in Greece. More specifically, further research will investigate users' reactions and attitudes towards the system. To that end, a causal research design will be employed in order to measure cause-and-effects relationships related to the system use. The long-term objective is to be able to use the platform during the Olympic Games that will be held in Athens in 2004.

References

Bloch, M., Steiner, T., & Pigneur, Y. (1996). Leveraging electronic commerce for competitive advantage: A business value framework. *Proceedings of the 9th International Conference on EDI – IOS:* June, Bled, Slovenia.

Coyne, K. (1989). Beyond service fads – meaningful strategies for the real world. *Sloan Management Review,* 30(4), 69-76.

Cronin, M. (1996). *The Internet Strategy Handbook.* Boston: Harvard Business School Press.

Earl, M. J. (1989). *Management Strategies for Information Technology,* Business Information Technology Series, Prentice Hall International (UK) Ltd.

Helm, L. (1995). Travel agents hit turbulence. *Los Angeles Times,* February 25, A1, A24.

Inskeep, E. (1991). *Tourism planning.* Canada: John Wiley & Sons, Inc.

Johnson, H.R., & Vitale, M.R. (1998). Creating competitive advantage with inter-organizational information systems. *MIS Quarterly,* 12(2), 153-165.

Kalakota, R., & Whinston, A. (1996). *Frontiers of Electronic Commerce.* Addison-Wesley.

466

Kotler, P., Heider, D.H., & Rein, I. (1993). *Marketing Places*. New York: The Free Press.

Malone, T., Yates, J., & Benjamin, R. (1987). Electronic markets and electronic hierarchies. *Communications of the ACM*, 30(6), 484-497.

Middleton, V. (1998). New market conditions and the strategic advantages of products similar to destinations. In AIEST (Ed.), *Destinations marketing*, St. Gallens: AIEST, 40, 153-165.

Poulymenakou, A., Drakos, W. Papazafeiropoulou, N., & Doukidis, G. (1998). Towards new web application development practices. *Australian Journal of Information Systems*, November, 107-113.

Reid, G.W. (1987). *Landscape Graphics*. New York: Whitney Library of Design.

Rockart, J., & Scott-Morton, P. (1991). Networked forms of organization. In S. P. Morton, (Ed.), *The Corporation of the 1990s*. New York : Oxford University Press.

The *Economist* (1995). Death of a Salesman. September 23 (54-55).

Travel Agents in the 'Information Age': New Zealand Experiences of Disintermediation.

Vladimir Garkavenko[a]
Hamish Bremner[a]
Simon Milne[a]

[a]New Zealand Tourism Research Institute,
Auckland University of Technology, New Zealand
{vladimir.garkavenko, hamish.bremner, simon.milne}@aut.ac.nz

Abstract

There is extensive discussion in the travel industry press about the inevitable disintermediation of travel agents as a result of the widespread availability of Information and Communication Technologies (ICTs). This study empirically investigates the effect of the evolving business environment on travel agents in New Zealand's largest city, Auckland. Our work is based on in-depth interviews conducted with agents, owners and travel sector experts in the greater Auckland region. We review the current pressures facing travel agents, emphasizing the pivotal role played by ICTs. The research highlights key points for the commercial survival of agents including: niche marketing, the provision of excellent service, marketing and purchasing alliances, and the capacity to keep pace with technological advances. The paper also reveals that the traditional boundaries existent between the principal and the consumer are open to negotiation and 'adding value' in the distribution chain is considered an essential strategy for survival.

Keywords: travel agents, information and communication technologies, disintermediation, Auckland, New Zealand.

1   Introduction

Information and communication technologies have dramatically changed the tourism industry and altered the competitiveness of tourism organisations and destinations. The rapid development of both tourism supply and demand makes ICTs an imperative tool for the industry, especially for marketing, distribution, promotion and coordination (Hall, 2000). The accessibility and relatively low cost of the Internet is bringing all players on-line - business and consumers, buyers and suppliers – resulting in new ways of doing business. According to Palmer and McCole (1999), the 'most important structural change that could be brought about by the Internet is disintermediation wherein principals bypass the intermediary and sell directly to end-users' (Palmer and McCole, 1999:37). This phenomenon is especially pronounced in the relationship between airlines and travel agents (principal and retailer). To minimise the risk of disintermediation, travel agents need to reduce their dependence on simple transaction processing and increase revenues from intelligent, knowledge based activities such as counselling, information brokering and package personalisation (O'Brien, 1999; Dull, 2001).

During the 1990s most airlines and other travel product providers introduced web sites which not only informed consumers, but also enabled itinerary building, fare construction and reservations (Buhalis, 2000). The resultant ability of airlines and others to communicate directly and cost-effectively with consumers, led to growing pressures on traditional distribution channels and the emergence of direct threats to travel agent survival (Bloch and Segev, 1996). Three major threats to the travel agent sector have been highlighted in the international literature: commission cuts by airlines, the launch of dynamic Internet sites by principals, and the growing willingness of consumers to purchase travel products on-line (Lewis et al. 1998; Milne et al. 2001).

Recent moves to cap the commissions received by travel agents represent an attempt by airlines to control what had been one of the major carriers' fastest growing cost items. In 2001 airline commissions to travel agents accounted for $3.4 billion of a total operating cost of $91 billion for the nine major airlines in the U.S. (De Lollis and Adams, 2002). This trend of commission reduction has only been intensified by the increasing financial uncertainty of many airlines due to harsh global competition and the ongoing ripple effects of the events of September 11 2001.

It is suggested that the Internet may account for one in every four travel purchases in the main generating markets within the next five years (WTOBC, 2001). The same source indicates that the use of the Internet by the tourism sector has already expanded at a dramatic rate and North America's initial predominance in the number of Internet users could be surpassed by the Asia-Pacific region as early as 2003. Internet sales currently account for about 15 per cent of travel purchases, but new studies indicate that Internet transactions may account for 20-25 per cent of all tourism sales in the main markets over the next four to five years (WTOBC, 2001).

This challenge becomes more pronounced with the launch of such web sites as Orbitz.com. Orbitz uses a new internet-based software search system that will allow consumers to view flight schedules and purchase tickets from 30 member airlines and 450 other airlines. Thus, the consumer has access to the prices of different airlines through one web site. Although the operation has been attacked as anti-competitive by the United States Travel Agents Association and others, it remains to be seen if the US courts will act and force it to discontinue.

The majority of research on travel agents has been conducted in nations with large domestic markets, including: Canada (Loverseed, 1999), Australia (O'Brien, 1999), the United States of America (Lewis et al., 1998) and Europe (Marcussen, 1999). There has been relatively little research conducted in peripheral developed nations. This paper addresses this imbalance by presenting the findings of 3 years of research into the changing competitive environment and organisational dynamics of travel agents in New Zealand's largest city, Auckland.

2   New Zealand context

The Travel Agents' Association of New Zealand has suggested that never before has the industry experienced such turbulence and change. Changes in airline ownership, the disappearance of some carriers from the New Zealand market and changing consumer demand and expectations have altered business dramatically (TAANZ, 2001). The collapse of Qantas New Zealand, Ansett Australia, the New Zealand government bail-out of Air New Zealand and the uncertainty surrounding the future of country's national carrier are evidence of an ever-changing market that is, at best, fluid in nature.

Tensions between agents and Air New Zealand have been growing. The airline has poured large sums of money into creating a new web-site which is designed to increase the number of Internet based bookings and sales (www.airnz.co.nz). Recently Air New Zealand decided to eliminate commission payment for its domestic flights and commissions on trans-Tasman flights have been reduced. The changing policies of the national carrier and its competitors are a major concern for New Zealand travel agents. According to TAANZ, 60% of processing by it members is airfare related, with a large proportion of this funding coming from Air New Zealand (TAANZ, 2001).

Air New Zealand, following the example of Lufthansa, has adopted the concept of paperless travel throughout its domestic network. The flexible reservation system was developed in late 1990s. The airline established three new e-commerce departments to market direct to the public: E-Direct, E-Agency, and E-Technologies. E-Direct drives the airline's direct Internet activities. It works closely with regional and retail teams to build a stronger on-line business. The E-Agency group establishes links with other sites, like America On-line, to grow the direct-to-customer international businesses. E-Technology is looking at technologies coming on stream such as smart cards and kiosks, which allow people to book and receive tickets at airports (Kennedy, 1997).

New Zealand agents also face challenges from principals other than Air New Zealand. In a move similar to the North American Orbitz initiative, Zuji.com launched its web site in the Asia Pacific region in 2001. Zuji.com has 16 members, including Japan Airlines, Qantas, Singapore Airlines, United and China Airlines. The company offers access to 56,000 hotels, 400 airlines and 50 car rental firms. Utilising the technology of the major GDS, Sabre, the portal allows the consumer direct access to systems that were previously the domain of travel agents alone. It should be noted however that Air New Zealand, while initially involved as an investor in Zuji.com, has now withdrawn investment and is only prepared to act as a supplier rather than a shareholder. It should also be noted that there are, like the Orbitz case, serious legal challenges to the operation (Griffin, 2002).

ICTs have dramatically changed the competitive environment for intermediaries in the New Zealand travel market in other ways. New players have entered the sector such as Internet Service Providers either through the establishment of new companies, taking shares in start-ups, or entering into arrangements with existing travel agents. Examples of this include the ISP 'ihug' part-owning the travel agency 'travelonline', and the nations dominant ISP 'xtra' (part of the major telecommunications company Telecom) linking with Harbour City United travel.

An indication of the growing importance of ICT stems from the fact that New Zealand is becoming increasingly 'wired'. Nearly 50% percent of the New Zealand population are now using the Internet (Nua, 2002) and over 40% of NZ households now own an Internet enabled computer. Indeed New Zealand has a global ranking of 15th in the recently introduced Technology Achievement Index (UNDP, 2001). These figures suggest that the adoption of ICTs should be a principle part of any strategy for travel agents to retain their position in the distribution chain.

The following objectives underpin this paper:

1. How is the competitive environment of Auckland travel agents evolving?
2. What strategies are being adopted by agents as they attempt to come to terms with the evolving competitive context? To what extent do ICTs represent a tool for survival as well as a potential threat?
3. To what extent are travel agents who use the Internet reaping fully its rewards as a tool for business survival and customer relationship development?
4. What is the future for the travel agent in New Zealand and what, if any, industry initiatives are likely to support the continued existence of this segment of the tourism industry?

3   Methods

The research revolves around in-depth structured interviews conducted with travel agents in 2000. Interviews were carried out with 25 retail travel agents, owners, wholesalers, and travel sector experts in the greater Auckland region to investigate their perceptions of the effects of the changing business environment (see Milne et al. 2001). The study consisted of semi-structured interviews, using open-ended questions. The duration of the interviews varied from 1 hour to 2 hours. The interviews were designed to produce a considered response to certain issues, including background about the firm and the individual, the characteristics of the agents' particular market and if/how these had changed over time, marketing, relationship with the airlines, the use of ICTs and associated issues, and the importance of business alliances and associations. A sampling method was adopted that allowed us to include the full range of travel agent operations. Corporate/business agents, franchised/chain retail agents, wholesalers, small independent operations, and leisure specialists were all represented in the interviews. Notes were taken during the interviews which were also tape recorded and then transcribed. Subsequent analysis of the interviews was employed to identify common themes.

The interviews have been supplemented with a detailed review of secondary data (published and unpublished) relating to ICTs and travel developments in New Zealand. The 2000 interviews are currently being repeated, with some additional elements added in order to ascertain perceptions of the changing business environment and travel agents responses to the recent development of the Air New Zealand web site and to the zero commission policy on domestic flights. A total of 30 agents are being interviewed. In many cases the same agents are being interviewed twice (2000, 2002) - offering a chance to ascertain how perceptions and performance have evolved over the period.

4   Results

Over 80% of the respondents emphasize that travel agencies have to compete not just with other agents, but with airlines, and a range of other providers. The establishment of Air New Zealand owned and operated travel centres was the first indication that Air New Zealand was prepared to act in direct competition with recognised travel agents. The creation of Air New Zealand's web-site in late 1990s reinforced the notion and led to such statements as 'in my opinion Air New Zealand is really aggressive in trying to get rid of travel agents'. Initially, the Air New Zealand web-site was regarded with disdain as it was unable to offer a serious threat to travel agents due to technical problems associated with it. However, the new site is fully functional and offers the consumer the ability to book flights directly at a price that is often the same that is available to the travel agents themselves. The cutting of commissions paid to agents for domestic flights is further proof that Air New Zealand is itself facing severe financial constraints and that one of the more controllable costs to the airline, can be curtailed, in the hope of providing increased profitability.

With regard to commissions some respondents, primarily corporate travel agents, are content with the new zero commission policy of Air New Zealand. The argument is that the Internet has made the cost of 'doing business' more transparent to the consumer. After searching the Internet and attempting to purchase all the required elements that make up a typical business trip it is soon realised that this requires time and effort and makes the consumer appreciate some of the costs incurred by the travel agency. One respondent mentioned that 'the client can see, for the first time, the true cost of what travel is … in that sense it becomes a very close business relationship'. A service fee is then seen as an acceptable recompense for the agencies operations. Others, particularly leisure retail agents, argue that a service fee will be detrimental to their profit margins as the consumer seeks the perceived cheapest price.

While there is a general level of optimism in the travel agency sector there is a strong realisation that 'change is the only constant' and that for agents to remain competitive they must adapt their operations to best utilise any opportunities. There is also a realisation that not all agents will survive. There is a general opinion that there are too many travel agents in the country and that this number will decrease through the attrition of poor performers, and perhaps more importantly through acquisitions and mergers. Indeed, TAANZ expect the number of travel agents within their organisation to drop to 470 in the next few years from the 1999 figure of 626 (NZ Herald, 2000; pers comm. TAANZ).

A part of this expected consolidation is clearly seen to be due to the increasingly competitive market and falling profit margins in the sector. As one respondent with international experience noted, 'the margins [in NZ] are just frightening'. Remaining price competitive is considered an essential factor in the equation for survival, as 'what attracts someone to walk into the shop in the first place is the price'. To this end, the ability to participate at appropriate economies of scale, either through alliances or through franchise ownership, is regarded as important. Small, independent travel agents, unless they have an established niche product, are unlikely to remain profitable in the near future. Likewise, those travel agents that rely on airline commissions for their profitability are already experiencing financial hardship.

Another perceived impact of ICTs on the travel agency sector is that the existing boundaries between the different operations within the distribution chain are open to negotiation (see also Buhalis, 2000) There is a perception that there are too many 'middlemen' and that a 'super middleman' between suppliers and consumers will emerge through the continued formation of alliances and company mergers. This is aptly demonstrated by the respondent who suggested that 'the lines are blurred and everybody at the moment in the travel industry is thinking how is my business going to be in 5 years time' and that 'we don't make a lot of dough but what we do make we throw away in distribution, wholesaler to retail head, retail head to agency, agency to customer .... it has to change, if it doesn't change then the industry will die a sad death.'

There is generally a positive attitude towards ICT use among those interviewed. The stated advantages of employing ICTs include: accuracy in making transactions, lowered costs of communication, increased speed in processing, improved access to information, and the potential of a broader market to sell to. Nevertheless new technologies are often not recognised as part of core development strategies among many smaller agents.

Many small tourism businesses in New Zealand are more concerned with Internet presence than its performance (Ministry of Economic Development, 2001). Information opportunities created via the Internet are often missed while the use of the Internet is not integrated into other parts of the organisation (Yoong and Huff,

2000). There is a similar scenario with travel agents in Auckland as one respondent suggested that 'sometimes I think we all have web-sites because everyone else does, they keep telling us that you have to have one'. Another agent suggested that 'a lot of companies have been stitched up by computer web developers and portal companies over selling the product.'

On the other hand there is also evidence of major investment into computerised operating and reporting systems of the larger corporate agencies. ICT has been a feature of some agencies and the ability to provide a business-to-business solution for corporate travel has clearly increased their share of the market.

While there is a perceived requirement for travel agents to keep pace with technological advancements there is a general consensus that acquiring staff with solid training in such areas can be difficult. These findings are consistent with the conclusions of recent research on ICT adoption by New Zealand businesses in general. It was pointed out that the main reason for slow ICT adoption is a lack of knowledge and training to design and maintain on-line businesses (Wilson, 2002).

There is a perception among agents that the New Zealand Internet consumer is not as sophisticated as their overseas counterparts and more unwilling to trust web based operations. Nevertheless New Zealand, per population, is considered one of the most 'wired' countries in the Asia/Pacific (Nua, 2002). At the same time interviewees recognised that the customers are getting more familiar with the Internet. Therefore it should be recognised that ICTs have an ever-increasing role to play and should be considered as an essential part of the travel agents' survival strategy.

There is a broad-based understanding that the relationship with customers is very important for the survival of travel agents. A common theme is that 'at the end of the day it comes down to service.' While many respondents indicated a concern for the needs of the client, there was limited knowledge of how best to maintain the relationship with their customers. Niche-market travel agents in New Zealand are seemingly more customer orientated and are able to recognise the international trend requiring travel agents to become an agent for the consumer rather than an agent of the principals. These agents maintain and utilise their customer database, communicate with customers by e-mail, and measure customer satisfaction to a much larger degree than the average retail travel agent.

In response to the number of on-line services it is recognised that 'the only way we can stay alive is to prove to everybody that we do add value in this chain of distribution.' In this regard the New Zealand travel agents are prepared to compete with on-line services, and their stated advantages include:

- Security of the Internet (credit-card theft and accountability for the product)
- Human reassurance for consumers decision-making

- Efficiency of the travel agent and their ability to act as an 'information broker'
- Perceived unbiased travel advice and access to wholesale prices

5 Conclusion

ICTs have changed the competitive environment for intermediaries in the New Zealand travel market. With the ability of airlines and other principals to market directly to consumers New Zealand travel agents, like their international counterparts, are facing increasing pressure in retaining their traditional role as intermediaries. The creation of alliances/networks and a focus on niche markets are the main perceived strategies for survival.

While airline policies regarding commissions are of particular concern, a singular focus on this aspect is not considered constructive. Instead of battling airlines, the core focus of travel agents should be on the provision of excellent service to the consumer. Although customer service was recognised by the majority of interviewees as important, the researchers consider travel agents could improve this aspect of their business.

There is generally a positive attitude towards ICT use among interviewees. However, ICT is rarely a core strategy of travel agencies. Although some travel agents perceived their customers as unsophisticated ICT users it is essential that travel agents utilise ICTs to strengthen their relationship with their customers. The fast-changing nature of ICT requires travel agents to keep pace with advancements and adopt policies that 'add value' in the distribution chain.

Therefore we suggest that future research be focussed on consumer attitudes towards, and response to, ICTs rather than the technology itself. Consumer abilities regarding ICTs need to be studied in order for agents to best utilise the available technologies. A web site 'usability' study is suggested as a tool for appreciating the manner in which the consumer responds to the agents presence on the Internet.

Travel agents need to re-position themselves from transaction processors, information providers and simple resellers to consultants, information brokers or 'infomediaries' and developers of personalised packages. As mentioned by the Economist (2002) travel agents 'need to become more like professional-service firms, rather than second-hand car dealers ... based on helping customers to buy what they want, not on helping suppliers to flog what they've got' (Economist, 2002).

References

Bloch, M. and Segev, A. (1996). *The impact of electronic commerce on the travel industry*. The Fisher Centre for Information Technology and Management, University of California. (http://groups.haas.berkeley.edu/citm/publications/papers/wp-1017.html [January 17, 2001]).

Buhalis, D. (2000). Tourism and information technologies: past, present and future. *Tourism Recreation Research, 25*(1), 41-58.

Dull, S. F. (2001). Customer Relationship Management: A breed apart. *Accenture Outlook Journal 2001, No 2.* (http://www.accenture.com/xd/xd.asp?it=enWeb&xd=ideas\outlook\7.2001\breed.xml [November 30, 2001]).

De Lollis, B. and Adams, M. (2002). Delta quits paying travel agent commissions. *USATODAY.com.* 14.03.2002. (http://usatoday.com/life/travel/leisure/2002/2002-03-15-commissions.htm [September 24, 2002]).

Economist. (2002). Business: Fit for DIY?; Travel Agents. *The Economist.* 1 June, 2002, 363 (8275), 63.

Hall, C. M. (2000). The future of tourism: a personal speculation. *Tourism Recreation Research, 25,* 1, 85-95.

Kennedy, G. (1997). Air NZ's electronic tickets are winning the paper war. *National Business Review,* May 30, p. 27.

Lewis, I, Semeijin, J. and Talalayevsky, A. (1998). The impact of information technology on travel agents. *Transportation Journal, Summer:* 20-25.

Loverseed, H. (1999). Travel agents in Canada. *Travel and Tourism Analyst, 1*: 71-86.

Marcussen, C. H. (1999). *Internet Distribution of European Travel and Tourism Services.* Nexo, Denmark: Research Centre of Bornholm.

Milne, S., Bremner, H. and Carter, P. (2001). Information technology and travel distribution channels: a future for travel agents? *Paper presented at the Annual Meeting of the Assoc. of American Geographers,* New York, March 2001.

Ministry of Economic Development. (2001). *Net Readiness in New Zealand Industries: Empirical Results, 2001.*(http://www.ecommerce.govt.nz/statistics/readiness/netreadiness-11.html [September 20, 2002]).

New Zealand Herald. (2000). *NZ Herald,* 17/12/00, C9.

Griffin, P. (2002). Travel giant casts shadow. *New Zealand Herald,* 10 September 2002. (http://www.nzherald.co.nz/storydisplay.cfm?thesection=technology&thesubsection=& storyID=2647048 [September 27, 2002]).

Nua Internet Surveys. (2002). *Nua Internet Surveys: How many Online. Asia.* (http://www.nua.com/surveys/how_many_online/asia.html [May 27, 2002]).

O'Brien, P. F. (1999). Intelligent Assistants for Retail Travel Agents. *Information Technology & Tourism, 2* (3/4), 213-28.

Palmer, A. and McCole, P. (1999). The virtual re-intermediation of travel services: A conceptual framework and empirical investigation. *Journal of Vacation Marketing, 6* (1), 33-47.

TAANZ. (2001*). TAANZ Actions... Industry News.* Issue 19, July 2001.

TAANZ Handbook. (2001). In *Travel Industry Directory & Information Guide 2001.*

UNDP, (2001). *Human Development Report 2001 "Making new technologies work for human development".* New York: Oxford University Press.

Wilson, H. (2002). *New Zealand small and medium-sized enterprises (SMEs):*

Annotated bibliography, key findings & recommendations for research & policy. Paper prepared for Auckland Regional Economic Development Strategies (AREDS).

(http://www.areds.co.nz/resources/New%20Zealand%20Small%20&%20Medium-sized%20Enterprises.doc [September 18, 2002]).

WTOBC (World Tourism Organisation Business Council). (2001). Internet poised to take a quarter of tourism sales. *WTO, Newsroom, New Release.* (http://www.world-tourism.org/newsroom/Releases/more_releases/October2001/011023.htm [December 18, 2001]).

Yoong, P. and Huff, S. (2000). Current issues and concerns regarding e-commerce: An exploratory study of SMEs in New Zealand. *In Proceedings of the 2000 ETEC Conference (Track 6), Kuala Lumpur, 2000.*

Springer Economics

Karl Wöber,
Andrew J. Frew, Martin Hitz (eds.)

Information and Communication Technologies in Tourism 2002

Proceedings of the International Conference in Innsbruck, Austria, 2002

2002. XIII, 526 pages. 108 figures.
Softcover EUR 98,–
(Recommended retail price)
Net-price subject to local VAT.
ISBN 3-211-83780-9

Consumers, who are increasingly mobile and better informed, are becoming more demanding in the eTourism world. However, what role will mobile devices play in the communication process with marketing, strategy and other functions? How do online consumers really behave in this globalizing, converging, and consolidating tourism industry? How does their behavior differentiate them from other customers? To what extent will new technologies, expert systems, new types of online-available consumer or business databases, and sophisticated models influence traditional research approaches in both the tourism and information technology arena? What is the contribution of information technology to organizational performance and social welfare when considering the microstructured characteristics of the tourism industry? What new forms of intermediation will emerge?
These questions have provoked a number of exciting responses. They are collected in this volume.

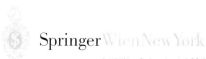

Springer Wien New York

A-1201 Wien, Sachsenplatz 4–6, P.O. Box 89, Fax +43.1.330 24 26, e-mail: books@springer.at, Internet: **www.springer.at**
D-69126 Heidelberg, Haberstraße 7, Fax +49.6221.345-229, e-mail: orders@springer.de
USA, Secaucus, NJ 07096-2485, P.O. Box 2485, Fax +1.201.348-4505, e-mail: orders@springer-ny.com
Eastern Book Service, Japan, Tokyo 113, 3–13, Hongo 3-chome, Bunkyo-ku, Fax +81.3.38 18 08 64, e-mail: orders@svt-ebs.co.jp

SpringerEconomics

Karl Wöber (ed.)

City Tourism 2002

Proceedings of European Cities Tourism's
International Conference in Vienna, Austria, 2002

2002. XII, 362 pages. 50 figures.
Softcover EUR 70,–
(Recommended retail price)
Net-price subject to local VAT.
ISBN 3-211-83831-7

This collection of papers presented at the International City Tourism
Conference (ICTC) concentrates on future challenges in city tourism
research and practices. The papers deal with visitor and tourist behavior in cities, concentration and dispersion effects in city tourism, the
effects of event management for cities, various city tourism policy
issues, the investigation of success factors for city tourism web sites,
and the impact and effectiveness of city tourism facilities and providers. They will enable tourism organizations and researchers to
explore innovative methods for serving the tourists of the future.

Please visit our website: **www.springer.at**

Springer Wien New York

A-1201 Wien, Sachsenplatz 4–6, P.O. Box 89, Fax +43.1.330 24 26, e-mail: books@springer.at, Internet: **www.springer.at**
D-69126 Heidelberg, Haberstraße 7, Fax +49.6221.345-229, e-mail: orders@springer.de
USA, Secaucus, NJ 07096-2485, P.O. Box 2485, Fax +1.201.348-4505, e-mail: orders@springer-ny.com
Eastern Book Service, Japan, Tokyo 113, 3–13, Hongo 3-chome, Bunkyo-ku, Fax +81.3.38 18 08 64, e-mail: orders@svt-ebs.co.jp

SpringerEconomics

Hannes Werthner,
Martin Bichler (eds.)

Lectures in E-Commerce

2001. IX, 216 pages. Numerous figures.
Softcover EUR 45,–
(Recommended retail price)
Net-price subject to local VAT.
ISBN 3-211-83623-3

Although only a few years old, electronic commerce offers new ways of doing business that no business can afford to ignore. This book is a collection of selected contributions from reknowned researchers who specialize in the various facets of electronic commerce, namely economics, finance, information technology, and education. The basic goal is to give an overview of some of the most relevant topics in E-Commerce.

Contents:

Measurement, Policy and Research Issues of a New Phenomena (P. Timmers) • Strategies for the Financial Services Industry in the Internet Age (H.U. Buhl, D. Kundisch, A. Leinfelder, W. Steck) • A Framework for Performance and Value Assessment of E-Business Systems in Corporate Travel Distribution (A.M. Chircu, R.J. Kauffman) • Defining Internet Readiness for the Tourism Industry: Concepts and Case Study (U. Gretzel, D.R. Fesenmaier) • Aggregation and Disaggregation of Information Goods: Implications for Bundling, Site Licensing, and Micropayment Systems (Y. Bakos, E. Brynjolfsson) • Auctions – the Big Winner Among Trading Mechanisms for the Internet Economy (R. Müller) • Security in E-Commerce (G. Müller) • XML@work: The Case of a Next-Generation EDI Solution (P. Buxmann) • E-Commerce/E-Business Education: Pedagogy or New Product Development? (P.M.C. Swatman, E.S.K. Chan)

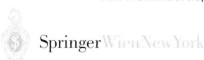

Springer Wien New York

A-1201 Wien, Sachsenplatz 4–6, P.O. Box 89, Fax +43.1.330 24 26, e-mail: books@springer.at, Internet: **www.springer.at**
D-69126 Heidelberg, Haberstraße 7, Fax +49.6221.345-229, e-mail: orders@springer.de
USA, Secaucus, NJ 07096-2485, P.O. Box 2485, Fax +1.201.348-4505, e-mail: orders@springer-ny.com
Eastern Book Service, Japan, Tokyo 113, 3–13, Hongo 3-chome, Bunkyo-ku, Fax +81.3.38 18 08 64, e-mail: orders@svt-ebs.co.jp

SpringerGeosciences

Yong-Qi Chen,
Yuk-Cheung Lee (eds.)

Geographical Data Acquisition

2001. XIV, 265 pages. 167 figures.
Softcover EUR 62,–
(Recommended retail price)
Net-price subject to local VAT.
ISBN 3-211-83472-9

This book is dedicated to the theory and methodology of geographical data acquisition, providing comprehensive coverage ranging from the definition of geo-referencing systems, transformation between these systems to the acquisition of geographical data using different methods. Emphasis is placed on conceptual aspects, and the book is written in a semi-technical style to enhance its readability. After reading this book, readers should have a rather good understanding of the nature of spatial data, the accuracy of spatial data, and the theory behind various data acquisition methodologies. This volume is a text book for GIS students in disciplines such as geography, environmental science, urban and town planning, natural resource management, computing and geomatics (surveying and mapping). Furthermore it is an essential reading for both GIS scientists and practitioners who need some background information on the technical aspects of geographical data acquisition.

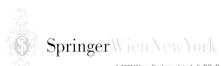

Springer Wien New York

A-1201 Wien, Sachsenplatz 4–6, P.O. Box 89, Fax +43.1.330 24 26, e-mail: books@springer.at, Internet: **www.springer.at**
D-69126 Heidelberg, Haberstraße 7, Fax +49.6221.345-229, e-mail: orders@springer.de
USA, Secaucus, NJ 07096-2485, P.O. Box 2485, Fax +1.201.348-4505, e-mail: orders@springer-ny.com
Eastern Book Service, Japan, Tokyo 113, 3–13, Hongo 3-chome, Bunkyo-ku, Fax +81.3.38 18 08 64, e-mail: orders@svt-ebs.co.jp

Springer-Verlag
and the Environment

WE AT SPRINGER-VERLAG FIRMLY BELIEVE THAT AN international science publisher has a special obligation to the environment, and our corporate policies consistently reflect this conviction.

WE ALSO EXPECT OUR BUSINESS PARTNERS – PRINTERS, paper mills, packaging manufacturers, etc. – to commit themselves to using environmentally friendly materials and production processes.

THE PAPER IN THIS BOOK IS MADE FROM NO-CHLORINE pulp and is acid free, in conformance with international standards for paper permanency.